THEORY OF SCHEDULING

Richard W. Conway, *Cornell University*
William L. Maxwell, *Cornell University*
Louis W. Miller, *RAND Corporation*

DOVER PUBLICATIONS, INC.
Mineola, New York

Bibliographical Note

This Dover edition, first published in 2003, is an unabridged republication of the work originally published by Addison-Wesley Publishing Company, Reading, Mass., in 1967.

Library of Congress Cataloging-in-Publication Data

Conway, Richard Walter, 1931–
    Theory of scheduling / Richard W. Conway, William L. Maxwell, and Louis W. Miller.
        p. cm.
    Reprint. Originally published: Reading, Mass. : Addison-Wesley, 1967.
    Includes bibliographical references and index.
    ISBN 0-486-42817-6 (pbk.)
    1. Production scheduling. I. Maxwell, William L., 1934– II. Miller, L. W. (Louis W.), 1935– III. Title.

TS155 .C637 2003
658.5'3—dc21

                                                        2003041047

Manufactured in the United States of America
Dover Publications, Inc., 31 East 2nd Street, Mineola, N.Y. 11501

PREFACE

Scheduling is a field in which there are some intriguing problems and some interesting answers. So far, however, the subject has not received the attention it deserves; work on it has been fragmented at best, and the published material on scheduling is to be found in a number of different disciplines. When one reads this published material, it is obvious both from its content and from the different papers the authors cite that investigators are often unaware of similar work along the same lines done by other people. This, then, is a first attempt to organize the work that has been done on scheduling. Also, since the field has long needed consistency of terminology and a reasonable taxonomy, we hope that this book will provide a start in that direction. We have tried to summarize much of what has been done in scheduling, to identify some of the interesting problems that have not yet been dealt with, and to place them in context with existing work. If this book encourages scholars to do enough research in scheduling to render the book itself obsolescent as a summary of accomplishment, it will have served well.

The theory of scheduling can be compared with inventory theory, although unquestionably it is less well developed and less coherent. The theory of scheduling has not yet attracted the attention of scholars comparable to Dvoretsky, Kiefer and Wolfowitz, or Arrow, Karlin, and Scarf; perhaps because the mathematical model which underlies the theory of scheduling probably cannot compare in elegance with that which underlies inventory theory. However, there is at least as much practical incentive to solve the problems of sequence as there is to solve those of inventory. We would also argue that there is in the theory of scheduling—even in its primitive state—as much of immediate practical value as there is in the much larger literature on inventories. Of course, few practical problems are explicitly solved by either theory (one-machine shops are no commoner than one-product inventories) but both theories contain implications and suggestions that are of value to industrial practitioners. Certainly the recurring themes of the effects of processing times and due-dates on sequence are useful.

We organized the book primarily according to the type of scheduling problem considered rather than the technique of its solution. Had we done otherwise, it would have been more readily apparent that there are basically three types of attack on scheduling problems: algebraic, probabilistic, and Monte Carlo simulation by computer. Even readers who do not have a substantial mathematical background can follow the algebraic work, since it is comparatively simple. The second approach treats the job-shop as a stochastic process. To fully appreciate the results obtained and the difficulties that face further progress, the reader should have had a course in advanced probability. The results obtained by using the third approach—computer simulation—are easily stated and readily understood, but the difficulty of obtaining significant results in this manner is underestimated. Much of the work described here was accomplished with machines and languages that were barely equal to the task. The work was done by people who were not only studying scheduling procedures but also pioneering in the art of simulation. Some of this experimental investigation of scheduling could hardly be considered strikingly successful, yet we considered it worth while to report the results in some detail. The reason we considered it so is that the basic set of procedures and experiments conceived by one would-be investigator is almost identical to that conceived by others. Until someone takes the trouble to report results fully, the pointless replication will continue.

We have used preliminary versions of this manuscript as the basis for a one-semester graduate course in Cornell's operations research program. Although these students have strong backgrounds in stochastic processes and simulation procedures, one semester is not enough time to fully explore all the material. However, it is appropriate to include some selected material in production courses for students majoring in either industrial engineering or business administration.

We began to be interested in problems of scheduling as early as 1958. Our inquiry has received support from a number of sources, among them the Office of Naval Research (Task NR 047-023), the National Science Foundation (GP 2729), the General Electric Company (Production Control Services), the Western Electric Company (Engineering Research Center), and Touche, Ross, Bailey, and Smart (Management Sciences Group). Each of us has pursued the work in both Ithaca and Santa Monica; both Cornell and The RAND Corporation have provided generous support and encouragement.

We are deeply indebted to a long procession of Cornell graduate students whose investigations represent a significant part of the scheduling literature. In something approaching chronological order, these graduate students are: B. M. Johnson, A. S. Ginsberg, J. P. Evans, C. E. Nugent, E. J. Ignall, L. E. Schrage, T. B. Crabill, A. Lalchandani, P. Everett, A. N. Bakhru, M. R. Rao, R. D. Wayson, J. W. Oldziey, G. L. Orkin, J. Soden, W. Walker, H. A. Neimeier, M. Mehra, and M. Naik. Much of the material in Chapters 8, 9, and 10 developed out of our association with Dr. Benjamin Avi-Itzhak of the Israel Institute of Technology during his stay as a visiting professor at Cornell.

The appearance of this book has long been delayed, since, early in our work on it, we recognized the properties of sequencing according to processing-time, and all our other tasks were shorter than this one.

Ithaca, N.Y.   R. W. C.
April 1967   W. L. M.
                                                                        L. W. M.

CONTENTS

CHAPTER 1

PROBLEMS OF SEQUENCE

Sequencing problems are very common occurrences. They exist whenever there is a choice as to the order in which a number of tasks can be performed. A problem could involve: jobs in a manufacturing plant, aircraft waiting for landing clearances, bank customers at a row of tellers' windows, programs to be run at a computing center, or just Saturday afternoon chores at home. Our basic thesis is that, regardless of the character of the particular tasks to be ordered, there is a fundamental similarity to the problems of sequence.

Sequencing problems obviously get solved, since most of the tasks are performed: the aircraft land, the bank customers transact their business and at least some of the Saturday chores are completed. However, most of these problems are solved quite casually or automatically without explicit recognition that a problem even existed, much less that a solution was obtained. Sometimes an ordering is determined essentially by chance; more often tasks are performed in the order in which they arise. An inherent sense of fair play has elevated the "first-come, first-served" solution of sequencing problems to an eminence out of all proportion to its basic virtue. It may be appropriate for patrons in a box-office line, but it need not necessarily be applied to inanimate jobs on a factory floor.

The formal statement and solution of a sequencing problem is probably a rare occurrence. It would read as follows:

1) α is the aggregate consequence of performing task A first, followed by B.

2) β is the aggregate consequence of performing task B first, followed by A.

3) Since α is preferable to β the ordering "A, then B" is selected.

We are not, of course, arguing that this formal analysis is appropriate for the ubiquitous sequencing problems of everyday life, but there are problems in industry, transportation, and governmental and institutional activities in which the results of sequence are nontrivial and systematic consideration is worth while. Yet all too often these problems are solved by default rather than design. The factory machine operator who decides which of several waiting jobs to process next often uses criteria which have little to do with the company's objectives. Programs may be run on a computer in the order submitted, a procedure that is undeniably fair, but far from optimal from anyone's point of view. It will be clearly demonstrated in the following chapters, by algebraic arguments, numerical examples, and computer experiments, that there are significant differences between alternative sequences. At least in some real problems these differences must result in sufficient cost or value to be worth considering. It is not a question of deciding whether or not to solve a particular sequencing problem—one way or another a sequence will be selected. The question is to determine how this decision will be made, by whom, and against what criteria.

Questions of sequence are perhaps not as vital as the decisions that determine what tasks are to be sequenced or how the task is to be performed once its turn has arrived. Nevertheless, if proper selection of sequence can yield some incremental improvement, it seems pointless to neglect the opportunity. The evidence in the following chapters would seem to indicate that even in the simplest abstractions of multi-activity systems, there are interesting differences attributable only to sequence and that in many cases procedures which are equally usable and practical have vastly different performance characteristics.

1-1 QUESTIONS OF "PURE" SEQUENCE

The difficulty in studying questions of sequence in actual applications is that these questions are not often completely separable—they interact and are confounded with other types of decisions. Very often the order of execution will have some influence on what tasks have to be performed, on the precise character of the task, or on the method that will be used for performance. For example, a manufacturing task might initially involve the production of 100 units of a certain product. The decision to defer the manufacture of this product for a month, rather than to start today, could have any of the following effects:

1) The entire order might be canceled.

2) An additional customer order might be received so that the production quantity becomes 150.

3) An engineering change in design specifications might occur.

4) One type of raw material, available now, might be out of stock in a month and require substitution.

5) A key machine, available now, may be down for repairs in a month and substitute equipment may have to be used.

6) A new machine, or new tooling, now on order might have been received, and this may improve processing.

7) The workers may have had practice on similar work during the month and may have become more skilled.

If any one of these changes is a reasonable possibility, then the question of sequence is inextricably involved with the decisions as to *what* is to be done and *how* it is to be done. In such an instance, engineering or managerial judgment is required in making decisions of sequence. Unfortunately, each problem is unique to its particular circumstances, and hence there is nothing of general interest that can be studied out of context.

The solution is to extract a problem that ignores the possibility of such changes and considers only the questions of sequence. Such a problem is unrealistic in the sense that it does not exactly represent any individual real situation, but rather gains in generality, since it approximates many situations. The results obtained from the study of this abstract idealized model do not represent a solution to any real sequencing problem; they represent information that should be available along with judgment and data on other aspects of the real problem. The decision-maker should have some idea of what the consequences of alternative ordering would be if none of the changes listed above were to take place. This thought may not dominate the decision but it should at least be considered in making it.

We shall consider a problem in which all the decisions relating to *what* and *how* have previously been made. Given that these decisions were entirely unaffected by the choice of sequence, the following assumptions will obtain throughout this work:

1) The tasks to be performed (a task represents a prescribed sequence of activities) are well defined and completely known; that is, both the character and magnitude of the tasks have been determined by some exogenous mechanism that is not of direct concern to the present work. Of course, it is often the case that the decisions for all future tasks are not announced to the scheduling process simultaneously, but rather are revealed sequentially over time. This is permitted under this assumption.

It is further assumed that all the assigned tasks must be performed because the option of selecting a subset of tasks that are in some sense preferable is not considered part of the problem of pure sequence. In a manufacturing situation, for example, we do not consider which of the jobs might be most profitable. We assume that this selection and decision has already been made and that the given tasks represent firm commitments, all of which must be met.

2) The resources or facilities that may be employed in the performance of these tasks have been entirely specified. For example, we might be told that 10 machines of various types can be operated by a crew of 8 men up to a maximum of 44 hours per week. The option of adding capacity by either acquiring more equipment or expanding the hours of use may be strongly suggested by the result of attempts to schedule the existing facilities, but it is not part of the problem of pure sequence.

3) The sequence of elemental activities required to perform each of the tasks is known, including any restrictions on the order in which these must be performed. The method of performing each of these elemental activities is entirely known and there is at least one of the set of facilities capable of performing each of these activities.

Much of the research literature in sequencing refers to the *job-shop scheduling problem* and uses the terminology of manufacturing: job, machine, operation, routing, and processing-time. In fact, the work is based on this type of idealized pure-sequence abstraction of such a manufacturing shop and the results are equally applicable to problems in transportation, communication, services, etc. Actually one might say that the results are equally inapplicable, since this idealized model is not an exact representation of any real job-shop. If the present work is more appropriate to manufacturing than to other areas it is because our background in that field has guided our selection of conditions for study, and not because we have chosen to continue the use of manufacturing vocabulary.

We considered drawing a fine distinction between sequencing and scheduling; to hold that the former is concerned only with the ordering of operations on a single machine, while the latter is a simultaneous and synchronized sequence on several machines. However, we found that no greater clarity resulted from such a distinction, and the two terms are used essentially as synonyms in the following chapters.

In the next section we shall define more carefully the assumptions of the idealized model that is the basis for the entire work.

1–2 THE JOB-SHOP PROCESS

The basic unit of the job-shop process is the *operation*. One can envision an operation as an elemental task to be performed, but insofar as a theory of scheduling is concerned the operation need not be defined. The theory is concerned only with what can be done with operations and not with what operations really are. Three primary attributes (others are described in Chapter 2) of each operation are given as part of the description of a particular problem:

1) A symbol identifying the operation with a particular *job*.

2) A symbol identifying the operation with a particular *machine*.

3) A real number representing the *processing-time* of the operation.

The set of all operations can be partitioned by the first identification symbol into disjoint, exhaustive, mutually exclusive subsets called *jobs*. Similarly, the set can be partitioned by the second symbol into exhaustive, mutually exclusive subsets associated with *machines*.

There is also given for each job a partial ordering of the operations that comprise the job. (Envision technological constraints on the order in which tasks must be accomplished.) This partial ordering between operations is given by a binary relationship called *precedence*. If x and y are two operations of the same job, and if for some reason the processing of x must begin before the processing of y, then x is said to *precede* y and is written

$$x > y.$$

Equivalently, if $x > y$, we may also write $y < x$ and say that operation y *follows* operation x. The precedence relationship is transitive:

$$x > y \quad \text{and} \quad y > z \quad \text{implies} \quad x > z.$$

A special case of this relationship exists when there are no intervening operations. We say operation x *directly precedes* operation y (or operation y *directly follows* operation x) and write

$$x \gg y \quad \text{or} \quad y \ll x \quad \text{if} \quad x > y,$$

and if there is no operation, z, such that $x > z > y$.

It is often convenient to display these relationships on a precedence graph such as the following:

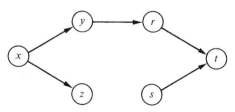

The nodes of the graph represent the operations and the directed arcs (arrows) represent "*directly-precedes*" relationships. The precedes relationship exists between two nodes if there is a path of directed arcs between them.

A *machine* in this process is intuitively a device or facility capable of performing whatever it is that has to be done in an operation, but abstractly, a machine is just a time scale with certain intervals available. A *job-shop* is the set of all the machines that are identified with a particular set of operations; a *job-shop process* consists of the machines, the jobs (operations), and a statement of the disciplines that restrict the manner in which operations can be assigned to specific points on the time scale of the appropriate machine.

Scheduling a job-shop process is the task of assigning each operation to a specific position on the time scale of the specified machine. For each operation of the process, this means the determination of one or more intervals (b_1, c_1), (b_2, c_2), . . . in such a manner that:

1) $(c_1 - b_1) + (c_2 - b_2) + \cdots$ is greater than or equal to the processing-time for the operation,

2) b_{1x}, the value of b_1 assigned to operation x, is not less than b_{1y} for every operation y of the same job as x such that $y > x$,

3) each of the intervals (b_i, c_i) lies wholly within one of the intervals available for assignment on the appropriate machine.

Alternatively, scheduling can be regarded as the task of constructing *an ordering of the operations associated with each machine*, somewhat analogous to the ordering that is given for the operations of each job.

There are many different disciplines that can be specified to constrain this assignment or ordering. Almost all the theory that has been developed to date has been concerned with a highly restrictive discipline that can be said to define a *simple job-shop process*. Additional restrictions are placed on the definitions of the job set and the machines, as well as on the manner in which a schedule may be constructed. These restrictions are:

1) Each machine is continuously available for assignment, without significant division of the time scale into shifts or days, and without consideration of temporary unavailability for causes such as breakdown or maintenance.

Each machine is simply a single interval $(0, T)$, where T is an arbitrarily large number.

2) Jobs are strictly-ordered sequences of operations, without assembly or partition.

For a given operation x, there is at most one operation y such that

$$y \gg x,$$

and at most one operation z such that

$$x \gg z.$$

3) Each operation can be performed by only one machine in the shop.

4) There is only one machine of each type in the shop.

The machine identification numbers are unique so that for each operation in the job set they identify one and only one machine in the shop.

5) Preemption is not allowed—once an operation is started on a machine, it must be completed before another operation can begin on that machine.

Only a single interval (b, c) is to be assigned to each operation with $(c - b)$ equal to the processing-time of the operation.

6) The processing-times of successive operations of a particular job may not be overlapped. A job can be in process on at most one operation at a time.

b_x, the value of b assigned to operation x, must be not less than c_y, for any operation y of the job set such that

$$y \succ x.$$

7) Each machine can handle at most one operation at a time.

Consider the interval (b_x, c_x), the assignment of operation x to a particular machine. For every other assignment (b_y, c_y) to that machine, either

$$b_x \geqq c_y \quad \text{or} \quad c_x \leqq b_y.$$

These restrictions are primarily for simplicity of structure, but at the same time they increase the generality of the model. Most of the different applications that can be based on this model require relaxing one or several of these restrictions, so that it is not an entirely realistic model for any application. But it appears that this simple job-shop process is the core of many applications, and insight that is useful in many different fields can be obtained by its investigation. In any event it is the model that has been used in most of the research to date and only scattered results exist for various different types of complex job-shop processes.

We should make one very important restriction explicit by noting that the simple job-shop process has *only one limiting resource.* We have assumed that an operation requires only one machine for its processing. Alternatively, one could assume that several machines, or a *machine*, a *man*, and a *tool*, were simultaneously required to perform the process. Attributes would be added to each operation to identify the particular type of man and type of tool required, and assignment of an operation to a point in time could not be made until the appropriate machine, man, and tool were all simultaneously available.

1-3 A CLASSIFICATION OF SCHEDULING PROBLEMS

A specific scheduling problem is described by four types of information:

1) The jobs and operations to be processed.

2) The number and types of machines that comprise the shop.

3) Disciplines that restrict the manner in which assignments can be made.

4) The criteria by which a schedule will be evaluated.

The previous section identified attributes of operations and the partial ordering of the operations of a job that are present in every scheduling problem. Problems differ in the number of jobs that are to be processed, the manner in which the jobs arrive at the shop, and in the order in which the different machine numbers appear in the operations of individual jobs. The nature of the job arrivals provides the distinction between *static* and *dynamic* problems. In a static problem a certain number of jobs arrive simultaneously in a shop that is idle and immediately available for work. No further jobs will arrive, so attention can be focused on scheduling this completely known and available set of jobs. In a dynamic problem the shop is a continuous process. Jobs arrive intermittently, at times that are predictable only in a statistical sense, and arrivals will continue indefinitely into the future. As might be expected, entirely different methods are required to analyze questions of sequence in these two problems. This distinction between static and dynamic problems provides the basis for the organization of this book—static problems are treated in Chapters 3 through 7, and dynamic problems are considered in Chapters 8 through 11.

The order in which the machine numbers appear in the operations of individual jobs determines whether a shop is a *flow-shop*. A flow-shop is one in which all the jobs follow essentially the same path from one machine to another. In terms of the model, this means that there is a numbering of the machines such that the machine number for operation y is greater than the machine number for operation x of the same job, if x precedes y.

At the opposite extreme is the *randomly routed job-shop*, in which there is no common pattern of movement from machine to machine. Each machine number is equally likely to appear at each operation of each job. Either of these extremes is undoubtedly rare in practice, with most real shops falling somewhere between these two limits, but almost all the research in scheduling has assumed one or the other of these two extreme cases.

The discipline question is largely covered by the assumptions of the simple job-shop process—unless clearly stated otherwise, the restrictions of Section 1-2 are assumed to hold.

A four-parameter notation will be used to identify individual scheduling problems—written as $A/B/C/D$.

A describes the job-arrival process. For dynamic problems, A will identify the probability distribution of the times between arrivals. For static problems, it will specify the number of jobs—assumed to arrive simultaneously unless stated otherwise. When n is given as the first term, it denotes an arbitrary, but finite, number of jobs in a static problem.

B describes the number of machines in the shop. A second term of m denotes an arbitrary number of machines.

C describes the flow pattern in the shop. The principal symbols are F for the flow-shop limiting case, R for the randomly routed job-shop limiting case, and G for completely general or arbitrary flow pattern. In the case of a shop with a single machine there is no flow pattern, and the third parameter of the description is omitted.

D describes the criterion by which the schedule is to be evaluated. The alternatives and notation for this fourth term are described in Chapter 2.

As an example of this notation, Johnson's problem ([103], Section 5-2) is described as

$n/2/F/F_{max}$: sequence an arbitrary number of jobs in a two-machine flow-shop so as to minimize the maximum flow-time (or schedule time).

The general job-shop problem—still unsolved—is

$n/m/G/F_{max}$: schedule n jobs in an arbitrary shop of m machines so that the last is finished as soon as possible.

CHAPTER 2

MEASURES FOR
SCHEDULE EVALUATION

The variety of different criteria that have been employed in theoretical studies of scheduling partly reflects the variety of different circumstances in which interesting scheduling problems arise and the different costs and values that are relevant in each case. However, the choice of criteria has undeniably also been influenced by the prospects of obtaining a solution. In some models it has been possible to find optimal procedures only by departing from what would be considered the most natural and realistic criteria. The purpose of the present chapter is to define and discuss a number of the measures that have been used and to show some of the interrelations between them. In the process we shall introduce the notation that will be required in later chapters.

It is important to clearly distinguish between the *variables that define the problem*, which in this consideration of "pure sequence" are assumed to be given by some external agency, and those *variables that describe the solution* produced by the scheduling process. To emphasize this distinction we have adopted the convention that lower-case letters denote the *given* variables and capital letters denote those that are *determined by scheduling*. The symbols h, x, y, z and Q, X, Y, Z will be used for variables which apply to an individual section.

2–1 VARIABLES THAT DEFINE A SCHEDULING PROBLEM

The problem description starts with a shop and a set of jobs. For the simple job-shop process the shop is completely described by giving the number of machines. We shall speak of an "m-machine shop" and assume that the individual machines may be identified by the integers $1, 2, \ldots, m$.

The jobs are similarly identified by integers $1, 2, \ldots, n$. The relevant attributes of job i that are given as part of the problem description are denoted by the following variables:

r_i is the *ready-time, release-time*, or *arrival-time*. This is the time at which the job is released to the shop by some external job-generation process. It is significant as the earliest time that processing of the first operation of the job could begin.

d_i is the *due-date*. This is the time at which some external agency would like to have the job leave the shop. It is the time by which the processing of the last operation *should* be completed.

We shall refer to the difference between these values sufficiently often to make it worth defining:

$a_i = d_i - r_i$,   the total *allowance* for time in the shop.

One can readily imagine situations in which different pairs of these three variables are actually given. In these theoretical exercises it doesn't matter which two are original and which derived; all three are simply given values for a job.

The job consists of a set of g_i operations. These are described by g_i pairs of values:

$$m_{i,1}, \qquad p_{i,1},$$
$$m_{i,2}, \qquad p_{i,2},$$
$$\vdots \qquad \vdots$$
$$m_{i,g_i}, \qquad p_{i,g_i},$$

where $m_{i,j}$ is the identification number of the machine that is required to perform the jth operation, $1 \leqq m_{i,j} \leqq m$; $p_{i,j}$ is the processing-time, the amount of time that will be required for machine $m_{i,j}$ to perform the operation.

Let $p_i = \sum_{j=1}^{g_i} p_{i,j}$ be the total processing-time for the job. There are various ways of regarding processing-time. One can ignore the variability that would almost always be present in practice and simply consider processing-time a variable given in the same sense as, say, the due-date. Alternatively, one can recognize the variability and say that the actual value will not be known until after the processing takes place; at best an estimate of $p_{i,j}$ is available before the work is done. However, the important distinction, and the reason that a lower-case letter is used, is that the value of the processing-time is assumed to be entirely independent of the scheduling decision. The value may not be known until after the operation is assigned a position in sequence, but it is assumed that the assignment has absolutely no effect on the value that is revealed.

In many cases in practice a job will consist of a "lot" of identical pieces each of which is to be processed in the same manner. The lot size is the number of pieces in the lot and the cycle-time for an operation is the time required to perform the operation on one piece. We shall use processing-time, $p_{i,j}$, to refer to the total time needed to perform an operation—the cycle-time multiplied by the lot size. In general, we shall assume that the lot size is determined before scheduling and that it is not changed by scheduling so that we shall not explicitly refer to either cycle-time or lot size.

Finally, we assume that $p_{i,j}$ includes whatever *setup* is necessary to prepare the machine to perform the operation and whatever *teardown* is required after the operation. This implies the assumption that the amount of such *changeover-time* is independent of the sequence of operations—that the length of time required to prepare the machine for a particular operation does not depend on what the machine did last. In a majority of cases this is probably a reasonable approximation to the fact and it greatly simplifies the analysis. However, there are situations in which it is untenable and an explicit setup-time, possibly sequence-dependent, must be recognized. This is discussed in Section 4–1 and Chapter 9.

2-2 VARIABLES THAT DESCRIBE
THE SOLUTION TO A SCHEDULING PROBLEM

Questions of pure sequence are considered simply to determine when each job should be done, or to be more specific, when each operation of each job should be done. This way of looking at the matter is equivalent to determining how long each operation of each job should *wait before processing begins*, and is denoted by

$W_{i,j}$, the *waiting-time* preceding the jth operation of job i,

the time that the job must wait after the completion of the $(j-1)$th operation before beginning the jth operation. The total waiting-time for a job is the sum of the waiting-times for all operations of the job:

$$W_i = \sum_{j=1}^{g_i} W_{i,j}.$$

The result of the scheduling process for a particular problem, the schedule, *is completely specified by giving a set of $W_{i,j}$*. Many other variables will be introduced for convenience and compactness of notation, but all these are functions of the $W_{i,j}$. In the final analysis every comparison of schedules is based on a comparison of different sets of $W_{i,j}$; the goodness of a schedule is completely a consequence of the values of the $W_{i,j}$.

The most important of the measures that may be derived from the $W_{i,j}$ are concerned with: (1) the time at which particular jobs leave the shop, (2) the length of time that particular jobs spend in the shop, and (3) the difference between the times when jobs leave the shop and when they were supposed to leave. These are defined as follows:

C_i, the *completion-time* of job i: the time at which processing of the last operation of the job is completed. Thus

$$C_i = r_i + W_{i,1} + p_{i,1} + W_{i,2} + p_{i,2} + \cdots + W_{i,g_i} + p_{i,g_i}$$

$$= r_i + \sum_{j=1}^{g_i} p_{i,j} + \sum_{j=1}^{g_i} W_{i,j} = r_i + p_i + W_i.$$

F_i, the *flow-time* of job i: the total time that the job spends in the shop, or

$$F_i = W_{i,1} + p_{i,1} + W_{i,2} + p_{i,2} + \cdots + W_{i,g_i} + p_{i,g_i}$$

$$= \sum_{j=1}^{g_i} p_{i,j} + \sum_{j=1}^{g_i} W_{i,j} = p_i + W_i = C_i - r_i.$$

The flow-time F_i is also called the *manufacturing interval* and the *shop-time*.
There is also L_i, the *lateness* of job i:

$$L_i = C_i - d_i = F_i - a_i.$$

In addition, there is T_i, the *tardiness* of job i:

$$T_i = \max(0, L_i),$$

and there is E_i, the *earliness* of job i:

$$E_i = \max(0, -L_i).$$

Lateness, tardiness, and earliness are three different ways of comparing the actual completion-time with the desired completion-time. *Lateness* considers the algebraic difference for each job, regardless of the sign of the difference. *Tardiness* considers only positive differences—jobs which are completed after their due-dates; and *earliness* considers only negative differences—jobs completed ahead of their due-dates.

A schedule is completely described by a set of $W_{i,j}$, and no less information is sufficient for description. Two schedules for a particular problem are identical if, and only if, they have identical sets of $W_{i,j}$. However, there are many situations in which one would wish to consider as equivalent certain schedules which are not strictly identical. For example, one might consider as equivalent all schedules for a particular problem that have a certain set of values of completion-time, or even consider as equivalent all schedules for a particular problem that have a certain average value of completion-time. In general, specifying a *measure of performance* in a scheduling problem is specifying a set of equivalence classes of schedules and a preference-ordering among these classes. For example, specifying the mean completion-time as the measure of performance for a particular problem means that:

1) All schedules that have a mean completion-time of $\bar{C}'$ are equivalent, so one is indifferent as to which is selected.

2) A schedule with mean completion-time of $\bar{C}'$ is preferred to a schedule with mean completion-time $\bar{C}''$ if and only if $\bar{C}' < \bar{C}''$.

A schedule S is *optimal* with respect to a certain measure of performance if it belongs to an equivalence class $\{S'\}$ such that there do not exist any (nonempty) classes which are preferred over $\{S'\}$.

In almost all the theoretical work on scheduling, very simple measures of performance have been employed. These have been the *average* or the *maximum* of the values of *completion-time, flow-time, lateness,* or *tardiness.* Certain of the results apply to all these criteria, since all are examples of what we call *regular measures of performance.* A regular measure is a value to be minimized that can be expressed as a function of the job completion-times:

$$M = f(C_1, C_2, \ldots, C_n),$$

and which increases only if at least one of the completion-times increases. That is, if $M' = f(C_1', C_2', \ldots, C_n')$, then

$$M' > M \quad \text{only if} \quad C_i' > C_i \quad \text{for at least one } i, \ 1 \leqq i \leqq n.$$

It is easily verified that the average or maximum of completion-times, flow-times, lateness, or tardiness satisfies these conditions. More complex measures with weighted averages, combinations of average and maximum, or functions of fractiles would also be included. But average or maximum earliness is not a regular measure, and measures such as the difference between the largest and the second-largest completion-times are not regular.

The averages of many of these derived variables are very closely related to each other. For each individual job the following relationship holds (by definition):

$$L_i = F_i - a_i = C_i - r_i - a_i = C_i - d_i.$$

For a problem consisting of n jobs, the n corresponding equations can be added:

$$\sum_{i=1}^{n} L_i = \sum_{i=1}^{n} F_i - \sum_{i=1}^{n} a_i = \sum_{i=1}^{n} C_i - \sum_{i=1}^{n} r_i - \sum_{i=1}^{n} a_i = \sum_{i=1}^{n} C_i - \sum_{i=1}^{n} d_i.$$

Dividing each term by n yields

$$\bar{L} = \bar{F} - \bar{a} = \bar{C} - \bar{r} - \bar{a} = \bar{C} - \bar{d}.$$

For each particular problem, $\bar{a}, \bar{r}$, and $\bar{d}$ are given constants, unaffected by the decisions of sequence. Therefore, for every schedule for this problem, $\bar{L}$ differs from $\bar{F}$ by precisely $\bar{a}$; $\bar{L}$ differs from $\bar{C}$ by $\bar{d}$; etc. This obviously means that the schedule that is optimal with respect to $\bar{F}$ is also optimal for $\bar{C}$ and $\bar{L}$.

Since each of these quantities is derived from the waiting-times, the average waiting-time is also related to these measures:

$$C_i = r_i + W_i + p_i, \qquad \bar{C} = \bar{r} + \frac{\sum_{i=1}^{n} W_i}{n} + \frac{\sum_{i=1}^{n} p_i}{n}.$$

Since $\bar{r}$ and the term involving the p_i are constants for a given problem, the average waiting-time is also minimized by the schedule that minimizes average completion-time.

Average tardiness and average earliness are related to average lateness, although not in a very useful way:

$$T_i - E_i = L_i, \qquad \bar{T} - \bar{E} = \bar{L}.$$

In general, this does not imply that a schedule that achieves minimum average lateness will also have minimum average tardiness. However, there are special circumstances in which this is true. For example, if the due-dates of a set of jobs are all so difficult or tight that all the jobs have positive values of lateness, then lateness and tardiness are equivalent measures. Alternatively, if the due-dates are set so loosely that all jobs are completed on time, then earliness is simply the negative of lateness.

These relationships between averages are important in both theoretical work and practical application. They are simply consequences of the definition of the variables and require no assumption as to the conditions of the shop or the type of scheduling procedure employed. However, one must always bear in mind that the relationship is strictly concerned with averages and implies nothing more. For example, if one is assured by these equations that two procedures have equal mean completion-times, he cannot conclude that they are equivalent procedures, for different distributions of completion-time can have the same mean. In particular, these relationships imply nothing at all with respect to the maximum of the values of the variables. A schedule that is optimum with respect to $\bar{F}$ is also optimum with respect to $\bar{L}$, but one knows nothing about $L_{\max}$ for a schedule that is optimum with respect to $F_{\max}$.

2–3 PERFORMANCE MEASURES FOR THE SHOP

It is sometimes convenient to evaluate a schedule with respect to measures that relate to the shop, rather than to the individual jobs. The most obvious and important of these are *facility utilization* and *work-in-process inventory*. Utilization is the fraction of available machine-capacity that is employed in the required processing—that is, the ratio of processing-time to available time. Since it is assumed throughout that the total amount of processing-time is predetermined and not affected by the scheduling decision, it can only be the denominator of the fraction (i.e., the available-time) that is of interest. In the continuous-process models of Chapters 8 through 11, the machines are always available so that the average utilization is simply a given parameter of the problem and not affected by the schedule. It is true that the schedule determines just when and in what pattern the idle-time occurs on each machine, and different patterns might have different utility—longer, connected intervals of idle-time might be preferred to frequent short periods—but no work has been done on the problem with this preference as a criterion.

In the finite problems of Chapters 3 through 7 one can assume that all the machines are available for the total time to process the n jobs—that is, from $r_{\min}$ to $C_{\max}$. Most of the problems assume that all the jobs become available simultaneously, in which case the average utilization is inversely proportional to the maximum flow-time:

$$\overline{U} = \frac{\sum_{i=1}^{n} p_i}{mF_{\max}}.$$

As a result, although utilization is obviously important in practical problems, it is not often mentioned explicitly in work on scheduling, since the models are either structured so that utilization is independent of schedule or so that it is a direct consequence of flow-time, in which case it would not need separate consideration.

There are exceptions in models in which setup-time is sequence-dependent or when some forms of preemption are used, since in these cases the total amount of work does depend on the schedule (Sections 4–1 and 8–7 and Chapter 9).

There are two principal ways of measuring work-in-process inventory, and many variations on these. Basically one can either count the number of jobs in the shop or sum the processing-times of those jobs. It is easy to suggest situations in which one or the other of these is appropriate. For example, in an Air Force base maintenance operation for a single type of aircraft, one would be interested in scheduling so as to minimize the number of jobs (aircraft) in the shop, since this would presumably maximize the number of aircraft which are available for service. On the other hand, in a manufacturing situation the costs of inventory are more likely to be related to the amount of work that the inventory represents rather than a simple count of jobs.

The following notation is used for work-in-process inventory:

$N(t)$, the number of jobs in the shop at time t,

$\overline{N}(t_1, t_2)$, the average number of jobs in the shop during the interval from t_1 to t_2:

$$\overline{N}(t_1, t_2) = \frac{1}{t_2 - t_1} \int_{t_1}^{t_2} N(t)\, dt,$$

$P(t)$, the sum of the processing-times of all operations on all jobs in the shop at time t, called the *work content* or *total work*, and

$\bar{P}(t_1, t_2)$, the average work content during the interval from t_1 to t_2:

$$\bar{P}(t_1, t_2) = \frac{1}{t_2 - t_1} \int_{t_1}^{t_2} P(t)\, dt.$$

It is sometimes interesting and useful to partition these measures. For example, the jobs in the shop at time t could be divided into two groups:

$N_q(t)$, the number of jobs which are in queue waiting to be processed, and

$N_p(t)$, the number of jobs which are actually on a machine being processed, where

$$N_p(t) + N_q(t) = N(t).$$

The averages, $\bar{N}_q(t_1, t_2)$ and $\bar{N}_p(t_1, t_2)$, could be defined in a manner analogous to $\bar{N}(t_1, t_2)$. These would be interesting because, as the length of the interval (t_1, t_2) increases, $\bar{N}_p(t_1, t_2)$ becomes independent of the schedule. It is simply a function of the amount of work that must be done and the number of machines that are available. Then the goodness of the schedule is reflected directly in $\bar{N}_q(t_1, t_2)$, a term that ideally could be reduced to zero.

Similarly, work content could be partitioned into the sums of processing-times of (a) operations already completed, (b) operations in progress, and (c) operations not yet started. In the long run, the average of (b) is again independent of schedule. The average of (a) would in many situations be the best estimate of the dollar value of work-in-process inventory, while the average of (c) would give a measure of the actual work backlog in the shop. The results in Chapter 11 clearly indicate that one must be precise and thoughtful in selecting an inventory measure, since different scheduling procedures will be appropriate for each choice.

2–4 THE RELATIONSHIP BETWEEN FLOW-TIME AND INVENTORY

It seems intuitively obvious that there is a relationship between flow-time and inventory—if individual jobs, on the average, spend more time in the shop, then the average number of jobs in the shop must be increased. Nevertheless it is surprising on first encounter to discover just how strong and direct this relationship is and how general its applicability is.

Different relationships apply in the *static* situation, in which all the jobs that are to be considered are available simultaneously at the beginning of the scheduling period, and the *dynamic* case, in which the jobs arrive intermittently during the period. In the static case we are interested in $\bar{N}(r_1, C_{max})$, the average number of jobs in the shop during the schedule-time for these n jobs. It is convenient, and causes no loss in generality, to assume that the jobs all arrive at time 0 so that this average is equivalent to $\bar{N}(0, F_{max})$:

$$\bar{N}(0, F_{max}) = \frac{1}{F_{max}} \int_{0}^{F_{max}} N(t)\, dt.$$

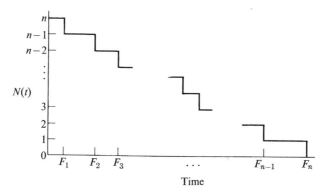

Fig. 2–1. Number of jobs versus time in a problem involving simultaneous arrivals.

The function $N(t)$ in this case looks like the graph in Fig. 2–1. With $N(t)$ changing value only at $n + 1$ points in time, the integral is evaluated as a sum of rectangles:

$$\int_0^{F_n} N(t)\, dt = nF_1 + (n - 1)(F_2 - F_1) + (n - 2)(F_3 - F_2)$$

$$+ \cdots + (n - (n - 1))(F_n - F_{n-1}) = \sum_{i=1}^{n} F_i,$$

$$\overline{N}(0, F_{\max}) = \frac{\sum_{i=1}^{n} F_i}{F_{\max}} = n\,\frac{\overline{F}}{F_{\max}}, \qquad \frac{\overline{N}(0, F_{\max})}{n} = \frac{\overline{F}}{F_{\max}}.$$

This result asserts that the *ratio of the average number of jobs in the shop to the maximum number of jobs in the shop is equal to the ratio of the average flow-time to the maximum flow-time.* It also says that for a given schedule-time $F_{\max}$ the average number of jobs in the shop is directly proportional to the average flow-time. Note that this relationship *holds for any set of n jobs ($n \geq 1$) which begin at the same time.* It does not depend on:

a) the number, kind, or arrangement of machines in any way;

b) the nature of the jobs, their routing over the machines, the processing-times or the *a priori* state of knowledge about processing-times;

c) other activities of the machines; they can have other jobs to process before, during, and after the arrival of these n jobs; or

d) the scheduling procedure.

Although the argument was based for convenience on the finite n/m problem, it is clear that the only part of the hypothesis that was actually used was the assumption that $r_1 = r_2 = \cdots = r_n$. The result holds equally well for any finite subset of the jobs of a continuous process that happens to start at the same time. ("Start" is used in the sense of being available for processing, and has nothing to do with the time at which work actually begins.)

Consider now the more general dynamic case in which the jobs *need not* arrive simultaneously, and assume initially that the jobs are numbered in the order of arrival,

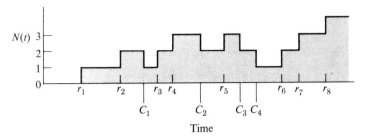

Fig. 2–2. Number of jobs versus time in a problem involving intermittent arrivals.

and that *they are completed in that same order:*

$$r_1 \leqq r_2 \leqq \cdots \leqq r_n, \qquad C_1 \leqq C_2 \leqq \cdots \leqq C_n.$$

We wish to find a relationship between (a) the mean number of jobs in the shop during the interval in which these n jobs are being processed and (b) the mean flow-time for the n jobs. By definition,

$$\overline{N}(0, C_n) = \frac{1}{C_n} \int_0^{C_n} N(t)\, dt.$$

A plot of $N(t)$ against time would look something like the graph in Fig. 2–2, with unit changes at $2n$ points in time as the n jobs arrive and depart. Alternatively, one could separately plot the cumulative number of arrivals and the cumulative number of completions against time, as in Figs. 2–3 and 2–4, and combine them on the same axes, as in Fig. 2–5. The graph in Fig. 2–5 has the property that the vertical distance

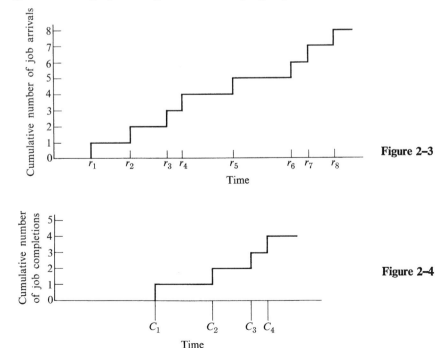

Figure 2–3

Figure 2–4

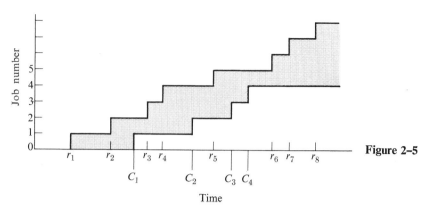

Figure 2–5

between the two lines is everywhere equal to $N(t)$ and that the shaded area between the lines is equal to the area under the $N(t)$ graph (Fig. 2–2) or the integral of $N(t)$. It is also interesting to note that each horizontal section of the graph represents a particular job. It has length F_i, beginning at r_i and ending at C_i. It is easier to give an expression for the area between the lines if one considers the horizontal sections:

$$\int_0^{C_n} N(t)\, dt = \sum_{i=1}^n (C_i - r_i) = \sum_{i=1}^n F_i,$$

$$\overline{N}(0, C_n) = \frac{1}{C_n}\sum_{i=1}^n F_i = \frac{n}{r_n + F_n}\frac{\sum_{i=1}^n F_i}{n},$$

$$\frac{\overline{F}}{\overline{N}(0, C_n)} = \frac{r_n}{n} + \frac{F_n}{n}.$$

The term r_n/n represents the mean time between job arrivals. If one takes larger and larger values of n and the situation approaches a continuous process, and if the process is in a "steady-state" such that the expected flow-time of jobs does not increase with time, then the term F_n/n becomes negligible. The result is that for a steady-state process (in which jobs are completed in the same order as they arrive) the mean flow-time is equal to the mean time between arrivals multiplied by the mean number of jobs in the system.

Relaxing the assumption that the jobs are completed in arrival order makes the justification slightly more complicated, but it does not alter the final expression. The graph would look something like Fig. 2–6. Some difficulty is encountered because at C_n, there is no assurance that all the $n - 1$ prior jobs have been completed. If we let $Y_i(t)$ represent the fraction of flow-time, F_i, still remaining after time t, then

$$\int_0^{C_n} N(t)\, dt = \sum_{i=1}^n (1 - Y_i(C_n))F_i = \sum_{i=1}^n F_i - \sum_{i=1}^n Y_i(C_n)F_i,$$

$$\overline{N}(0, C_n)\left[\frac{r_n}{n} + \frac{F_n}{n}\right] = \overline{F} - \frac{\sum_{i=1}^n Y_i(C_n)F_i}{n}.$$

Again, as the number of jobs n increases, if the process is in a steady-state the term F_n/n will become negligible. The second term on the right will also become negligibly

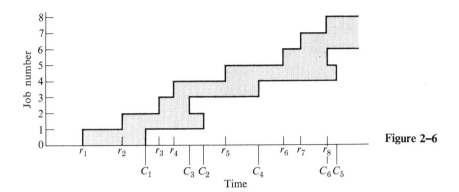

Figure 2–6

small since, for most of the terms in the summation in the numerator, Y_i will be zero—the fraction of the total number of jobs that are still uncompleted at time C_n will decrease. One can bound the left-hand expression:

$$\overline{N}(0, C_n)\left[\frac{r_n}{n} + \frac{F_n}{n}\right] \leq \overline{F},$$

$$\overline{N}(0, C_n)\left[\frac{r_n}{n} + \frac{F_n}{n}\right] \geq \overline{F} - \frac{N(C_n)F_{\max}}{n}.$$

If the process is in a steady-state, then $N(C_n)$, the number of jobs in the system at time C_n, and $F_{\max}$ are both finite, and the ratio of their product to n can be made arbitrarily close to zero by taking n sufficiently large.

The result* is that for any steady-state process the *mean flow-time is equal to the product of the mean number in the system and the mean time between arrivals*, or

$$\overline{F} = \frac{\overline{N}}{\lambda}, \quad \text{where } \lambda \text{ is the mean arrival rate.}$$

This is a most important result, for both theory and practice. It is intuitively appealing, and one can give many heuristic arguments in its defense, but neither intuition nor heuristic logic suggests the truly catholic nature of its applicability. It asks only that the system be in a weak sort of steady-state—that the long-run average rate of completion be equal to the average rate of arrival. It does not depend on:

a) the type of system or where the system boundaries are drawn,

b) any characteristic of the arriving stream of work, or how it arrives (except that in the long run it must not arrive in such quantities as to saturate the system), or

c) how work is scheduled within the system.

This relationship applies to an entire plant, to a department within the plant, and to an individual machine. It applies to the complete product mix of a company, or to one product line, or to an individual product. One must be careful only to count and measure in consistent units.

---

* Little [122] published a formal proof of this relationship, in a queuing context, in 1961, but the result was certainly well known, and it is possible that prior proofs exist.

The relationship can be expressed in terms of facility utilization instead of mean job-arrival rate. In a shop of m machines, the average utilization $\overline{U}$ is given by

$$\overline{U} = \frac{\lambda \overline{p}}{m},$$

where λ is the mean arrival rate (number of jobs per unit time), and $\overline{p}$ is the mean amount of work per job. Substituting for λ in the flow equation, we have

$$\overline{F} = \frac{\overline{N}\overline{p}}{m\overline{U}}.$$

Including earlier relationships between mean flow-time, mean lateness, and mean waiting-time, we have

$$\overline{F} = \frac{\overline{N}\overline{p}}{m\overline{U}} = \overline{L} + \overline{a} = \overline{W} + \overline{p},$$

where $\overline{F}$ is the mean flow-time, $\overline{L}$ is the mean lateness, $\overline{W}$ is the mean waiting-time, $\overline{a}$ is the mean allowable flow-time, $\overline{p}$ is the mean total processing time per job, $\overline{N}$ is the mean number of jobs in the system, $\overline{U}$ is the mean utilization of the machines in the system, and m is the number of machines in the system.

This interdependence of measures is important in the comparison of scheduling procedures. For any given situation, described by a particular set of values for $\overline{p}$, $\overline{a}$, m, and $\overline{U}$,

a) the mean flow-time is directly proportional to the mean work-in-process inventory (as measured by the number of jobs),

b) the mean flow-time differs from the mean lateness and the mean waiting-time by a constant amount,

c) a scheduling procedure which is relatively good with respect to any one of these mean values is comparably good with respect to each of the others, and

d) a scheduling procedure which minimizes mean flow-time also minimizes mean lateness, mean waiting-time, and the mean number of jobs in the system.

2–5 COSTS ASSOCIATED WITH SCHEDULING DECISIONS

Although in practice the questions of when and in what order tasks are performed would have some effect on many of the costs of performance, and indeed, on whether the task is performed at all, the costs that can be directly associated with the limited question of pure sequence are restricted.

The assumption that the set of tasks is determined beforehand and that it is unaffected by scheduling decisions means that the total revenue of the enterprise is fixed, or at least unaffected by, and irrelevant to, scheduling. The assumption that the method and equipment to be used and the efficiency with which they will be employed are also unaffected by scheduling decisions means that all of the costs that are normally classified as *direct costs* are irrelevant for our purposes. In fact, the costs that may be attributed to decisions of pure sequence are entirely what would be classified as *facility costs* rather than *product costs*. Even cost items such as overtime wage premiums or penalties exacted for late delivery that could be identified with a particular job are really consequences of decisions made on many other jobs and it would be unfair to assign them to the unfortunate jobs with which they are identified. There are three principal types of costs that can be affected by the decisions of pure sequence. These are the costs of inventory, utilization, and lateness. Much has been written on the costs of inventory, and there are still some areas of debate. There is at least agreement that these costs are real, nontrivial (estimates of 2%–3% of value per month are not unusual), in some way proportional to the physical level of inventory, and very difficult to quantify. In different circumstances, costs might be related to the number of jobs, work content, or work completed, but in general we can conclude that there are economic reasons for our interest in reducing average inventories. Under the restrictions assumed here, these costs also provide the incentive for minimizing average flow-time, since they are directly related to average inventory. In practice, reduced flow-time would probably also provide a competitive sales advantage that would be even more significant.

Facility utilization is a very important economic consequence of sequencing decisions. The ability to compact the busy intervals and produce a short schedule-time or a low mean flow-time simply implies a procedure that will permit a given facility to do more work. Conversely, an efficient scheduling procedure will permit a given work load to be accomplished with a smaller aggregate demand on facilities. In the long run, this will be reflected in either the amount of plant required or the amount of business that can be accommodated. In the short run, it is reflected in the costs of overtime, additional shift operation, and overload subcontracts.

In some situations, especially construction projects, the cost of lateness is obvious, explicit, and unequivocal. For example, a penalty of X dollars per day may be deducted from payment for each day that completion is delayed beyond a specified due-date. In manufacturing situations the penalty for lateness is seldom this obvious and immediate but presumably no less real, since the customer's displeasure will have implications for future business. One scheduler described the relative importance of this consideration to the authors by saying that he might be admonished for high inventories, but that he would be fired for excessive lateness.

Whether one considers the limited questions of pure sequence or a broader and more realistic view of scheduling, it is still true that the costs and values are difficult to identify and harder still to measure. A modern costing system includes procedures specifically designed to call attention to inefficient use of labor or material or to indirect items that exceed budget allowances, but there are no accounts or procedures that clearly signal inefficiency in scheduling. The costs are nonetheless real, and judging from the demonstrated differences in performance between alternative scheduling procedures there must be some situations in which this choice is vital.

CHAPTER 3

FINITE SEQUENCING
FOR A SINGLE MACHINE

This chapter is concerned with the special case of the scheduling problem in which each job consists of a single operation. Since the set of jobs can be partitioned depending on the machine required to perform the operation, each machine in the shop is independent of the others and can be scheduled separately. Therefore we can limit our attention to a single machine, and to the set of jobs that is to be processed on that machine. We shall assume that the number of jobs n is finite, that n is known in advance of scheduling, and that all the n jobs must be processed. Furthermore, we shall assume that the machine is to have no other obligations and that it will be continuously available, without breakdown, until all the jobs are completed. We shall also assume that all the n jobs are available for processing simultaneously so that the schedule *could* begin with any one of them, and that there is either no setup-time required for the jobs, or that the setup-time does not depend on the nature of the preceding job on the machine. In the latter case the setup-time for a particular job depends only on the characteristics of that job and may, for present purposes, be included in the processing-time for the job. The notation that will be used is the following:

$m = 1$ There is a single relevant machine.

$g_i = 1$ By assumption, each job has one operation.

$r_i = 0$ By assumption, all the jobs are available simultaneously. One can, without loss of generality, translate the time scale so that the origin is at the moment of availability.

$p_{i,1} = p_i$ The processing-times are arbitrary, determined by some process independent of the scheduling procedure, and are assumed known at the time of scheduling. Since each job has a single processing-time, the second subscript (operation) will be omitted.

$a_i = d_i$ Allowance for waiting and processing, also due-date, since $d_i = r_i + a_i = 0 + a_i$.

$W_{i,1} = W_i$ The waiting-time of the ith job before processing begins. Since each job waits for a single operation, the second subscript (operation) will be omitted.

$C_i = F_i = p_i + W_i$ The time at which the job is completed. It is equal to the flow-time, since the time scale begins as the job becomes available.

There are both practical and theoretical reasons for examining this special case in detail. This is the simplest scheduling problem and it is by far the best understood. It serves to illustrate the differences between different scheduling procedures and the consequences of choosing different measures of performance. Some of the solutions to this relatively simple problem are useful in more complex cases both in suggesting directions for research and in providing bases for approximate practical procedures. One can really appreciate how limited the results for a general job-shop are when one has a rather complete understanding of the one-operation case, and can compare the two.

From a practical point of view direct applications of this model are both more frequent and more important than one might initially suspect. In addition to shops that are actually a single machine, there are many situations in which a large complex of equipment behaves as if it were a single machine. In the chemical and processing industries an entire plant often represents an integrated process that operates on one product at a time, but on many distinctly different products in sequence. Two specific examples are the manufacture of detergents, in which many different types and brands of detergents are processed separately and in sequence on the same processor, and the manufacture of paints, in which paints of many different colors are processed as separate batches on the same line of equipment. There are also situations in which, although each job actually consists of several operations to be performed on different machines, there is one particular machine whose value and function so dominate the process that it is scheduled as if the other machines did not exist. An example is found in modern papermaking in which one integrated machine, completely dominating the other steps in the process, works on a sequence of jobs representing different colors, weights, and textures of paper. While at the moment examples are easier to cite in the processing industries, the mechanical manufacturing industries, with increasing automation, are headed in this direction. As the use of automatic-control equipment and complex transfer machines increases, the number of separate "scheduleable" units decreases.

There are also cases of temporary dominance in which a short-run bottleneck situation on one machine is so restrictive that the work for this machine should be scheduled as if the other machines did not exist. For example, if while scheduling air traffic for a network of terminals one learns that the capacity of an airport has been reduced by weather conditions, he might have to allow the scheduling of this one terminal to completely dominate the entire network.

3-1 PERMUTATION SCHEDULES

Two schedule tactics that have some utility in more complex situations are *preemption*, in which processing is interrupted and a job is removed from a machine before an operation is completed, and *inserted idle-time*, in which a machine is held idle although there is work ready and waiting to be done. If these tactics are to be allowed in the present case it means that for a problem with two jobs, shown in Fig. 3-1, one would, in addition to schedules 1 and 2, consider such variations as are shown in schedules 3 through 6.

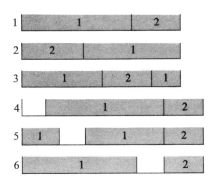

Figure 3-1

However, one is relieved of this necessity by the following result.

Theorem 3-1. When one schedules an $n/1$ problem with respect to a regular measure of performance, it is not necessary to consider schedules which involve either preemption or inserted idle-time.

Proof (for inserted idle-time). Consider a schedule **S** that has an inserted idle-period from t_1 to t_2,

$$0 \leqq t_1 < t_2 \leqq C_{max}.$$

There is a corresponding schedule **S'** which differs from **S** only in that the machine is not idle from t_1 to t_2 and all the actions and events after t_2 in **S** have been advanced by an amount $t_2 - t_1$. This obviously does not change the completion-time of any job completed before t_1, and it reduces the completion-time of each job completed after t_2 in **S** by the amount $t_2 - t_1$. It is clear that

$$C_i' \leqq C_i, \quad i = 1, 2, \ldots, n,$$

so that the value of a regular measure of performance (Section 2-2) could not be less for **S** than for **S'**. It is therefore sufficient to consider only schedules similar to **S'** in that they do not contain any periods of inserted idle-time.

Proof (for preemption). Consider a schedule **S** in which job I is started at time t_1, and at time t_2, before it is completed, job I is preempted by job J. (It is convenient, but not necessary, to assume that J is not in turn preempted.) At some later time t_3 work resumes on job I and it is ultimately completed.

Schedule S

	I		J				I	
t_1		t_2			C_J	t_3		C_I

Now consider schedule **S'** which differs from **S** only in that the positions of job J and the first portion of job I have been interchanged.

Schedule S′

The result is simply that job J is completed earlier under S′ and no other job completion-time is changed. Therefore the value of any regular measure of performance cannot be less for S than for S′. One could repeat this operation on S′ by advancing any other jobs that might still come between the portions of job I. One would eventually connect the portions of job I through a sequence of changes, none of which would increase any regular measure. Any preemption can be eliminated from any schedule in this manner without increasing any regular measure. It follows that it is sufficient to consider only schedules without preemption.

The significance of the fact that neither preemption nor inserted idle-time need be considered is that one can restrict attention in this problem to the class of *permutation schedules*. These are schedules that are completely specified by giving the order in which the jobs will be processed. Such an ordering is described by giving one of the $n!$ possible permutations of the job identification numbers $1, 2, \ldots, n$. Given this specification of sequence and the job processing-times, one can calculate the values of the W_i and from them any other property of the schedule.

In general, square brackets will be used to denote position in sequence in a permutation schedule. The symbol [1] means the job which is in first position in sequence; $p_{[i]}$ means the processing-time of whichever job occupies the ith position in sequence; and [3] = 7 means that job number 7 has been assigned to the third position in sequence. For example, using this notation, it is always true in a permutation schedule that

$$C_{[1]} \le C_{[2]} \le \cdots \le C_{[n]}.$$

However, sometimes it will be convenient, after a sequence has been specified, to imagine a renumbering of the jobs in the order in which they appear in sequence so that the square brackets may be omitted. This means that jobs are reassigned identification numbers so that

$$[i] = i \quad \text{for} \quad i = 1, 2, \ldots, n.$$

There will be ample warning in the text whenever such a renumbering is employed.

One should note that for permutation schedules in this $n/1$ problem the maximum flow-time is simply the sum of the n processing-times. Its value depends only on this given information, and is exactly the same for each of the $n!$ possible sequences. In the more complex models of Chapters 4, 5, and 6, the maximum flow-time is the primary measure of performance, but in this case it is independent of sequence and the selection of a sequence must be based on other measures. It is trivially obvious that the maximum number of jobs in the shop, the minimum number of jobs in the shop, the maximum work-content in the shop, the minimum waiting-time, and machine utilization are also measures that are, in this case, independent of sequence.

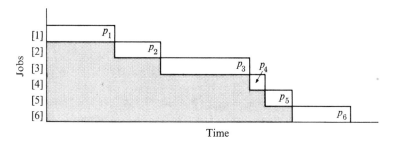

Fig. 3–2. Graph of an $n/1$ schedule.

3-2 SEQUENCING ACCORDING TO PROCESSING-TIME

A sequence for an $n/1$ problem can be represented graphically in the manner of Fig. 3–2. The total area of this graph—both labeled blocks and shaded portion—represents the sum of the job flow-times. Observe that the area represented by the labeled blocks is independent of the sequence. Although rearranged depending on sequence, precisely these same blocks will be present in any graph of a schedule for this problem. However, the magnitude of the shaded area is not constant and depends on the manner in which the labeled blocks are arranged. Each labeled block can be characterized by a vector on its diagonal, with slope $-1/p_i$, as shown in Fig. 3–3.

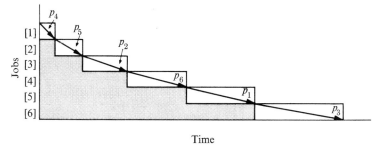

Fig. 3–3. Graph of an $n/1$ SPT schedule.

Each schedule graph corresponds to a head-to-tail chain of these vectors. It seems intuitively clear that the shaded area of such a graph is minimized when the blocks are arranged so that the vectors form a convex curve. We do this by starting with the block whose vector has the most negative slope and arranging the others in order of increasing (toward zero) slope, as shown in Fig. 3–3. This is equivalent to sequencing the jobs so that

$$p_{[1]} \leqq p_{[2]} \leqq \cdots \leqq p_{[n]}.$$

Following this procedure results in *shortest-processing-time sequencing* (abbreviated SPT) which is easily the most important concept in the entire subject of sequencing. It recurs, with many variations, and performs with surprising efficacy throughout the following chapters.

The optimality of SPT sequencing can be formally established by the following:

Theorem 3–2.* When one schedules an $n/1//\bar{F}$ problem, the mean flow-time is minimized by sequencing the jobs in order of nondecreasing processing-time,

$$p_{[1]} \leqq p_{[2]} \leqq \cdots \leqq p_{[n]},$$

and maximized by sequencing the jobs in order of nonincreasing processing-time,

$$p_{[1]} \geqq p_{[2]} \geqq \cdots \geqq p_{[n]}.$$

Proof. Since only permutation schedules need be considered, the flow-time of the job in kth position of an arbitrary sequence is simply

$$F_{[k]} = \sum_{i=1}^{k} p_{[i]}.$$

The mean of the flow-times of the n jobs is given by

$$\bar{F} = \frac{\sum_{k=1}^{n} F_{[k]}}{n} = \frac{\sum_{k=1}^{n} \sum_{i=1}^{k} p_{[i]}}{n}$$

$$= \frac{1}{n}(p_{[1]} + p_{[1]} + p_{[2]} + p_{[1]} + p_{[2]} + p_{[3]}$$

$$+ \cdots + p_{[1]} + \cdots + p_{[n]})$$

$$= \frac{\sum_{i=1}^{n} (n - i + 1)p_{[i]}}{n}.$$

It is well known† that such a sum of pairwise products of two sequences of numbers will be minimized if one sequence is arranged in increasing order and the other in decreasing order. Since the $(n - i + 1)$ coefficients are already in decreasing order, the minimization is accomplished by sequencing the jobs so that the processing-times are in increasing (or at least nondecreasing) order. The sum of products is maximized if both sequences have the same ordering, which means sequencing the jobs so that the processing-times are in decreasing (or at least nonincreasing) order.

A slightly stronger version of Theorem 3–2 can be given, and an alternative proof will be given that illustrates an approach used frequently in later sections. The measure of performance of Theorem 3–2 can be generalized to the minimization of

$$\frac{\sum_{i=1}^{n} F_i^{\alpha}}{n} \quad \text{for } \alpha > 0.$$

The previous statement was for the special case with $\alpha = 1$.

Proof. Consider a schedule **S** that is not identical to a schedule produced by SPT. (Note that because of the possibility of subsets of jobs with equal processing-times

---

* We are somewhat at a loss to identify the original author of this result. It was given by Smith [187] in 1956 but was surely well known before that.

† Hardy, Littlewood, and Polya, *Inequalities*, second edition, New York: Cambridge University Press, 1952, page 261.

many SPT sequences can exist, all with the same values of mean flow-time. S is not any of these SPT sequences.) There is some position k in S such that

$$p_{[k]} > p_{[k+1]}.$$

Let K and K' be the numbers of the jobs in the kth and $(k + 1)$th positions in S, respectively.

Now consider a schedule S' that differs from S only in that job K' is in kth position and job K in $(k + 1)$th position. The sequence is the same for the first $k - 1$ jobs and the last $(n - k - 1)$ jobs under either S or S' and it is obvious that each of these $n - 2$ jobs begins and is completed at exactly the same moment under either schedule. S and S' differ only in the flow-times of jobs K and K'. If

$$t = \sum_{i=1}^{k-1} p_{[i]}$$

(same under S and S') then these flow-times, raised to power α, are the following.

Under S: $F_K^\alpha = F_{[k]}^\alpha = (t + p_K)^\alpha,$

$F_{K'}^\alpha = F_{[k+1]}^\alpha = (t + p_K + p_{K'})^\alpha.$

Under S': $F_K^\alpha = F_{[k+1]}^\alpha = (t + p_{K'} + p_K)^\alpha,$

$F_{K'}^\alpha = F_{[k]}^\alpha = (t + p_{K'})^\alpha.$

The term $F_{[k+1]}$ is the same under both schedules, so that S and S' differ only in $F_{[k]}$. For any positive α,

$$(t + p_K)^\alpha > (t + p_{K'})^\alpha,$$

since position k was chosen so that $p_K > p_{K'}$.

This means that S' is better than S. One could continue to improve the schedule in this way so long as such interchanges were possible, that is, until an SPT sequence was achieved. In summary:

1) If a sequence is not SPT there exists a position k such that $p_{[k]} > p_{[k+1]}$.

2) The schedule will be improved by interchanging the jobs in positions k and $k + 1$.

3) Any of the $n!$ possible sequences can be transformed to an SPT sequence by a series of such interchanges, with an improvement being obtained at each step.

The analysis of schedules by considering the interchange of all possible adjacent pairs of jobs is a useful approach, but it is not sufficient under all circumstances. It quickly breaks down in the more complex situations of later chapters (see Section 5–4, for example) and there are even $n/1$ problems for which it is inadequate. For example, consider the following $3/1//\overline{T}$ problem:

Job	Processing-time	Due-date
a	1	4
b	2	2
c	3	3

The measure of performance is

$$\bar{T} = \tfrac{1}{3}\big(\max(C_a - d_a, 0) + \max(C_b - d_b, 0) + \max(C_c - d_c, 0)\big).$$

Note that this is a regular measure. The six possible sequences for this problem are shown in Fig. 3–4, with the arrows indicating the transformations possible by the interchange of adjacent-job pairs. Each sequence can reach two others by this means. The sum of tardinesses for each schedule is given below the description of sequence. If one were to start with schedule S(*cab*) there would be no adjacent-job interchange that would improve the measure, and one would presumably not find the optimal schedule which is S(*bac*). The difficulty lies in the transitivity of the ordering under this measure of performance. In comparing S(*acb*) with S(*cab*) one would conclude that *c* should precede *a*. In comparing S(*cab*) with S(*cba*) one would conclude that *a* should precede *b*. However, this does not establish that *c* should precede *b*. Moreover the pairwise choices are dependent on the position of the third job. For example, with *b* in third position, *c* should precede *a* (S(*acb*) vs. S(*cab*)); with *b* in either first or second position, *a* should precede *c* (S(*bac*) vs. S(*bca*)); (S(*abc*) vs. S(*cba*)).

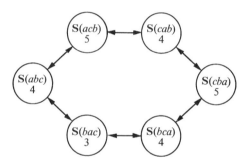

Fig. 3–4. Adjacent-job interchanges in a 3/1 problem.

One can show, either by similar proofs or by simply appealing to the definitions and interrelationships of Section 2–2, that SPT sequencing also minimizes the mean completion-time, mean waiting-time and mean lateness. There are trivial properties of a schedule that are also obviously minimized, such as minimum flow-time, minimum completion-time, maximum waiting-time, and minimum nonzero waiting-time. More important, the mean number of jobs in the shop is also minimized. This follows from the relationship in Section 2–4:

$$\frac{\bar{N}(0, F_{\max})}{n} = \frac{\bar{F}}{F_{\max}}.$$

Since *n* and $F_{\max}$ are both constants in this problem, $\bar{N}(0, F_{\max})$ is minimized along with $\bar{F}$.

It is, in general, true that *longest-processing-time sequencing* (LPT) maximizes whatever SPT minimizes, so that if for some reason we wished to maximize one of these properties we would know the method.

There are, of course, many reasonable and interesting measures of performance in an $n/1$ problem that are not optimized by SPT sequencing. The most important are functions of due-date. Consider the following 2/1 problem:

Job	p	d
a	1	3
b	2	2

There are only two sequences to be considered, ab and ba. SPT selects sequence ab, but ba has smaller maximum lateness and maximum tardiness, smaller mean tardiness, and fewer jobs tardy. In the special case in which the due-dates are so tight that all the jobs are tardy, SPT does minimize the mean tardiness, simply because this is identical to mean lateness under these circumstances. In the special case in which

$$d_1 = d_2 = \cdots = d_n,$$

SPT also minimizes the number of jobs tardy,

The 2/1 example above also demonstrates that SPT sequencing does not minimize the variance of job flow-time,

$$\text{var } F = \frac{1}{n} \sum_{i=1}^{n} (F_i - \bar{F})^2 = \frac{1}{n} \sum_{i=1}^{n} F_i^2 - \bar{F}^2.$$

SPT sequencing minimizes both terms on the right, but not the difference between these terms.

3–3 SEQUENCING ACCORDING TO DUE-DATE

Of the various measures of performance that have been considered in research on sequencing, certainly the measure that arouses the most interest in those who actually face practical problems of sequencing is the satisfaction of preassigned job due-dates. Equipment utilization, work-in-process inventory, and job flow-time are all interesting and more or less important, but the ability to fulfill delivery promises on time undoubtedly dominates these other considerations.

Since the job due-dates are involved in the measure of performance, it would seem only natural that the information they represent be employed in the decision rule that generates a schedule. It is therefore somewhat surprising on first encounter to discover that mean lateness is minimized by SPT sequencing, which does not consider due-date information in any way. The obvious use of due-dates is simply to sequence the jobs so that

$$d_{[1]} \leqq d_{[2]} \leqq \cdots \leqq d_{[n]}.$$

What is accomplished by this ordering is not altogether obvious, but is shown by the following theorem:

Theorem 3–3. (Jackson 1955, [90]). When one schedules an $n/1//L_{\max}$ problem, the maximum job lateness and maximum job tardiness are minimized by sequencing the jobs in order of nondecreasing due-dates.

Proof. Consider a schedule **S** that is not identical to one produced by the due-date rule. There is some position k in **S** such that

$$d_{[k]} > d_{[k+1]}.$$

Let K and K' be the numbers of the jobs in the kth and $(k + 1)$th positions in **S**, respectively.

Now consider a schedule **S**′ that differs from **S** only in that job K' is in kth position and job K in $(k + 1)$th position. The sequence is the same for the first $k - 1$ jobs and the last $(n - k - 1)$ jobs under either **S** or **S**′ and it is obvious that each of these $n - 2$ jobs begins and is completed at exactly the same moment; therefore, the lateness is exactly the same for these jobs under either schedule. Let L be the maximum of these $n - 2$ latenesses, obviously the same for both **S** and **S**′. The two schedules differ only in the latenesses of jobs K and K'. Our task is to show that

$$\max(L, L_K, L_{K'})$$

under **S**′ is less than or equal to the corresponding expression under **S**. The values will be the same if L is the dominating term in each expression; it is to be established that the value under **S**′ will be smaller if L_K or $L_{K'}$ dominates either expression. If $t = \sum_{i=1}^{k-1} p_{[i]}$ (same under **S** and **S**′), then the relevant latenesses are the following.

Under **S**: $L_K(\mathbf{S}) = L_{[k]} = t + p_K - d_K,$

                $L_{K'}(\mathbf{S}) = L_{[k+1]} = t + p_K + p_{K'} - d_{K'}.$

Under **S**′: $L_K(\mathbf{S}') = L_{[k+1]} = t + p_{K'} + p_K - d_K,$

                $L_{K'}(\mathbf{S}') = L_{[k]} = t + p_{K'} - d_{K'}.$

The lateness $L_{K'}(\mathbf{S})$ dominates the relationship:

$$L_{K'}(\mathbf{S}) > L_K(\mathbf{S}') \quad \text{since } d_K > d_{K'},$$
$$L_{K'}(\mathbf{S}) > L_{K'}(\mathbf{S}') \quad \text{since } p_K > 0,$$
$$L_{K'}(\mathbf{S}) > \max(L_K(\mathbf{S}'), L_{K'}(\mathbf{S}')),$$
$$\max(L, L_K(\mathbf{S}), L_{K'}(\mathbf{S})) \geqq \max(L, L_K(\mathbf{S}'), L_{K'}(\mathbf{S}')).$$

Thus for any schedule **S** there is a potential improvement by interchanging adjacent pairs of jobs into due-date order.

The minimization of maximum tardiness follows directly from the lateness result:

$$T_{\max}(\mathbf{S}) = \max(0, L_{\max}(\mathbf{S})) \geqq \max(0, L_{\max}(\mathbf{S}')) = T_{\max}(\mathbf{S}').$$

A second method of using due-date information in sequencing is to consider *slack-time*. The slack-time of job i at moment t is the amount of time remaining before this job must be started if it is to be completed on time: $d_i - p_i - t$. The job with minimum slack-time presumably offers the greatest risk of being late, and should therefore be given precedence in the schedule. Since at any decision point the term t is common to each job, we must sequence so that

$$d_{[1]} - p_{[1]} \leqq d_{[2]} - p_{[2]} \leqq \cdots \leqq d_{[n]} - p_{[n]}.$$

Sequencing of this type has an intuitive appeal as a refinement of simple due-date sequencing and in more complex situations (Section 11–3) it appears to give superior performance. Therefore it usually comes as something of a surprise to discover what slack-time sequencing actually accomplishes in an $n/1$ problem:

Theorem 3–4. When one schedules an $n/1$ problem, the minimum job lateness and minimum job tardiness are maximized by sequencing the jobs in order of nondecreasing slack-time.

Proof. The proof exactly follows that of Theorem 3–3:

$$L_K(S) < L_{K'}(S') \quad \text{since} \quad (d_K - p_K) > (d_{K'} - p_{K'}),$$

$$L_K(S) < L_K(S') \quad \text{since} \quad p_{K'} > 0,$$

$$L_K(S) < \min(L_{K'}(S'), L_K(S')).$$

Now, letting L be the minimum of the latenesses of the $n - 2$ jobs (excluding K and K'), we obtain

$$\min(L, L_K(S), L_{K'}(S)) \leq \min(L, L_{K'}(S'), L_K(S')).$$

In aggregate, these results provide substantial control over the distribution of latenesses. One can minimize the mean (SPT), minimize the maximum value (due-date) or maximize the minimum value (slack-time). With respect to tardiness one can minimize the maximum value or maximize the minimum value. However, none of these rules in general minimizes the mean tardiness, as shown by Table 3–1.

Table 3–1

		Job						
		1	2	3	4	$\overline{T}$	T_{max}	T_{min}
Due-date	d_i	1	2	4	6			
processing-time	p_i	2	4	3	1			
SPT sequence	C_i	3	10	6	1			
4, 1, 3, 2	T_i	2	8	2	0	3.0	8	0
Due-date sequence	C_i	2	6	9	10			
1, 2, 3, 4	T_i	1	4	5	4	3.5	5	1
Slack-time sequence	C_i	6	4	9	10			
2, 1, 3, 4	T_i	5	2	5	4	4.0	5	2
X sequence	C_i	2	10	5	6			
1, 3, 4, 2	T_i	1	8	1	0	2.5	8	0

If one is fortunate enough to encounter a problem in which due-date sequencing can satisfy all the jobs' due-dates (the maximum tardiness equals 0), then it may be that there are other sequences that also satisfy due-dates but that are in some other sense preferable to the due-date sequence. If this secondary criterion is the minimization of mean flow-time, then a sequencing procedure can be based on the following:

Theorem 3–5. (Smith 1956, [187].) When one schedules an $n/1$ problem, if there exists a sequence such that the maximum job tardiness is zero, then there is an ordering of the jobs with job K in the last position which minimizes mean flow-time (subject to the condition that the maximum tardiness remain zero), if and only if

(a) $d_K \geq \sum_{i=1}^{n} p_i,$

(b) $p_K \geq p_i$ for all i with $d_i \geq \sum_{j=1}^{n} p_j.$

Theorem 3–5 says that a job can be in last position only if this does not cause it to have nonzero tardiness and if the job has the greatest processing-time of the set of all jobs that could be in the last position without being tardy.

Proof. Consider a schedule that has the following properties:

1) the maximum tardiness is zero;

2) job K, in position n, satisfies conditions (a) and (b) of the theorem.

Now show that any modification of this schedule that displaces K from position n either makes a job late, and/or increases the mean flow-time. Observe that no other job J, in an earlier position, say $n - v$, satisfies both conditions (a) and (b), unless $p_J = p_K$, in which case interchange of J and K is immaterial. If J does not satisfy (a), then the interchange of J and K will make J have nonzero tardiness, which is prohibited. If J does not satisfy (b), since $p_J < p_K$, then the interchange of J and K will increase the sum of flow-times by $v(p_K - p_J)$.

The procedure is quite clearly SPT sequencing subject to due-date constraints. The complete schedule that minimizes the mean flow-time, subject to the satisfaction of due-dates, is developed recursively, by selecting a job for position n, then $n - 1$, $n - 2$, etc. The job with the greatest processing-time from the set of jobs with due-dates not smaller than the completion-time of the job in position n (known in advance to be the sum of the processing-times of all jobs) is selected for position n. Once a person makes this selection, he knows the completion-time for the job in position $n - 1$:

$$C_{[n-1]} = \sum_{i=1}^{n} p_i - p_{[n]}.$$

Among the remaining jobs, the job with the largest processing-time from the set of jobs with due-dates not smaller than $C_{[n-1]}$ is selected for position $n - 1$.

The procedure is much easier to use than it is to describe. For example, consider the following 6/1 problem:

Job

	1	2	3	4	5	6
d_i	24	21	8	5	10	23
p_i	4	7	1	3	2	5

Due-date sequencing of these jobs results in a maximum tardiness of zero, so that it is an appropriate problem for improvement by the secondary measure. The sum of the processing-times is 22. Only jobs 1 and 6 have due-dates not less than 22, so one of these must go in position 6. Job 6 is selected since $p_6 > p_1$. Excluding job 6, the sum of the processing-times for the 5/1 problem that remains is 17. Jobs 1 and 2 are candidates for last position; job 2 is selected. Continuing in this manner, one obtains the sequence 3, 4, 5, 1, 2, 6.

3–4 RANDOM SEQUENCING

A useful standard in $n/1$ sequencing is provided by *random scheduling of the jobs* (abbreviated RANDOM). As the term will be used here the randomness pertains to the processing-times. The sequencing may be deliberate and systematic, using any attribute of the job that is independent of the processing-time, and the process is still characterized as random sequencing. For example, assuming that the due-dates in a problem are independent of processing-times, due-date sequencing is, for present purposes, effectively random sequencing, while slack-time sequencing is not.

Although previously the situation has been entirely deterministic, so that repeated application of a given rule to a given problem would produce identically the same schedule (assuming that a complete rule contains sufficient provision for breaking ties), we must now expect different schedules from repeated application of the rule. The flow-times of individual jobs and the mean flow-time for all n jobs are now random variables. A distribution of values of mean flow-time is associated with random sequencing; in particular, we are concerned with the expected value of the mean flow-time.

Theorem 3–6. In an $n/1$ problem, the expected value of the mean flow-time for procedures that determine sequence independent of the processing-times of the jobs is

$$E(\overline{F}) = \frac{n+1}{2}\,\overline{p},$$

if a specific set of n jobs is known and $\overline{p}$ is the average of their processing-times. It is

$$E(\overline{F}) = \frac{n+1}{2}\,\hat{p},$$

if the processing-times are considered independent, identically distributed random variables, not yet observed, and $\hat{p}$ is their expected value.

Proof. Mean flow-time in an $n/1$ problem is given by

$$\overline{F} = \frac{1}{n}\sum_{i=1}^{n}(n-i+1)p_{[i]}.$$

If we regard the processing-times as random variables, then

$$E(\overline{F}) = \frac{1}{n}\sum_{i=1}^{n}(n-i+1)E(p_{[i]}).$$

Since the individual processing-times are identically distributed and jobs are assigned to a position in sequence in a manner that does not depend on the value of processing-time,

$$E(p_{[1]}) = E(p_{[2]}) = \cdots = E(p_{[n]}) = \hat{p},$$

$$E(\overline{F}) = \frac{\hat{p}}{n} \sum_{i=1}^{n} (n - i + 1) = \frac{n+1}{2} \hat{p}.$$

If a particular set of n jobs has been selected and the processing-times are known, then the situation can be regarded somewhat differently. The mean flow-time is a sum of n terms, each term being the product of one of the given processing-times and a coefficient which is determined by sequence. Thus

$$\overline{F} = \frac{1}{n} (Y_1 p_1 + Y_2 p_2 + \cdots + Y_n p_n).$$

The coefficient assigned to a particular job depends on the random process of sequencing and may be considered a random variable, so that

$$E(\overline{F}) = \frac{1}{n} (p_1 E(Y_1) + p_2 E(Y_2) + \cdots + p_n E(Y_n)).$$

Consider job 1. If it is assigned to position 1, then $Y_1 = n$; if assigned to position 2, then $Y_1 = n - 1$; ... ; if assigned to position n, then $Y_1 = 1$. One can interpret random sequencing as meaning that job 1 (or any other job) is equally likely to be assigned to any one of the n positions, so that Y_1 will take on one of the values $1, 2, \ldots, n$, each value having equal probability:

$$E(Y_1) = \frac{1}{n} 1 + \frac{1}{n} 2 + \frac{1}{n} 3 + \cdots + \frac{1}{n} n = \frac{n+1}{2}.$$

The same argument applies to each of the jobs and each of the Y_i, so that

$$E(\overline{F}) = E(Y_i) \frac{1}{n} (p_1 + p_2 + \cdots + p_n) = \frac{n+1}{2} \overline{p}.$$

Trivial variations of this argument will show that for random sequencing:

a) the expected value of mean waiting-time,

$$E(\overline{W}) = \frac{n-1}{2} \overline{p},$$

b) the expected value of mean lateness,

$$E(\overline{L}) = \frac{n+1}{2} \overline{p} - \overline{a},$$

c) the expected value of the mean number of jobs in the shop,

$$E(\overline{N}(0, F_{\max})) = \frac{n+1}{2}.$$

3–5 PROPERTIES OF ANTITHETICAL RULES

We call a pair of scheduling procedures *antithetical* if they produce opposite sequences in an $n/1$ problem; that is, $\mathbf{R}$ and $\mathbf{R'}$ are antithetical if the job that is assigned to position i by $\mathbf{R}$ is assigned to $(n - i + 1)$ by $\mathbf{R'}$. The obvious and most important example of a pair of antithetical rules is the shortest-processing-time rule and the longest-processing-time rule. What makes antithetical rules worth considering is the interesting property that for the primary measures of performance they are symmetric about the expected value for random sequencing:

Theorem 3–7. In an $n/1//\bar{F}$ problem, if $\mathbf{R}$ and $\mathbf{R'}$ are antithetical rules, then the mean flow-time under $\mathbf{R}$ is related to the mean flow-time under $\mathbf{R'}$ in the following way:

$$\bar{F}_{\mathbf{R}} + \bar{F}_{\mathbf{R'}} = (n + 1)\bar{p}.$$

Proof. If $[i]$ represents position under $\mathbf{R}$ and $[i]'$ position under $\mathbf{R'}$, then

$$\bar{F}_{\mathbf{R}} = \frac{1}{n}\sum_{i=1}^{n}(n - i + 1)p_{[i]} = \frac{1}{n}\sum_{i=1}^{n}(n + 1)p_{[i]} - \frac{1}{n}\sum_{i=1}^{n}ip_{[i]},$$

$$\bar{F}_{\mathbf{R'}} = \frac{1}{n}\sum_{i=1}^{n}(n - i + 1)p_{[i]'} = \frac{1}{n}\sum_{i=1}^{n}(n - i + 1)p_{[n-i+1]} = \frac{1}{n}\sum_{i=1}^{n}ip_{[i]},$$

$$\bar{F}_{\mathbf{R}} + \bar{F}_{\mathbf{R'}} = \frac{1}{n}\sum_{i=1}^{n}(n + 1)p_{[i]} = (n + 1)\bar{p}.$$

Similar argument will establish the same symmetry for mean completion-time, mean waiting-time, mean lateness, and the mean number of jobs in the shop.

The result is illustrated in Fig. 3–5. For any pair of antithetical rules, $\mathbf{R}$ and $\mathbf{R'}$, the distances x and x' are equal. For the particular pair, shortest-processing-time/longest-processing-time, x and x' take on the maximum possible value for the given problem (Theorem 3–2).

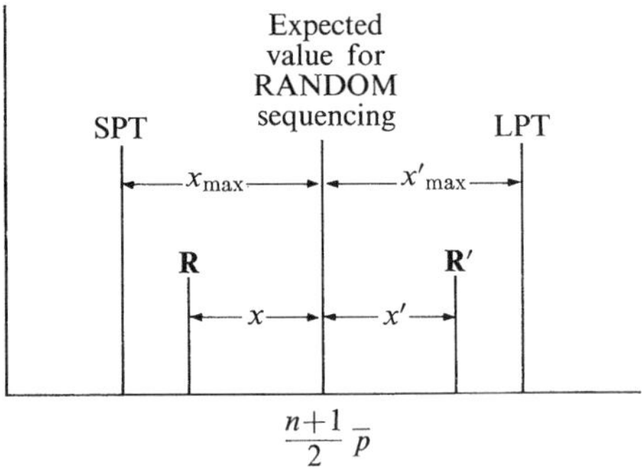

Fig. 3–5. Mean flow-time under antithetical rules in an $n/1$ problem.

The next question is of course the magnitude of x and $x_{\max}$. These values are peculiar to a particular problem and depend on the differences among the n values of processing-times. If the processing-times of the n jobs are all equal, then the mean flow-time is identically the same for each of the $n!$ possible sequences. If there is at least one pair of jobs with different processing-times, then the selection of sequence can affect the mean flow-time. A relationship between mean flow-time under SPT and the distribution of processing-times is given by Theorem 3–8.

Theorem 3–8. In an $n/1//\overline{F}$ problem, if the processing-times are assumed to be n independent, identically distributed random variables, with distribution function $G(p)$ and expected value $\hat{p}$, then the expected mean flow-times for SPT and LPT sequencing are given by

$$E(\overline{F}_{\text{SPT}}) = n\hat{p} - (n - 1)Q, \qquad E(\overline{F}_{\text{LPT}}) = \hat{p} + (n - 1)Q,$$

where $Q = \int_0^\infty pG(p)\,dG(p)$.

Proof. It is simpler to consider LPT sequencing first, and then to use Theorem 3–7 to obtain the SPT result.

We shall depart from our standard notation and use a capital P for processing-time, to denote its character as a random variable. Further, we shall assume that the n values of P that constitute a particular problem have been renumbered so that $P_1 \leq P_2 \leq \cdots \leq P_n$. The LPT sequence will take the jobs in reversed order:

$$\overline{F}_{\text{LPT}} = \frac{1}{n} \sum_{i=1}^n (n - i + 1)P_{[i]} = \frac{1}{n} \sum_{i=1}^n iP_i,$$

$$E(\overline{F}_{\text{LPT}}) = \frac{1}{n} \sum_{i=1}^n iE(P_i).$$

From the theory of order statistics,* the differential of the distribution function of the ith ordered processing-time is given by

$$dG_i(p) = \frac{n!}{(i - 1)!(n - i)!} (G(p))^{i-1}(1 - G(p))^{n-i}\,dG(p).$$

The expected value of the ith job processing-time is

$$E(P_i) = \int_0^\infty p\,dG_i(p),$$

$$E(\overline{F}_{\text{LPT}}) = \frac{1}{n} \sum_{i=1}^n i \int_0^\infty p \frac{n!}{(i - 1)!(n - 1)!} (G(p))^{i-1}(1 - G(p))^{n-i}\,dG(p).$$

Reversing the order of integration and summation, we have

$$E(\overline{F}_{\text{LPT}}) = \frac{1}{n} \int_0^\infty p \sum_{i=1}^n i \frac{n!}{(i - 1)!(n - 1)!} (G(p))^{i-1}(1 - G(p))^{n-i}\,dG(p).$$

When we let $j = i - 1$ and factor out n, the inner sum is

$$n\sum_{j=0}^{n-1} (j + 1) \frac{(n - 1)!}{j!(n - 1 - j)!} (G(p))^j(1 - G(p))^{n-1-j} = n(n - 1)G(p) + n,$$

$$E(\overline{F}_{\text{LPT}}) = \int_0^\infty p\big((n - 1)G(p) + 1\big)\,dG(p) = \hat{p} + (n - 1) \int_0^\infty pG(p)\,dG(p)$$

$$= \hat{p} + (n - 1)Q.$$

---

* H. L. Johnson, and S. C. Leone, *Statistics and Experimental Design* (New York: Wiley, 1964), II, Chapter 6.

Theorem 3-7 gives $\overline{F}_{SPT} + \overline{F}_{LPT} = (n + 1)\hat{p}$. Taking expected values does not alter this relationship:

$$E(\overline{F}_{SPT}) + E(\overline{F}_{LPT}) = (n + 1)\hat{p},$$

$$E(\overline{F}_{SPT}) = (n + 1)\hat{p} - \hat{p} - (n - 1)Q = n\hat{p} - (n - 1)Q.$$

In general, Q cannot be conveniently expressed in terms of the moments of the distribution of processing-times, $G(p)$. However, values can be obtained for specific distributions. For example, the Erlang distribution is often used as a model for a source of processing-times. The density function is of the form

$$g(p) = \frac{(k/\hat{p})^k p^{k-1}}{(k - 1)!} e^{-kp/\hat{p}}, \qquad p > 0,$$

where $\hat{p}$ is the mean and $\hat{p}^2/k$ is the variance of the distribution. With $k = 1$, the Erlang distribution corresponds to the exponential distribution. As k increases the variance decreases, retaining a right-skewness. With infinite k, the Erlang distribution is a point distribution with value $\hat{p}$ for each variable. Values of Q for different Erlang distributions with $p = 1.00$ are given in Table 3-2. Expected mean flow-times for an 11/1 problem with Erlang-distributed processing-times are given in Table 3-2 and Fig. 3-6.

Table 3-2

Expected Mean Flow-Times with Erlang Processing-Times
$\hat{p} = 1, n = 11$

k	Variance, $\hat{p}^2/k$	Q	$E(\overline{F}_{SPT})$	$E(\overline{F}_{LPT})$	$E(\overline{F}_{RANDOM})$
1	1.000	0.750	3.50	8.50	6.00
2	0.500	0.688	4.13	7.88	
3	0.333	0.656	4.44	7.56	
4	0.250	0.637	4.63	7.37	
5	0.200	0.623	4.77	7.23	
6	0.167	0.613	4.87	7.13	
7	0.143	0.605	4.95	7.05	
8	0.125	0.598	5.02	6.98	
9	0.111	0.593	5.07	6.93	
10	0.100	0.588	5.12	6.88	
15	0.067	0.572	5.28	6.72	
20	0.050	0.563	5.37	6.63	
30	0.033	0.551	5.49	6.51	
40	0.025	0.544	5.56	6.44	
50	0.020	0.540	5.60	6.40	
75	0.013	0.532	5.67	6.33	
100	0.010	0.528	5.72	6.28	
∞	0	0.500	6.00	6.00	

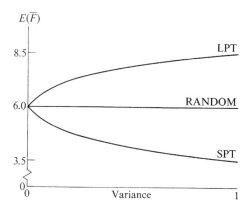

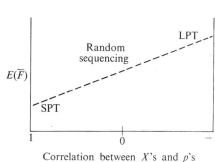

Fig. 3–6. Expected mean flow-time for an 11/1 problem with Erlang processing-times; $\hat{p} = 1.00$.

Fig. 3–7. Conceptual relationship between quality of estimates of processing-times and expected mean flow-time for SPT sequencing.

3–6 SPT SEQUENCING WITH INCOMPLETE INFORMATION

So far we have been assuming that the job processing-times are known before processing takes place and that this information is available to the scheduling process. However, in a majority of practical scheduling problems the processing-times are not known precisely until after the event, and scheduling decisions are at best based on *a priori* estimates of the processing-times.

Suppose that the jobs of an $n/1$ problem all have different processing-times and are renumbered so that

$$p_1 < p_2 < \cdots < p_n.$$

Now suppose that $X_1, X_2, \ldots, X_n$ are a set of *a priori* estimates of the processing-times on which SPT sequencing is to be based. That is, the jobs will be sequenced so that

$$X_{[1]} \leqq X_{[2]} \leqq \cdots \leqq X_{[n]}.$$

At least in concept the situation is as shown in Fig. 3–7. If the X's are perfectly correlated with the p's, then the optimum SPT sequence is obtained. If the X's are independent of the p's, the result is random sequencing; if the X's are inversely correlated, the result is LPT sequencing. In general, it should not be unreasonable to expect estimating procedures to fall somewhere between perfect correlation and independence, since there is usually an opportunity to obtain the p's after the fact and refine the process. Unfortunately we have not yet been able to obtain the actual relationship suggested in Fig. 3–7 for any interesting model.

Note that the necessary condition to obtain the optimal sequence is that $[i] = i$ and not that $X_i = p_i$. That is, the estimates do not have to be perfect; they need only be sufficient to obtain the appropriate relative ordering. A consistent bias in the estimates has no effect whatever on the quality of the schedule, and errors of estimate that are small relative to the differences between the p's are of no consequence. Estimating processes with reasonable degrees of precision will produce X's that will

Table 3–3

Position in sequence	Coefficient in sum of flow-times	Probability of occurrence
1	n	$\displaystyle\prod_{\substack{i=1\\i\neq I}}^{n} (1 - G_i(p))$
2	$n - 1$	$\displaystyle\sum_{\substack{j=1\\j\neq I}}^{n}\left(G_j(p)\prod_{\substack{i=1\\i\neq I,j}}^{n}(1 - G_i(p))\right)$
3	$n - 2$	$\displaystyle\sum_{\substack{k=1\\k\neq I}}^{n}\sum_{\substack{j>k\\j\neq I}}\left(G_k(p)G_j(p)\prod_{\substack{i=1\\i\neq I,j,k}}^{n}(1 - G_i(p))\right)$
$\vdots$		
n	1	$\displaystyle\prod_{\substack{k=1\\k\neq I}}^{n} G_k(p)$

correctly order the jobs when the differences between the p's are large and ordering is important, and will yield X's that result in misordering only when the differences between the p's are small and order is less important.

One model that can be used to study this problem is obtained by assuming that each job i draws its processing-time from a separate distribution with distribution function $G_i(p)$ and expected value $\hat{p}_i$. Suppose that the $\hat{p}_i$ are known and that the jobs are so numbered that

$$\hat{p}_1 \leqq \hat{p}_2 \leqq \cdots \leqq \hat{p}_n.$$

Suppose that the actual processing-times, P_i, are not known and that the $\hat{p}_i$ are used to order the jobs. (This is not an unrealistic situation: lacking the actual processing-time of a specific job, one uses the historical average processing-time for similar jobs.)

Let **N** be the sequence with the jobs in numbered order (SPT using $\hat{p}_i$),

$$\overline{F}_N = \frac{1}{n} \sum_{i=1}^{n} (n - i + 1)P_i.$$

Let **S** be the optimal SPT sequence with

$$P_{[1]} \leqq P_{[2]} \leqq \cdots \leqq P_{[n]}, \qquad \overline{F}_S = \frac{1}{n} \sum_{i=1}^{n} (n - i + 1)P_{[i]}.$$

When we seek an expression for $E(\overline{F}_N - \overline{F}_S)$, $E(\overline{F}_N)$ is obvious:

$$E(\overline{F}_N) = \frac{1}{n} \sum_{i=1}^{n} (n - i + 1)\hat{p}_i,$$

but $E(\overline{F}_S)$ is much more difficult. Since each processing-time is a random variable, under **S** it is possible for any job to appear in any position, although some configu-

Table 3–4

Position in sequence	Coefficient in sum of flow-times	Probability of occurrence
1	4	$(1 - G_1(p))(1 - G_2(p))(1 - G_4(p))$
2	3	$G_1(p)(1 - G_2(p))(1 - G_4(p))$ $+ G_2(p)(1 - G_1(p))(1 - G_4(p))$ $+ G_4(p)(1 - G_1(p))(1 - G_2(p))$
3	2	$G_1(p)G_2(p)(1 - G_4(p))$ $+ G_1(p)G_4(p)(1 - G_2(p))$ $+ G_2(p)G_4(p)(1 - G_1(p))$
4	1	$G_1(p)G_2(p)G_4(p)$

rations would be highly unlikely. Consider a particular job I with processing-time $P_I = p$. In position i, job I has coefficient $n - i + 1$ in the sum of flow-times. The probability associated with each possible value of the coefficient for job I is given in Table 3–3. For example, with $n = 4$ and $I = 3$, these expressions become the values shown in Table 3–4. The expected value of the contribution of job I to the expected value of the mean flow-time under S is obtained by taking the expected value of the probability distribution in Table 3–4 for a given value of p, and then taking the expected value with respect to the distribution $G_I(p)$. The value of this contribution is given by the following:

Theorem 3–9. In an $n/1//\bar{F}$ problem, if each processing-time, P_i, is a random variable with distribution $G_i(p)$, and the jobs are sequenced in order of increasing processing-time, then the expected value of the contribution of job I to the expected value of the mean flow-time is:

$$Q_I = \hat{p}_I - \frac{1}{n} \int_0^\infty p \left(\sum_{\substack{i=1 \\ i \neq I}}^n G_i(p) \right) dG_I(p),$$

where $\hat{p}_I = E(P_I)$.

Proof (by induction). For $n = 1$, the expression given is $Q_I = \hat{p}_I$. Since job I is the only job, the expected mean flow-time is the expected processing-time of job I, so the expression holds for $n = 1$.

Assume that the expression holds for n jobs, one of which is job I, and consider the case in which another job is added to the set of jobs. For a given value of p (the processing-time of job I) let $h_{i,n}(p)$ be the probability that job I occupied position i in the ordered sequence when there were n jobs in total, and $h_{i,n+1}(p)$ be the probability that this job will be in position i when there are $n + 1$ jobs in total. The probability that job I would occupy position i with $n + 1$ jobs can be found by multiplying the probability that job I was in this position with n jobs and the probability that the additional job has a processing-time larger than p, and adding to this figure the product of the probability that job I was in position $i - 1$ with n jobs and the probability that the additional job has a processing-time less than p. Thus

$$h_{i,n+1}(p) = h_{i,n}(p)\big(1 - G_{n+1}(p)\big) + h_{i-1,n}(p)G_{n+1}(p) \qquad \text{for } i = 2, 3, \ldots, n.$$

For the special cases at the beginning and end of the sequence,

$$h_{1,n+1}(p) = h_{1,n}(p)\big(1 - G_{n+1}(p)\big),$$
$$h_{n+1,n+1}(p) = h_{n,n}(p)G_{n+1}(p).$$

For a given value of p (the processing-time of job I), the expected value of the coefficient for job I in the expression for mean flow-time for $n + 1$ jobs is

$$\sum_{i=1}^{n+1} (n + 1 - i + 1)h_{i,n+1}(p).$$

Substituting for $h_{i,n+1}(p)$ in terms of $h_{i,n}(p)$, we have

$$\sum_{i=1}^{n+1} (n + 1 - i + 1)h_{i,n+1}(p) = \sum_{i=1}^{n} (n - i + 1)h_{i,n}(p) + \big(1 - G_{n+1}(p)\big).$$

Let $Q_I^{(n)}$ and $Q_I^{(n+1)}$ represent the expected values of the contributions of job I to the mean flow-time with n jobs and $n + 1$ jobs, respectively:

$$Q_I^{(n+1)} = \frac{1}{n + 1} \int_0^{\infty} p \sum_{i=1}^{n+1} (n + 1 - i + 1)h_{i,n+1}(p)\, dG_I(p)$$

$$= \frac{1}{n + 1} Q_I^{(n)} + \frac{1}{n + 1} \int_0^{\infty} p\big(1 - G_{n+1}(p)\big)\, dG_I(p)$$

$$= \frac{n}{n + 1} Q_I^{(n)} + \frac{\hat{p}_I}{n + 1} - \frac{1}{n + 1} \int_0^{\infty} pG_{n+1}(p)\, dG_I(p).$$

Recall the induction assumption that the expression of the theorem holds for $Q_I^{(n)}$:

$$Q_I^{(n+1)} = \hat{p}_I - \frac{1}{n + 1} \int_0^{\infty} p \left(\sum_{\substack{i=1 \\ i \neq I}}^{n+1} G_i(p) \right) dG_I(p).$$

The expected mean flow-time under $\mathbf{S}$ is

$$E(\overline{F}_{\mathbf{S}}) = \sum_{i=1}^{n} Q_i = \sum_{i=1}^{n} \hat{p}_i - \frac{1}{n} \sum_{i=1}^{n} p \sum_{\substack{j=1 \\ j \neq i}}^{n} G_j(p)\, dG_i(p)$$

$$= \sum_{i=1}^{n} \hat{p}_i + \frac{1}{n} \sum_{i=1}^{n} \int_0^{\infty} pG_i(p)\, dG_i(p) - \frac{1}{n} \int_0^{\infty} p \left(\sum_{i=1}^{n} G_i(p) \right) d\left(\sum_{i=1}^{n} G_i(p) \right).$$

The expected value of the effect of imperfect knowledge of the processing-times on the mean flow-time—that is, the expected increase in mean flow-time if one sequences by the expected values of the processing-times rather than the actual processing-times—is

$$E(\overline{F}_N) - E(\overline{F}_S) = \frac{1}{n}\left(\sum_{i=1}^{n}(1 - i)p_i - \sum_{i=1}^{n}\int_0^\infty pG_i(p)\,dG_i(p)\right.$$
$$\left. + \int_0^\infty p\left(\sum_{i=1}^{n}G_i(p)\right)d\left(\sum_{i=1}^{n}G_i(p)\right)\right).$$

This difference depends entirely on the distributions from which the processing-times are obtained, and although the difference does not reduce to a simple expression, it can be computed from the distribution functions of a particular problem.

3–7 SEQUENCING AGAINST WEIGHTED MEASURES OF PERFORMANCE

In most cases, jobs are not all equally important, a fact that is reflected in the measures by which schedules are evaluated. We shall assume that a value u_i is given for each job to describe relative importance, and use these values as coefficients in performance measures. Mean weighted flow-time is given by

$$\overline{F}_u = \frac{1}{n}\sum_{i=1}^{n}u_iF_i,$$

but one could just as well use

$$\overline{F}_{u'} = \frac{\sum_{i=1}^{n}u_iF_i}{\sum_{i=1}^{n}u_i},$$

since the divisor in either case is a constant independent of sequence; however, we shall use $\overline{F}_u$ for convenience.

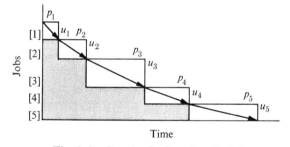

Time

Fig. 3–8. Graph of $n/1$ ratio schedule.

Figure 3–8 shows a schedule similar to that of Fig. 3–3, except that the rectangle for each job has height proportional to u_i. The diagonal vector for each job has slope $-u_i/p_i$. The total weighted flow-time is the sum of the area of the blocks and the shaded region under the blocks. The area of the job-blocks is again a constant, and the shaded area is minimized by arranging the jobs so that the job-vectors form a convex curve, as shown. This is accomplished by ordering the jobs so that

$$\frac{p_{[1]}}{u_{[1]}} \leq \frac{p_{[2]}}{u_{[2]}} \leq \cdots \leq \frac{p_{[n]}}{u_{[n]}}.$$

The optimality of this generalization of SPT sequencing is shown by the following theorem, a generalization of Theorem 3–2.

Theorem 3–10. (Smith 1956, [187]). In an $n/1//\overline{F}_u$ problem, the total weighted flow-time $\sum_{i=1}^{n} u_i F_i$ is minimized by sequencing the jobs so that

$$\frac{p_{[1]}}{u_{[1]}} \leqq \frac{p_{[2]}}{u_{[2]}} \leqq \cdots \leqq \frac{p_{[n]}}{u_{[n]}},$$

and it is maximized by the antithetical procedure.

Proof. Consider a schedule **S** that is not ordered by increasing p/u ratio. There is some position k such that

$$\frac{p_{[k]}}{u_{[k]}} > \frac{p_{[k+1]}}{u_{[k+1]}}.$$

Let K and K' be the numbers of the jobs in the kth and $(k + 1)$th positions in **S**, respectively. Let **S**′ be a schedule that differs from **S** only in that job K' is in position k and job K in position $k + 1$. The first $(k - 1)$ and the last $(n - k - 1)$ jobs have exactly the same completion-time under **S** and **S**′ and make the same contribution to total weighted flow-time. Schedules **S** and **S**′ differ only in the flow-times of jobs K and K'. If $t = \sum_{i=1}^{k-1} p_{[i]}$, then these flow-times, appropriately weighted, are the following:

Under **S**: $u_K F_K = u_{[k]} F_{[k]} = u_K(t + p_K),$

            $u_{K'} F_{K'} = u_{[k+1]} F_{[k+1]} = u_{K'}(t + p_K + p_{K'}).$

Under **S**′: $u_K F_K = u_{[k+1]} F_{[k+1]} = u_K(t + p_{K'} + p_K),$

            $u_{K'} F_{K'} = u_{[k]} F_{[k]} = u_{K'}(t + p_{K'}).$

Eliminating common terms, this reduces to:

Under **S**: $u_{K'} p_K.$

Under **S**′: $u_K p_{K'}.$

Jobs K and K' were selected initially so that $u_K p_{K'} < u_{K'} p_K$; hence **S**′ represents an improvement over **S**.

Many of the earlier results can also be generalized to include the u_i coefficients:

1) Mean weighted lateness and mean weighted waiting-time are minimized by p/u ratio sequencing.

2) For sequencing that is independent of both the processing-time and the weight u_i, the expected value of mean weighted flow-time is

$$\frac{\sum_{i=1}^{n} u_i \sum_{i=1}^{n} p_i + \sum_{i=1}^{n} u_i p_i}{2n},$$

if a specific set of n jobs is known, and

$$\frac{n + 1}{2} \hat{u}\hat{p},$$

if the processing-times are considered independent identically distributed random variables with expected value $\hat{p}$, and the weights are considered independent identically distributed random variables with expected value $\hat{u}$.

3) If $\mathbf{R}$ and $\mathbf{R}'$ are antithetical procedures, their mean weighted flow-times are related:

$$\bar{F}_{u\mathbf{R}} + \bar{F}_{u\mathbf{R}'} = \frac{\sum_{i=1}^{n} u_i \sum_{i=1}^{n} p_i + \sum_{i=1}^{n} u_i p_i}{n}.$$

An obvious method of sequencing in this case (and probably one often followed in practice) is to consider the weights alone; that is, to sequence so that

$$u_{[1]} \geqq u_{[2]} \geqq \cdots \geqq u_{[n]}.$$

It is interesting to determine what the results of this procedure are with respect to mean weighted flow-time. Several special cases are obvious from Theorem 3–10:

1) If all the processing-times are equal, then we have the same ordering as the p/u ratio and hence this sequence is optimal.

2) If the weight is directly proportional to the processing-time ($u_i = kp_i$), then the p/u ratio is the same for each job and all orderings are equivalent.

3) It is possible for such a sequence to be antithetical to the optimal sequence. (For example, let $u_i = k\sqrt{p_i}$.)

In general, suppose that the p_i are independent identically distributed random variables with expected value $\hat{p}$, and the u_i are independent identically distributed random variables with expected value $\hat{u}$ (with u_i and p_i uncorrelated). Suppose that the jobs are numbered so that

$$u_1 \geqq u_2 \geqq \cdots \geqq u_n.$$

Let $\mathbf{N}$ be the sequence in which the jobs are in numbered order. Then

$$\bar{F}_{u\mathbf{N}} = \frac{1}{n} \sum_{i=1}^{n} u_i F_i = \frac{1}{n} \sum_{i=1}^{n} u_i \sum_{j=1}^{i} p_j,$$

$$E(\bar{F}_{u\mathbf{N}}) = \frac{\hat{p}}{n} E\left(\sum_{i=1}^{n} i u_i\right).$$

We can use Theorem 3–8 to give an expression for $E(\sum_{i=1}^{n} i u_i)$ by substituting u for p:

$$E\left(\sum_{i=1}^{n} i u_i\right) = n(n\hat{u} - (n-1)Q_u), \quad \text{where } Q_u = \int_0^{\infty} u G(u) \, dG(u),$$

$$E(\bar{F}_{u\mathbf{N}}) = n\hat{p}\hat{u} - \hat{p}(n-1)Q_u = \frac{(n+1)\hat{p}\hat{u}}{2} - \hat{p}(n-1)\left(Q_u - \frac{\hat{u}}{2}\right).$$

The first term on the right is the expected mean weighted flow-time for a random sequencing of the jobs. The second term on the right is always nonnegative, since it can be shown that $Q_u \geqq (\hat{u}/2)$ (with equality only for a degenerate distribution in which all values are equal). This shows that taking the jobs in order of decreasing u will result, in general, in a smaller mean weighted flow-time than would be obtained in

a random sequence. Of course, if the criterion is mean *unweighted* flow-time, and the p_i and u_i are uncorrelated, then ordering the jobs by u is random sequencing.
The problem of minimizing mean weighted tardiness,

$$\bar{T}_u = \frac{1}{n} \sum_{i=1}^{n} u_i T_i = \frac{1}{n} \sum_{i=1}^{n} u_i \max(0, L_i),$$

has been the subject of considerable recent work [51, 70, 118, 133, 171, 181, 182]; certain special cases have been solved, and some interesting bounds and limits have been obtained, but the general problem remains open.

For a given problem of this sort, one might first try sequencing according to due-date. This will, of course, minimize the maximum tardiness and if this maximum happens to be zero, then $\bar{T}_u$ is also zero and the sequence is clearly optimal. Failing this, one might try sequencing by increasing p/u ratio, which will minimize the mean weighted lateness. If the result is that every job has nonzero tardiness, then tardiness and lateness are equivalent and the sequence is optimal with respect to $\bar{T}_u$. Unfortunately, for the interesting problems in which the due-dates are neither so generous that all the jobs have zero tardiness nor so difficult that all the jobs have nonzero tardiness, no generally optimal procedure is known.

Schild and Fredman [181] have given an adjusting procedure that begins with a p/u sequence and seeks interchanges to reduce the value of $\bar{T}_u$. This appears to produce good schedules but it cannot in general guarantee an optimum. The following is a restatement and explanation of their procedure.

Assume that a set of jobs is so numbered that

$$\frac{p_1}{u_1} \leqq \frac{p_2}{u_2} \leqq \cdots \leqq \frac{p_n}{u_n},$$

and that it is initially sequenced by job number. Let I be the first job that has zero tardiness. (If there is no such job the sequence is optimal and the procedure is terminated.) Let T_u be the sum of the weighted tardinesses. T_u will not be improved by interchanging job I with a job i earlier in the sequence ($i < I$) or by simply moving job I up to an earlier position in the sequence (since job I makes no contribution to T_u, further reduction in F_I is of no value, and one or more other jobs will be delayed with possible increase in their contribution to T_u). However, it is possible that I could be moved to a later position so that one or more other jobs could be advanced with a net reduction in T_u. The strategy is to seek an appropriate job J, which is after I in the initial sequence, and insert it between $I - 1$ and I. The flow-times in the revised sequence would be

$$F'_i = F_i + p_J, \qquad I \leqq i < J,$$

$$F'_J = F_J - \sum_{i=I}^{J-1} p_i,$$

$$F'_i = F_i, \qquad i < I \quad \text{and} \quad i > J.$$

The amount by which job I can be postponed without increasing its contribution to T_u is $(a_I - F_I)$.

The contribution to T_u of the jobs between I and J in the revised sequence is

$$\sum_{i=I}^{J} u_i \max(0, F_i' - a_i)$$

$$= \sum_{i=I}^{J-1} u_i \max(F_i + p_J - a_i, 0) + u_J \max\left(F_J - \sum_{i=I}^{J-1} p_i - a_J, 0\right).$$

The change in T_u is given by

$$Z_I(J) = \sum_{i=I}^{J-1} u_i \max(F_i + p_J - a_i, 0) + u_J \max\left(F_J - \sum_{i=I}^{J-1} p_i - a_J, 0\right)$$

$$- \sum_{i=I}^{J-1} u_i \max(F_i - a_i, 0) - u_J \max(F_J - a_J, 0),$$

which can be rewritten as

$$Z_I(J) = \sum_{i=I}^{J-1} \left(\max(F_i + p_J - a_i, 0) - \max(F_i - a_i, 0)\right)$$

$$+ u_j \left(\max\left(F_J - \sum_{i=I}^{J-1} p_i - a_J, 0\right) - \max(F_J - a_J, 0)\right),$$

$$Z_I(J) = \sum_{i=I}^{J-1} u_i p_J + \sum_{i=I}^{J-1} u_i\left(\max(F_i - a_i, -p_J) - \max(F_i - a_i, 0)\right)$$

$$- \sum_{i=I}^{J-1} u_J p_i + u_J \left(\max\left(F_J - a_J, \sum_{i=I}^{J-1} p_i\right) - \max(F_J - a_J, 0)\right),$$

$$Z_I(J) = \sum_{i=I}^{J-1} u_i p_J - \sum_{i=I}^{J-1} u_i \min\left(p_J, \max(a_i - F_i, 0)\right) - u_J \sum_{i=I}^{J-1} p_i$$

$$+ u_J \max(a_J - F_J', 0).$$

The individual terms in this final form may be interpreted as follows:

$u_i p_J$ — Each job from I to $J - 1$ finishes p_J time units later under the revised sequence. The T_u is increased by this amount, assuming that each of these jobs was already late in the original sequence. If not, the correction is given by:

$u_i \min\left(p_J, \max(a_i - F_i, 0)\right)$. The penalty $u_i p_J$ is not fully incurred if job i was not late in the original sequence:

If $F_i' = (F_i + p_J) \leqq a_i$, then $u_i p_J$ is not incurred at all. If $F_i \leqq a_i < (F_i + p_J)$, then the portion of $u_i p_J$ not incurred is $u_i(a_i - F_i)$.

$u_J \sum_{i=I}^{J-1} p_i$ — The contribution of job J to T_u is decreased by the product of u_J and the processing-time of each job that it displaces.

$u_J \max(a_J - F_J', 0)$ — The decrease in the contribution of job J to T_u continues only to the point where this job becomes early, if this occurs.

This expression is exact, it gives the actual change in T_u that will result if a change in sequence is made, but as a working procedure it is convenient to use a simpler expression that provides only an upper bound on the change by assuming that all the jobs between I and J are late in the original sequence. Then

$$\min\big(p_J, \max(a_i - F_i, 0)\big) = 0 \quad \text{for } I < i < J.$$

Also, since $a_I > F_I$ by the way that job I was chosen,

$$\min\big(p_J, \max(a_I - F_I, 0)\big) = \min(p_J, a_I - F_I).$$

The resulting expression is

$$Z_I'(J) = p_J \sum_{i=I}^{J-1} u_i - u_I \min(p_J, a_I - F_I) - u_J \sum_{i=I}^{J-1} p_i$$
$$+ u_J \max(a_J - F_J', 0) \leqq Z_I(J).$$

The Schild and Fredman procedure is as follows:

1) Initially order the jobs by p/u ratio.

2) Find the first job in the resulting sequence with $F_i < a_i$. Call this job I.

3) Evaluate $Z_I'(k)$ successively for jobs following I in sequence—$k = I + 1$, $I + 2, \ldots$ If a k is encounted such that $Z_I'(k) < 0$, then move job k in front of job I in sequence.

4) Repeat steps 2 and 3 as long as changes are encountered.

While this procedure can sometimes effect an improvement in a sequence, it does not guarantee an optimal sequence, as shown by the following example.

Job	1	2	3
p	1	3	7
u	2	5	11
$a = d$	6	5	10

Sequencing by due-date (2, 1, 3) does not complete all jobs on time: $T_3 = 1; T_u = 11$. Sequencing by p/u ratio (1, 2, 3) does not complete all jobs late: $T_1 = 0$, $T_2 = 0$; $T_u = 11$. The Schild and Fredman procedure cannot find an improvement in the p/u ratio sequence:

$$Z_1'(2) = 6 - 6 - 5 + 10 = 5,$$
$$Z_1'(3) = 49 - 10 - 44 + 33 = 27,$$
$$Z_2'(3) = 35 - 5 - 33 + 22 = 19.$$

However, a better sequence exists: (2, 3, 1) gives $T_u = 10$. We should note that the failure of the procedure is not due to the use of $Z_I'(k)$ as an approximation for $Z_I(k)$, but is due to the inadequacy of pairwise comparison for this measure of performance, as discussed earlier in Section 3-2.

3–8 SEQUENCING WITH MULTIPLE CLASSES

A class structure can be superimposed on the simple sequencing rules considered up to this point by requiring that a certain subset of the jobs be placed first, in some order; the jobs of another subset next, in some order; etc. A discipline of this type has three components:

1) a rule for partitioning the n jobs into distinct, mutually exhaustive subsets called classes,

2) a rule for specifying sequence among the classes,

3) a rule (or rules) for specifying the sequence of jobs within each class.

For example, a manufacturer might have some jobs for specific customer orders and some jobs for stock replenishment. He might decide to do the customer jobs first, say in SPT order, and then the stock jobs, also in SPT order. It is clear that this discipline will have greater mean flow-time than simple SPT sequencing, but it also seems likely that it will be better than random sequencing. This type of overriding class structure is common, and it is interesting and useful to determine how its performance compares with the class-free optimal procedures.

This can be done by studying a two-class model in which k of the n jobs are selected at random to be members of the preferred class, and in which the same procedure is used to determine the order of jobs within each of the classes. The overall (both classes) mean flow-time in this case is a random variable, owing to the method of class determination, and its expected value is given by Theorem 3–11.

Theorem 3–11. In an $n/1//\overline{F}$ problem, if **R** and **R′** are any pair of antithetical rules, and $\overline{F}_\mathbf{R}$ and $\overline{F}_{\mathbf{R}'}$ are the respective mean flow-times, then the expected mean flow-time for a procedure in which k of the jobs are selected at random to be processed first (in order according to **R**) followed by the remaining $(n - k)$ jobs (also in order according to **R**) is given by

$$E(\overline{F}_k) = (1 - \alpha)\overline{F}_\mathbf{R} + \alpha\overline{F}_{\mathbf{R}'},$$

where

$$\alpha = \frac{k(n - k)}{n(n - 1)}.$$

Proof. Mean flow-time can be written as

$$\overline{F}_k = \frac{1}{n} \sum_{i=1}^{n} Y_i p_i,$$

where the coefficient Y_i is one of the numbers $1, 2, \ldots, n$ determined by the position job i takes in sequence. This is a random variable in this case so that

$$E(\overline{F}_k) = \frac{1}{n} \sum_{i=1}^{n} E(Y_i) p_i.$$

We shall find an expression for $E(Y_i)$.

Assume that the jobs are renumbered in the order in which they would be processed under **R**. Since **R** sequencing is used in both classes, it is clear that the sequence is completely determined by giving the result of the selection; that is, jobs in each class are simply processed in number order.

There are $\binom{n}{k}$ ways of selecting k out of n jobs, and by the assumption of random selection, each of these is equally likely to occur. We shall compute z_{ij} (the number of ways in which job number i can appear in position j in the schedule) and use $z_{ij}\binom{n}{k}^{-1}$ as the probability that job i will be in position j in the computation of $E(Y_i)$. There are n jobs and n positions in sequence:

Jobs: 1   2   3   ... i ... n

Positions: 1   2   ... j ... k | $k+1$ ... $k+(n-k)$

            Selected jobs | Remaining jobs

Consider first the case in which job i is selected, meaning that it goes into one of the first k positions. Out of all possible sequences job i will go into position j each time that there are also selected $j-1$ jobs with number smaller than i, and $k-j$ jobs with number larger than i. There are $\binom{i-1}{j-1}$ ways to do the former and $\binom{n-i}{k-j}$ ways to do the latter, so that

$$z_{ij} = \binom{i-1}{j-1}\binom{n-i}{k-j}.$$

For $i \leq k$, if i is selected, it cannot appear later than position i, so that in this expression $j = 1, 2, \ldots, i$. For $i > k$, if i is selected, then $j = k-(n-i), \ldots, k$. It is convenient to adopt the usual combinatorial convention that $\binom{x}{y} = 0$ if either $y < 0$ or $(x-y) < 0$. Then satisfactory limits on position are simply $j = 1, \ldots, i$.

If job i is not selected it occupies position $k+j$ when $j-1$ jobs of lower number are also "not selected" and $n-k-j$ jobs of higher number are not selected, so that

$$z_{ij} = \binom{i-1}{j-1}\binom{n-i}{n-k-j}.$$

With the same convention, limits on j are again $j = 1, \ldots, i$:

$$E(Y_i) = \sum_{j=1}^{n} (n-j+1)\,\text{Prob (job } i \text{ in position } j),$$

$$\frac{E(Y_i)}{n} = \sum_{j=1}^{i} \frac{(n-j+1)\binom{i-1}{j-1}\binom{n-i}{k-j}}{n\binom{n}{k}}$$

$$+ \sum_{j=1}^{i} \frac{(n-k-j+1)\binom{i-1}{j-1}\binom{n-1}{n-k-j}}{n\binom{n}{k}}.$$

Let $J = j - 1$ and separate the summations:

$$\frac{E(Y_i)}{n} = \sum_{J=0}^{i-1} \frac{\binom{i-1}{J}\binom{n-i}{k-1-J}}{\binom{n}{k}} - \frac{1}{n}\sum_{J=0}^{i-1} J\frac{\binom{i-1}{J}\binom{n-i}{k-1-J}}{\binom{n}{k}}$$

$$+ \frac{n-k}{n}\sum_{J=0}^{i-1} \frac{\binom{i-1}{J}\binom{n-i}{n-k-I-J}}{\binom{n}{k}}$$

$$- \frac{1}{n}\sum_{J=0}^{i-1} J\frac{\binom{i-1}{J}\binom{n-i}{n-k-1-J}}{\binom{n}{k}}.$$

The summations can be eliminated by the use of two identities that are obtained from properties of the hypergeometric distribution:*

$$\sum_{j=0}^{a}\binom{a}{j}\binom{b}{c-j} = \binom{a+b}{c}, \qquad \sum_{j=0}^{a} j\binom{a}{j}\binom{b}{c-j} = \frac{ca}{a+b}\binom{a+b}{c}.$$

$$\frac{E(Y_i)}{n} = \frac{\binom{n-1}{k-1}}{\binom{n}{k}} + \frac{(n-k)\binom{n-1}{n-k-1}}{n\binom{n}{k}} - \frac{(k-1)(i-1)\binom{n-1}{k-1}}{n(n-1)\binom{n}{k}}$$

$$- \frac{(n-k-1)(i-1)\binom{n-1}{n-k-1}}{n(n-1)\binom{n}{k}}$$

$$= \frac{k}{n} + \frac{(n-k)^2}{n^2} - \frac{k(k-1)(i-1)}{n^2(n-1)} - \frac{(n-k)(n-k-1)(i-1)}{n^2(n-1)}.$$

$$E(Y_i) = (n-i+1)\left(1 - \frac{k(n-k)}{n(n-1)}\right) + i\frac{k(n-k)}{n(n-1)}$$

$$= (n-i+1)(1-\alpha) + i\alpha, \qquad \text{where } \alpha = \frac{k(n-k)}{n(n-1)};$$

$$E(\overline{F}_k) = \frac{(1-\alpha)}{n}\sum_{i=1}^{n}(n-i+1)p_i + \frac{\alpha}{n}\sum_{i=1}^{n} ip_i = (1-\alpha)\overline{F}_\mathbf{R} + \alpha\overline{F}_{\mathbf{R}'}.$$

---

* E. Parzen, *Modern Probability Theory* (New York: Wiley, 1964), page 164.

Recall from Theorems 3–6 and 3–7 that the performance of antithetical rules is symmetric about the expected mean flow-time for random sequencing:

$$\overline{F}_R + \overline{F}_{R'} = 2E(\overline{F}_{RANDOM}).$$

From this and Theorem 3–11 the performance of a two-class rule can be given in terms of the performance of the rule used to sequence within classes and simple random sequencing:

$$E(\overline{F}_k) = (1 - 2\alpha)\overline{F}_R + 2\alpha E(\overline{F}_{RANDOM}).$$

The selection number, k, enters the result only in α. This term has the following characteristics:

a) α is symmetric in k. This means that a situation with 10% preferred jobs has the same performance as one with 90% preferred jobs.

b) α achieves its maximum for $k = n/2$; that is, the two-class rule gives its worst performance when the two classes are of equal size.

c) The maximum value of α ranges from $\frac{1}{2}$ down to approach $\frac{1}{4}$ as n becomes large. This means that for large n the performance of the worst two-class rule is halfway between the performance of the intra-class rule and random sequencing.

d) The graph of α vs. k is strictly convex; the introduction of a preferred class causes more than proportional deterioration in overall performance.

e) When either all of the jobs or none of the jobs are selected into the preferred class, only the intraclass rule is effective and $E(\overline{F}_0) = E(\overline{F}_n) = \overline{F}_R$.

A typical graph relating performance to class proportion is shown in Fig. 3–9.

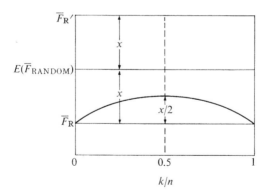

Fig. 3–9. Performance of a two-class sequencing rule in an $n/1$ problem.

We have not tried to obtain comparable results for more than two classes, but presumably these results would fall between the two-class rule and random sequencing, since an increase in the number of classes further interferes with the ordering under **R**. The worst three-class rule should have poorer performance (in the sense of $E(\overline{F})$) than the worst two-class rule.

CHAPTER 4

FURTHER PROBLEMS
WITH ONE OPERATION PER JOB

There are several interesting and important problems which are similar to the $n/1$ problems of Chapter 3, in that each job has only a single operation, but which violate one or more of the other restrictions of Chapter 3. The following sections consider cases in which the setup-time is sequence dependent, the jobs do not arrive simultaneously, and the jobs have required precedence relationships. The final section of the chapter considers the case in which two or more machines are used in parallel to process the n jobs, and it serves as a bridge to the general n/m problems of the following chapters.

4–1 SEQUENCE-DEPENDENT SETUP-TIMES

There are some situations in which it is simply not acceptable to assume that the time required to set up the facility for the next task is independent of the task that was the immediate predecessor on the facility. In fact, in some of these cases the variation of setup-time with sequence provides the dominant criterion for evaluating a schedule—far overshadowing concern with average flow-time or lateness.

Table 4–1

Hours to Clean Equipment Between Colors

Preceding color	Following color			
	White	Yellow	Red	Blue
White	0	1	2	3
Yellow	6	0	1	2
Red	8	6	0	1
Blue	10	8	6	0

A good example is found in the manufacture of paint. Different colors of paint are produced in sequence on the same pieces of equipment, which must be cleaned when the manufacturer changes from one color to another. Quite obviously the thoroughness of the cleaning is heavily dependent on the color being removed and the color for which the machine is being prepared. For example, suppose that only four colors were involved, and that the cleaning times were as given in Table 4–1. There are six different sequences for these four colors, and the total time to make the four changes required per cycle varies from 13 to 21 hours, as shown below.

Sequence	Total changeover time per cycle, hr
White-yellow-red-blue-white . . .	13
White-red-blue-yellow-white . . .	17
White-blue-red-yellow-white . . .	21
White-yellow-blue-red-white . . .	17
White-red-yellow-blue-white . . .	20
White-blue-yellow-red-white . . .	20

There are many similar examples in the chemical and processing industries in which a facility operates on one product at a time, but on many distinctly different products in sequence. An interesting variation called the *come-down problem* occurs in the rolling of steel strips. The rollers are slightly scored by the edges of the strip being rolled. This causes no difficulty if the next strip in sequence is of narrower width, but this scoring would mar the surface of a wider strip. The result is that changeover time is small so long as one comes down in width, but an increase in width requires removal and regrinding of the rollers. The problem can be equally important with general-purpose machine tools. Certain jobs can have similar setups so that changing from one to another is simply a matter of adjusting stops and perhaps changing tools. Other jobs on the same machine could require an entirely different setup. For example, changing a punch press from a setup for continuous-strip operation to a setup for individual-piece single-step operation is much more time-consuming than changes within these classes. There is evidence to suggest that setup similarity is given more weight than any other factor in current industrial practice of sequence determination.

Where setup is assumed to be independent of sequence, setup-time can be included in the processing-time p_i. In this section we must separate the setup-time and introduce a matrix of setup-times, S (similar to Table 4-1) in which s_{ij} represents the time to change over from job i to job j. It is often convenient to add an imaginary $(n + 1)$-job to the given n jobs to represent the idle condition of the facility, and a 0th position in the sequence to represent the preliminary idle state. The term $s_{[0],[1]}$ would be the time required to bring the facility from idleness to a state ready to process the first job in sequence.

The circumstances are considerably altered by this change in assumption. Where previously the total amount of time required of the machine was a constant, independent of sequence, it now depends on how the jobs are ordered. The flow-times now are given by

$$F_{[1]} = s_{[0],[1]} + p_{[1]},$$
$$F_{[2]} = F_{[1]} + s_{[1],[2]} + p_{[2]},$$
$$F_{[i]} = F_{[i-1]} + s_{[i-1],[i]} + p_{[i]}.$$

Where previously the maximum flow-time was a constant and attention was directed to aggregate measures such as mean flow-time, now even the reduction of maximum flow-time is a formidable problem:

$$F_{\max} = F_{[n]} = \sum_{i=1}^{n} s_{[i-1],[i]} + \sum_{i=1}^{n} p_{[i]}.$$

The sum of the processing-times is still a constant, and may be ignored; the maximum flow-time is minimized by minimizing the sum of the n setup-times.

The task of minimizing the sum of the setup-times corresponds to what is usually called the *traveling-salesman problem*. This problem is presented as a situation in which a salesman must visit each of n cities once and only once and return to his point of origin, and do so in a way that minimizes the total distance traveled (or total time, or cost, etc.). Each city corresponds to a job; and the distance between cities corresponds to the time required to change over from one job to another. (The scheduling problem stated above actually corresponds to an $(n + 1)$-city problem, since the idle or preparatory state must be treated as a job.) In general, one is given an n-by-n matrix S (distances, times, etc.). The object is to find an n-by-n matrix X, with

$$x_{ik} = 0 \text{ or } 1, \quad i = 1, 2, \ldots, n, \quad k = 1, 2, \ldots, n,$$

$$\sum_{i=1}^{n} x_{ik} = 1, \quad k = 1, 2, \ldots, n,$$

$$\sum_{k=1}^{n} x_{ik} = 1, \quad i = 1, 2, \ldots, n,$$

and such that $\sum_{i=1}^{n} \sum_{k=1}^{n} x_{ik} s_{ik}$ is a minimum. Up to this point the problem is the *assignment problem*.* But for the traveling-salesman problem there is the additional restraint that no tour can return to its starting point until all n cities have been visited. This means that X matrices such as the following, each of which would be a valid solution to an assignment problem, are not allowed:

$$\begin{bmatrix} 1 & 0 & 0 & 0 \\ 0 & 1 & 0 & 0 \\ 0 & 0 & 1 & 0 \\ 0 & 0 & 0 & 1 \end{bmatrix}, \begin{bmatrix} 0 & 1 & 0 & 0 \\ 1 & 0 & 0 & 0 \\ 0 & 0 & 0 & 1 \\ 0 & 0 & 1 & 0 \end{bmatrix}, \begin{bmatrix} 1 & 0 & 0 & 0 \\ 0 & 0 & 1 & 0 \\ 0 & 0 & 0 & 1 \\ 0 & 1 & 0 & 0 \end{bmatrix}, \begin{bmatrix} 1 & 0 & 0 & 0 \\ 0 & 0 & 0 & 1 \\ 0 & 0 & 1 & 0 \\ 0 & 1 & 0 & 0 \end{bmatrix}, \begin{bmatrix} 0 & 0 & 0 & 1 \\ 0 & 0 & 1 & 0 \\ 0 & 1 & 0 & 0 \\ 1 & 0 & 0 & 0 \end{bmatrix}.$$

The seemingly innocent necessity of excluding solutions of this type differentiates between what is perhaps the simplest of all mathematical programming forms and one of the more difficult problems. Although no algorithm of comparable directness to, say, the simplex method for linear programming is known for the traveling-salesman problem, two procedures that will obtain optimum solutions to problems of modest size and approximate solutions to larger problems have recently been offered. These are described in the following sections.

The problem of sequencing jobs corresponds to a very general form of the traveling-salesman problem. The matrix S can be nonsymmetric, for the time to change over from job i to job j can, in general, be different from the time to change over from job j

---

* G. Hadley, *Linear Programming* (Reading, Mass.: Addison-Wesley, 1962), page 367.

to job i. In the classical traveling-salesman form, the distance from city i to city j will be equal to the distance from city j to city i, although the problem may be stated in terms of travel time or cost, and these matrices need not be symmetric. In addition, distances are subject to further restrictions if they are to be mutually consistent and plottable on a Euclidean plane. For example, the total distance along a path between two given cities, with one or more intermediate stops, should be greater than or equal to the distance of a direct trip between the two cities. (This corresponds to the statement that the sum of the length of any two sides of a proper Euclidean triangle must be greater than the length of the third side.) There is no corresponding constraint on the times to change over from one job to another. However, this additional generality does not appear to make the problem any more difficult to solve. In fact, the procedure described in Section 4–1.1 appears to be slightly more effective for asymmetric problems.

The difficulty encountered in solving the problem for the simple criterion of minimizing the maximum flow-time has apparently discouraged any work involving more complex measures, such as minimizing the mean flow-time. This problem can be shown to correspond to a quadratic assignment problem, but again an algorithm for its solution is not known.

4–1.1 A "Branch-and-Bound" Algorithm for the Traveling-Salesman Problem

An ingenious recursive computational procedure for the traveling-salesman problem has recently been offered by Little, Murty, Sweeney, and Karel [123]. The procedure involves the maintenance of a list of unsolved, closely related traveling-salesman problems, and three processing routines to apply to the problems on that list. Initially, one places the original problem on the list. The original problem is then processed (and hence removed from the list) by one of the routines which will sometimes create other problems to be placed on the list. One simply continues, applying these same routines to each problem until there are no more on the list, at which point the solution to the original problem has been revealed. Roughly speaking, the three routines are:

1) A *solution routine*, which directly solves a problem from the list if the problem is easy (small) enough.

2) An *elimination routine*, which discards a problem from the list if it can show that this problem can make no contribution to the solution of the original problem.

3) A *partitioning routine* which replaces a problem too difficult to solve with two related subproblems.

One is given an n-city problem by the specification of an n-by-n matrix S, where s_{ij} gives the distance (or cost, or time, etc.) to travel from city i to city j. A solution to the problem, a tour of the cities, is given by a permutation of the integers $1, 2, \ldots, n$:

$$[1], [2], \ldots, [n].$$

The *value* of a solution is the sum of the n elements from the S-matrix specified by the permutation

$$s_{[1],[2]} + s_{[2],[3]} + \cdots + s_{[n-1],[n]} + s_{[n],[1]}.$$

The optimum solution is that permutation which has minimum value. A particular arc, say from i to j, may be *prohibited* if one makes s_{ij} infinite. Then if *any* finite solution exists, the permutation for the optimum solution will not include the ordered pair $... i, j, ...$. On the other hand, one can specify that a certain arc *must* be included in the tour, and be interested in the optimum solution, subject to the restriction that the permutation must contain the ordered pair (i, j).

Each unsolved problem on the list of this procedure involves n cities. Of the n steps in the tour (ordered pairs in the permutation) of a particular problem, k of the steps may already be specified. One has to choose the remaining $n - k$ steps in an optimum manner. It is also necessary to give, for each problem on the list, a value, Y, which is a *lower bound* on the value of all solutions, and hence on the optimum solution, to the problem. There are of course trivial lower bounds, such as the minimum element of S or the sum of the n row minima of S. The ingenuity of Little's procedure lies essentially in the manner in which he constructs usefully large lower bounds on the problems on the list.

Each unsolved problem on the list, then, is characterized by $n - k$, the number of unspecified steps on the tour, and Y, the lower bound on the solutions to the problem. One can also assume that at least one solution to the original problem (the permutation $1, 2, ..., n$ is a solution) is known at the outset, and can let Z be the value of the best solution known up to a given point in the procedure. (Initially Z may not be finite.) One attacks unsolved problems on the list in the following way:

a) If $n - k = 2$, there are only two more steps to be specified; the *solution routine* is used to solve the problem directly. If the value of this solution is less than Z, then Z is set equal to this new value, and the new solution is recorded as the incumbent "best."

b) If Y is greater than or equal to Z, this problem is incapable of contributing a solution any better than one that is already known, and the *elimination routine* removes the problem from the list.

c) If neither of these conditions is satisfied, then the *partitioning routine* replaces the problem with two other problems:

  i) One of these problems has an additional step (i, j) specified and may also have an increased lower bound Y.

  ii) The second problem has the step (i, j) *prohibited* and certainly has an increased lower bound Y.

So the problems appearing on the list have increasing lower bounds and/or an increasing number of steps specified. Eventually each problem is disposed of by using either the *solution* or the *elimination* routine; the list becomes empty, and the original problem is solved.

The *solution* and *elimination* routines are obvious; the essence of the Little procedure lies in the *partitioning* routine. This involves the notions of *reduction* and *selection*. Reduction is simply the process of obtaining at least one zero in each row and each column of S. One notes that, since every solution to a problem includes one and only one element from each row or column, then if a constant is either subtracted or added to each element in a row or column, all solutions will be equally affected; thus in particular the optimum solution will still be optimum. If one subtracts a constant, h, from each element of a row or column in a problem S, one obtains a problem S' whose optimum solution is the same permutation as the optimum solution for S. Also $Y' = h$ can be associated with S' to provide a lower bound on the values of solutions to the original problem that can be obtained by solving S'. One can continue subtracting an amount equal to the smallest element in each row or column until each column and row has at least one zero. The sum of these *reducing constants* provides an initial lower bound, Y, for the original problem.

Say that a problem S, fully reduced, is being considered for solution by the partitioning routine. One is concerned with the selection of a particular path, say from i to j, that will be the basis for creating two new problems:

1) S_{ij}, the problem of finding the best solution from among all the solutions to S that include the step (i, j).

2) $S_{n(ij)}$, the problem of selecting the best from among all the solutions to S that do not include the step (i, j).

Since in problem S_{ij} it has been decided to go from i to j, one can prohibit going from i to any other city, and prohibit arriving at j from any other city, by making all the entries in the ith row and the jth column, except s_{ij}, equal to infinity. One must also prohibit the future selection of the element s_{ji}, by making it infinite also, since a tour cannot include s_{ij} and s_{ji} and still visit all n cities before returning to the starting point. Since these prohibitions may have eliminated some of the zeros of S, one can possibly further reduce S_{ij}, and thus establish a new and greater lower bound on solutions of S obtained by solving S_{ij}.

In problem $S_{n(ij)}$ one prohibits travel from i to j by making $s_{ij} = \infty$. Again, a further reduction of this modified matrix may be possible which will give an increased lower bound on solutions obtained through problem $S_{n(ij)}$. Now the objective of the selection of (i, j) is to make the lower bound on $S_{n(ij)}$ as great as possible in hopes that it can be discarded from the list of unsolved problems by the elimination routine without further partition. To accomplish this, one looks ahead at the reduction that will be possible in $S_{n(ij)}$ for each possible (i, j) and makes the selection such that the sum of the two subsequent reducing constants will be a maximum. It should be obvious that it is necessary to consider only the zero elements of S as candidates, since if a nonzero element is selected no further reduction of $S_{n(ij)}$ will be possible.

The procedure is perhaps more easily described by means of an example than by stating algorithms. Suppose one has a six-city problem specified by the matrix S, given below, which can be reduced to S-(16) by subtracting a total of 16 from rows and columns (a dash indicates a prohibited, or infinite, entry).

		S						S-(16)			
−	1	7	3	14	2	−	0^2	3	2	13	1
3	−	6	9	1	24	2	−	2	8	0^6	23
6	14	−	3	7	3	3	11	−	0^0	4	0^1
2	3	5	−	9	11	0^2	1	0^2	−	7	9
15	7	11	2	−	4	13	5	6	0^2	−	2
20	5	13	4	18	−	16	1	6	0^1	14	−

At stage (1) of the procedure, the list consists of problem S-(16), the (16) indicating that any solution to the original problem obtained through S-(16) will have a value of at least 16. A solution 1–2–3–4–5–6 is, of course, known so that at this point $Z = 43$. The superscripts to each of the zeros of S-(16) indicate the reduction that would be possible if that zero were prohibited. The greatest of these is 6, for element s_{25}, so the partition is based on this element, yielding the problems shown below:

List, stage (2) ($Z = 43$)	S_{25}-(16)						$S_{n(25)}$-(22)					
S_{25}-(16)	–	0^2	3	2	–	1	–	0	3	2	9	1
$S_{n(25)}$-(22)	–	–	–	–	$\underline{0}$	–	0	–	0	6	–	21
3	11	–	0^0	–	0^1		3	11	–	0	0	0
0^3	1	0^3	–	–	9		0	1	0	–	3	9
13	–	6	0^2	–	2		13	5	6	0	–	2
16	1	6	0^1	–	–		16	1	6	0	10	–

At stage (2) the list consists of problems S_{25}-(16) and $S_{n(25)}$-(22). Note that $S_{n(25)}$ has been further reduced by 6 (4 in the 5th column and 2 in the 2nd row), giving a bound of 22. In problem S_{25} the element s_{25} has been underlined, indicating that this is *required* in the solution, and the other elements in that row and column have been prohibited. Note also that element s_{52} has been prohibited. Since solutions to this problem are known to contain s_{25}, it is clear that they cannot contain s_{52}, which would cause a cyclic subtour of fewer than n cities. Since both problems on the list have bounds of less than Z and more than two cities to specify, they cannot yet be either *solved* or *eliminated*, and one of them must be attacked with the *partitioning* routine. With more cities specified, and a lower value of Y, S_{25} is probably the more promising of the two problems. Again, the value of each zero as a basis for partition is given as a superscript. Selecting element s_{41} produces problems $S_{25,41}$-(19) and $S_{25,n(41)}$-(19); $S_{n(25)}$-(22) remains on the list from stage (2).

List, stage (3) ($Z = 43$)	$S_{25,41}$-(19)						$S_{25,n(41)}$-(19)					
$S_{n(25)}$-(22)	–	0^1	0^3	–	–	1	–	0^2	3	2	–	1
$S_{25,41}$-(19)	–	–	–	–	$\underline{0}$	–	–	–	–	–	$\underline{0}$	–
$S_{25,n(41)}$-(19)	–	11	–	0^0	–	0^1	0^{10}	11	–	0^0	$\underline{0}$	0^1
	$\underline{0}$	–	–	–	–	–	–	1	0^4	–	–	9
	–	–	3	0^2	–	2	10	–	6	0^2	–	2
	–	1	3	0^1	–	–	13	1	6	0^1	–	–

List, stage (4) ($Z = 43$)	$S_{25,41,13}$-(20)						$S_{25,41,n(13)}$-(22)					
$S_{n(25)}$-(22)	–	–	$\underline{0}$	–	–	–	–	0	–	–	–	1
$S_{25,n(41)}$-(19)	–	–	–	–	$\underline{0}$	–	–	–	–	–	$\underline{0}$	–
$S_{25,41,13}$-(20)	–	10	–	–	–	0^{12}	–	11	–	0	–	0
$S_{25,41,n(13)}$-(22)	$\underline{0}$	–	–	–	–	–	$\underline{0}$	–	–	–	–	–
	–	–	–	0^2	–	2	–	–	0	0	–	2
	–	0^{10}	–	0^0	–	–	–	1	0	0	–	–

List, stage (5) $(Z = 43)$ $S_{25,41,13,36}$-(20) $S_{25,41,13,n(36)}$-(32)

$S_{n(25)}$-(22)	–	–	0	–	–	–	–	–	0	–	–	–
$S_{25,n(41)}$-(19)	–	–	–	–	0	–	–	–	–	–	0	–
$S_{25,41,n(13)}$-(22)	–	–	–	–	–	0	–	0	–	–	–	–
$S_{25,41,13,36}$-(20)	0	–	–	–	–	–	0	–	–	–	–	–
$S_{25,41,13,n(36)}$-(32)	–	–	–	0	–	–	–	–	–	0	–	0
	–	0	–	0	–	–	–	0	–	0	–	–

Note that at stage (4) when $S_{25,41}$ is partitioned on the basis of element 13, the blocking of elements of $S_{25,41,13}$ which cannot be used includes the usual symmetric element 31 and also element 34. Since both 41 and 13 are required steps in the path, element 34 would close a three-city subtour 4–1–3–4.

Partitioning is continued until at stage (5) there is a solvable problem on the list—$S_{25,41,13,36}$-(20) has only two steps remaining to be specified. Four specified steps must involve a minimum of five cities; there is at most one city remaining, and only one way that it can be fitted in to give a valid tour. In this case the four steps happen to involve all six cities in two disconnected trips: 2–5 and 4–1–3–6. The solution routine connects these trips, specifying s_{54} and s_{62} (there is no choice) to produce a solution 6–2–5–4–1–3 with a value of 20.

This new value, $Z = 20$, allows the *elimination* routine to discard three problems from the list, so that all that remains is problem $S_{25,n(41)}$-(19). It is still possible that this problem could yield a solution with a value less than 20. This problem is partitioned on s_{31} to produce $S_{25,n(41),31}$-(20) and $S_{25,n(41),n(31)}$-(29). Both these problems can be discarded by the elimination routine so that the list becomes empty and the problem is solved.

List, stage (7) $(Z = 20)$ $S_{25,n(41),31}$-(20) $S_{25,n(41),n(31)}$-(29)

$S_{25,n(41),31}$-(20)	–	0	–	2	–	0	–	0	3	2	–	1
$S_{25,n(41),n(31)}$-(29)	–	–	–	–	0	–	–	–	–	–	0	–
	0	–	–	–	–	–	–	11	–	0	–	0
	–	1	0	–	–	8	–	1	0	–	–	9
	–	–	6	0	–	1	0	–	6	0	–	2
	–	1	6	0	–	–	3	1	6	0	–	–

One can conveniently show the partitioning of the various problems by means of a tree. Each node in Fig. 4–1 corresponds to a problem on the list.

There are several strategies that one might use in selecting problems from the list for partition. Selecting the problem with the most cities already specified tends to keep the number of problems on the list at any given time small. Selecting the problem with the smallest lower bound tends to reduce the total number of problems that must be partitioned. Which is preferable depends primarily on whether one expects to be limited by storage space or by computing time.

Little *et al.* have programmed the algorithm for the IBM 7090 and they cite computational experience ([123], page 986): "Problems up to 20 cities usually require

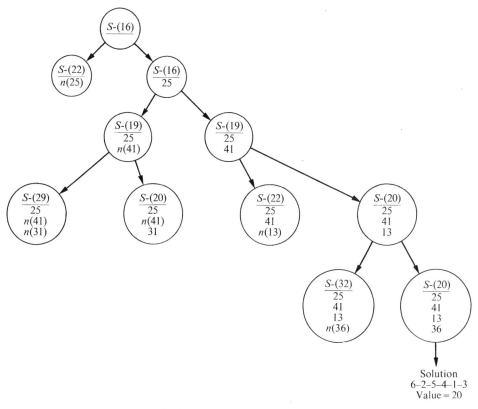

Solution
6–2–5–4–1–3
Value = 20

Fig. 4–1. Partitioning tree for the traveling-salesman problem.

only a few seconds. The time grows exponentially, however, and by 40 cities is beginning to be appreciable, averaging a little over 8 minutes. As a rule of thumb, adding 10 cities to the problem multiplies the time by a factor of 10."

This exponential growth of the computational task as the number of cities increases is characteristic of every algorithm for the traveling-salesman problem that has been offered to date. In fact, Little's algorithm retards this growth more effectively than any other known procedure for the problem, and is therefore capable of solving substantially larger problems.

4–1.2 Solution of the Traveling-Salesman Problem by Dynamic Programming

An alternative solution to the problem, by means of dynamic programming, has been offered by Bellman [19], and by Held and Karp [75]. This procedure is perhaps more general than the branch-and-bound technique—it is applicable to a larger class of sequencing problems—but its present computational limits are reached on a somewhat smaller problem than the branch-and-bound method.

Again, assume that one is given an n-city traveling-salesman problem with distances specified by the elements of a matrix, S. Without loss of generality one can choose any city, c_o, as the origin. Suppose that one partitions the set of n cities into four exhaustive, mutually exclusive subsets:

1) $\{c_o\}$, a set consisting of a single city, the origin.

2) $\{c_i\}$, a set consisting of a single city which is not the origin.

3) $\{C_k\}$, a set consisting of k cities, not c_o or c_i.

4) $\{C_{(n-k-2)}\}$, a set consisting of the remaining $(n - k - 2)$ cities.

In particular, suppose that one knows the ordering from an optimal tour beginning and ending at c_o. One could then choose a city c_i and a subset $\{C_k\}$ of k cities so as to make the following statement: Starting at c_o, on this optimal tour, one visits each of the cities of $\{C_{(n-k-2)}\}$, in some particular order, and arrives at c_i, with the cities of $\{C_k\}$ still to be visited, in some order, before returning to c_o.

Now considering only the portion of the tour from c_i through the cities of $\{C_k\}$ and back to c_o, one knows that this will be the shortest possible path from c_i to c_o, with intermediate stops at the k cities of $\{C_k\}$. If this were not the case, then without altering the route allowed before reaching c_i one could find a better path for completing the tour, and hence a shorter tour, which would contradict the assumption that the original tour was optimal to begin with.

Therefore let $f(c_i; \{C_k\})$ be the length of the shortest possible path from c_i back to c_o, with intermediate stops at the k cities of $\{C_k\}$. Note in particular that when $k = 0$, then

$$f(c_i; \{\ \}) = s_{io}$$

is just a specific element from S, and if one allowed $k = n - 1$ and c_i to coincide with the origin, then $f(c_o; \{C_{n-1}\})$ would be the length of an optimal tour of the original problem. The essence of the procedure is to begin with $k = 0$ and to increase k in steps of one. In traditional dynamic programming fashion, one begins at the destination (c_o) and works one's way backward, one city at a time, until the origin (c_o) is reached and the optimal solution is revealed.

For this problem the basic principle of optimality of dynamic programming states that

$$f(c_i; \{C_k\}) = \min_{c_j \in \{C_k\}} \left(s_{ij} + f(c_j; \{C_k\} - \{c_j\}) \right).$$

This says that to find the best way of starting at c_i and returning to c_o after visiting k cities one has to select the shortest of the k alternatives that start with the step from c_i to one of the k cities and *then follow the shortest path from there back to c_o, visiting the other $k - 1$ cities in some order*. The difficulty lies in the fact that one has to know the length of the shortest path (and the order of visiting) from each of these k cities back to c_o after visiting the other $k - 1$. Each of these k problems is in turn expressed as the minimum of $k - 1$ choices, by reapplication of the basic recursive relationship given above. Eventually one reaches the point at which the terms on the right are simply elements from S. Since the solutions to these one-step problems are known, one can begin here and build up solutions to the larger problems and eventually to the n-city problem.

For example, suppose that one is given a 5-city problem and city 5 is designated as the origin. Then $f(c_5; \{c_1, c_2, c_3, c_4\})$ is the length of the shortest tour and any ordering that achieves this length is an optimal solution.

At stage (0) one computes the solution to the four problems with $k = 0$:

Stage (0)
_____
$k = 0$
4 problems

$$f(c_1; \{\}) = s_{15}, \quad f(c_2; \{\}) = s_{25}, \quad f(c_3; \{\}) = s_{35}, \quad f(c_4; \{\}) = s_{45}.$$

At stage (1), one expresses the solution to the problems with $k = 1$ in terms of the $(k = 0)$ problems already solved:

Stage (1)
_____
$k = 1$
12 problems

$$f(c_1; \{c_2\}) = s_{12} + f(c_2; \{\}),$$
$$f(c_1; \{c_3\}) = s_{13} + f(c_3; \{\}),$$
$$f(c_1; \{c_4\}) = s_{14} + f(c_4; \{\}),$$
$$f(c_2; \{c_1\}) = s_{21} + f(c_1; \{\}),$$
$$f(c_2; \{c_3\}) = s_{23} + f(c_3; \{\}),$$
$$\vdots$$
$$f(c_4; \{c_3\}) = s_{43} + f(c_3; \{\}).$$

At stage (2) one expresses the solution to all problems with $k = 2$ in terms of the problems solved in stage (1):

Stage (2)
_____
$k = 2$
12 problems

$$f(c_1; \{c_2, c_3\}) = \min(s_{12} + f(c_2; \{c_3\}), s_{13} + f(c_3; \{c_2\})),$$
$$f(c_1; \{c_2, c_4\}) = \min(s_{12} + f(c_2; \{c_4\}), s_{14} + f(c_4; \{c_2\})),$$
$$f(c_1; \{c_3, c_4\}) = \min(s_{13} + f(c_3; \{c_4\}), s_{14} + f(c_4; \{c_3\})),$$
$$f(c_2; \{c_1, c_3\}) = \min(s_{21} + f(c_1; \{c_3\}), s_{23} + f(c_3; \{c_1\})),$$
$$f(c_2; \{c_1, c_4\}) = \min(s_{21} + f(c_1; \{c_4\}), s_{24} + f(c_4; \{c_1\})),$$
$$\vdots$$
$$f(c_4; \{c_2, c_3\}) = \min(s_{42} + f(c_2; \{c_3\}), s_{43} + f(c_3; \{c_2\})).$$

Note that stage (2) uses every one of the problems with $k = 1$ that was solved in stage (1). Note also that when the selection between the terms on the right is made one should record which term was selected, since this indicates the actual route to be followed. For example, if

$$s_{12} + f(c_2; \{c_3\}) < s_{13} + f(c_3; \{c_2\}),$$

then

$$f(c_1; \{c_2, c_3\}) = s_{12} + f(c_2; \{c_3\}).$$

We now know that whenever one is at c_1, with c_2 and c_3 still to visit, *regardless of how one arrived at* c_1, one would, from c_1, go first to c_2, then to c_3, and then back to c_5, traveling a distance $f(c_1; \{c_2, c_3\})$ after leaving c_1.

Continuing, the stage (3) problems use every one of the problems solved in stage (2) but none of the problems solved in stage (1):

$$\underline{\text{Stage (3)}}$$
$$k = 3$$
$$4 \text{ problems}$$

$f(c_1; \{c_2, c_3, c_4\})$
$$= \min\big(s_{12} + f(c_2; \{c_3, c_4\}), s_{13} + f(c_3; \{c_2, c_4\}), s_{14} + f(c_4; \{c_2, c_3\})\big),$$
$f(c_2; \{c_1, c_3, c_4\})$
$$= \min\big(s_{21} + f(c_1; \{c_3, c_4\}), s_{23} + f(c_3; \{c_1, c_4\}), s_{24} + f(c_4; \{c_1, c_3\})\big),$$
$f(c_3; \{c_1, c_2, c_4\})$
$$= \min\big(s_{31} + f(c_1; \{c_2, c_4\}), s_{32} + f(c_2; \{c_1, c_4\}), s_{34} + f(c_4; \{c_1, c_2\})\big),$$
$f(c_4; \{c_1, c_2, c_3\})$
$$= \min\big(s_{41} + f(c_1; \{c_2, c_3\}), s_{42} + f(c_2; \{c_1, c_3\}), s_{43} + f(c_3; \{c_1, c_2\})\big).$$

Finally, at stage (4), the solution to the original problem is obtained:

$$\underline{\text{Stage (4)}}$$
$$k = 4$$
$$1 \text{ problem}$$

$$f(c_5; \{c_1, c_2, c_3, c_4\}) = \min\big(s_{51} + f(c_1; \{c_2, c_3, c_4\}), s_{52} + f(c_2; \{c_1, c_3, c_4\}),$$
$$s_{53} + f(c_3; \{c_1, c_2, c_4\}), s_{54} + f(c_4; \{c_1, c_2, c_3\})\big).$$

To summarize the amount of work that was involved, consider the number and size of the problems solved at each stage, as shown in Table 4–2. In total, 33 problems were solved by looking up 108 terms from either the S-matrix or a list of previously solved problems. This can be compared with finding a solution by exhaustive enumeration, which would require generating each of the $(n - 1)!$ or 24 possible paths. Each of these would involve five terms from the S-matrix, or 120 terms in all. The comparison becomes increasingly favorable to dynamic programming as the number of cities increases.

Table 4–2

k	Number of problems	Number of alternatives in each problem	Number of s- and f-terms per problem
0	4	1	1
1	12	1	2
2	12	2	4
3	4	3	6
4	1	4	8

In general, at each stage there are $n - 1$ ways of selecting city c_i, and for each of these, there are $\binom{n-2}{k}$ ways of selecting the cities of $\{C_k\}$. This means that there are

$$(n - 1) \binom{n - 2}{k} = \frac{(n - 1)!}{k!(n - 2 - k)!}$$

problems. For $k \geq 1$ there are k alternatives to be compared in each of these, and this means finding $2k$ s- and f-terms for each. (For $k = 0$ there is only 1 term for each of the $n - 1$ problems.) The total number of terms at all stages is given by

$$2 \cdot \sum_{k=1}^{n-1} \frac{k(n - 1)!}{k!(n - 2 - k)!} + (n - 1)$$

$$= 2 \cdot \sum_{k=1}^{n-1} \binom{n - 1}{k - 1} (n - k)(n - k - 1) + (n - 1)$$

$$< (n - 1)(n - 2)(2^n - 2) + (n - 1)$$

$$< n^2 2^n.$$

This is not a very good upper bound, but it still compares favorably with the $n!$ terms required for enumeration. For example, when $n = 8$, the situation is as shown in Table 4–3. The product of $8^2 \cdot 2^8$ is 16,384. Enumeration involves $7! = 5040$ alternatives with a total of 40,320 terms.

Table 4–3

k	0	1	2	3	4	5	6	7	Total
Number of problems	7	42	105	140	105	42	7	1	449
Number of terms	7	84	420	840	840	420	84	14	2709

Actually, the practical limit on computation for this dynamic programming solution is probably storage space rather than the amount of computation. Since one cannot overwrite any of the problems solved at a given stage until all the problems at the following stage have been solved, one must be able to store all the problems at two consecutive stages. In particular, one must provide for the two stages, with k near $(n - 2)/2$, when there is the greatest number of problems. Each problem will require several computer words to store the necessary identification information, the length of the optimum tour, and the order of cities in the optimum tour. A storage of 32,768 words might just hold a 13-city problem (with 5544 problems for $k = 5$ and 5544 for $k = 6$), but recall (Section 4–1.1) that problems of 40 cities have been solved by branch and bound on a machine with this storage capacity. There are, of course, larger memories available, and one could resort to auxiliary storage devices, but note that the storage requirements for dynamic programming are more than doubled for each additional city.

4-1.3 The "Closest-Unvisited-City" Algorithm for the Traveling-Salesman Problem

An obvious procedure for the traveling-salesman problem (and one which is probably employed in practice if traveling salesmen are really concerned with such problems) is to always move next to the closest city which has not yet been visited. That is, given a present location corresponding to a specification of a row in the S-matrix, one chooses for the next city the one associated with the minimum element in that row (excluding cities already visited, which would prematurely close the tour). Not only is this the obvious commonsense heuristic for the problem, but its analogy to the shortest-processing-time sequencing commends it. It is easy to exhibit counter-examples to show that this procedure does not yield an optimal solution, and in fact, one can readily construct examples in which this procedure would lead one into a trap, and would result in disastrously bad solutions. Nevertheless one might still be interested in the application of this procedure for normal or random cases that were not deliberately contrived to defeat it. If one were offered assurance that the performance would not be too bad, one might choose to employ such an approximation, and forfeit some advantage in the value of the solution for the undeniable ease with which one was able to obtain it.

Table 4-4

			S		
–	1	7	3	14	2
3	–	6	9	1	24
6	14	–	3	7	3
2	3	5	–	9	11
15	7	11	2	–	4
20	5	13	4	18	–

Table 4-5		
Origin	Sequence	Value
1	1 – 2 – 5 – 4 – 3 – 6	32
2	2 – 5 – 4 1 – 6 – 3	34
3	3 – 4 – 1 – 2 – 5 – 6	24
	3 – 6 – 4 – 1 – 2 – 5	22
4	4 – 1 – 2 – 5 – 6 – 3	24
5	5 – 4 – 1 – 2 – 3 – 6	32
6	6 – 4 – 1 – 2 – 5 – 3	22

It is interesting to note that while optimal solutions are cyclic permutations and one can arbitrarily designate the origin, the particular solution obtained by the closest-unvisited-city algorithm does depend on the designation of the origin. For example, Table 4-4 restates the problem given in Section 4-1.1. Table 4-5 gives the closest-unvisited-city solutions to this problem. Origins 3 and 6 each yield the solution 3-6-4-1-2-5, which has a length of 22. This is not much greater than the optimum length of 20 (3-6-2-5-4-1) although this solution has only three of six steps in common with the optimum tour obtained in Section 4-1.1. For this particular problem, if one examined the closest-city solutions from each possible origin (or if he were lucky in his choice of an origin) he would do quite well relative to the optimum solution, but if he were unfortunate in the choice of origin the result would be rather poor. In general, even if one considers each possible origin, the computation is only increased by a factor of n and is still modest in comparison with known procedures that guarantee an optimum solution. In particular, the computation increases only as n^2 for this procedure, which means that it can be applied to very large problems.

Only if there were frequent ties in the selection of the closest city, such as for origin 3 in the example above, and only if one were to try to enumerate all possible closest-city solutions, would the computation become burdensome.

Gavett [60] has tested the closest-unvisited-city algorithm for a large number of problems in which the elements of the distance-matrix are independent identically distributed random variables—in some cases from a normal distribution, in others from a rectangular distribution. A number of examples of each type of problem were generated and tested. The *average* tour lengths for each type of problem are given in Tables 4–6 and 4–7. These results include: the expected tour length if a tour is selected at random, the length of the optimal tour (obtained by the Little procedure), the length of tour generated by the closest-unvisited-city algorithm with arbitrarily designated origin, and the length of the best tour obtained by the closest-unvisited-city algorithm, exploring each city as the origin. The difference between the average optimum tour and the expected tour length with random routing increases as the number of cities does and, of course, decreases as the standard deviation of the matrix elements decreases. The performance of the closest-city algorithm seems to weaken as the number of cities increases. For twenty cities, and normally distributed distances with standard deviation of 0.4, the average arbitrary-origin closest-city tour is 26% longer than the optimum in this sample; the average all-origin closest-city tour is 18% longer. For rectangularly distributed distances these values are 51% and 38%, respectively. Roughly speaking, for these data one might characterize the all-origin algorithm as midway between the arbitrary-origin algorithm and the optimum in performance.

4–2 INTERMITTENT JOB ARRIVALS

Throughout Chapter 3 each of the n jobs was assumed to arrive and be ready for processing at the same time. When this is not the case and $r_i \neq r_j$ for some i and j, then the questions of *preemption* and *inserted-idleness* must be considered.

Preemption is said to occur when the processing of a job that has been started is stopped before completion. If preemption must be considered, then a sequence of n integers is no longer sufficient to describe a schedule for n jobs, since individual jobs may appear two or more times in the schedule and one must specify how much time is to be allowed at each appearance. Different kinds of preemption are possible, depending on the treatment accorded the interrupted job when it returns to the machine for further work. At one extreme the processing may resume where it left off without any extra work or time being occasioned by the interruption. Under these circumstances the total processing-time for the job is a constant, independent of the number of interruptions that it may suffer. This is called a *preempt-resume* discipline. At the other extreme, called *preempt-repeat*, the benefit of any processing that has been done is forfeited with the interruption, so that processing must be repeated when the job returns to the machine. The stated processing-time for the job is then a minimum, achieved only if the schedule permits the job to be processed without interruption. One very reasonable and important discipline between these extremes involves preempt-resume operation for the processing-time with preempt-repeat for the setup-time.

Table 4-6   Average Tour Lengths for Traveling-Salesman Problems with Normally Distributed Distances   (*Gavett* [60])

Number of cities	Normal distribution		Number of problems	Average tour lengths			
	Mean	Standard deviation		Random	Optimum	Closest unvisited city	
						Arbitrary origin	All origin
5	1.0	0.4	100	5.0	3.28	3.64 (1.11)*	3.44 (1.05)
5	1.0	0.3	45	5.0	3.54	3.98 (1.12)	3.76 (1.06)
5	1.0	0.2	30	5.0	4.09	4.32 (1.06)	4.21 (1.03)
5	1.0	0.1	10	5.0	4.49	4.59 (1.02)	4.57 (1.02)
10	1.0	0.4	10	10.0	4.78	5.51 (1.15)	5.26 (1.10)
10	1.0	0.3	45	10.0	5.92	6.63 (1.12)	6.34 (1.07)
10	1.0	0.2	30	10.0	7.18	7.69 (1.07)	7.43 (1.04)
10	1.0	0.1	10	10.0	8.69	9.00 (1.04)	8.87 (1.02)
15	1.0	0.4	10	15.0	6.40	7.81 (1.22)	7.23 (1.13)
15	1.0	0.3	45	15.0	8.11	9.33 (1.15)	8.90 (1.10)
15	1.0	0.2	30	15.0	10.34	11.09 (1.07)	10.83 (1.05)
15	1.0	0.1	10	15.0	12.83	13.17 (1.03)	13.00 (1.01)
20	1.0	0.4	10	20.0	7.68	9.65 (1.26)	9.07 (1.18)
20	1.0	0.3	45	20.0	10.00	11.52 (1.15)	11.02 (1.10)
20	1.0	0.2	30	20.0	13.36	14.35 (1.07)	14.11 (1.06)
20	1.0	0.1	10	20.0	16.51	16.96 (1.03)	16.84 (1.02)

Table 4-7   Average Tour Lengths for Traveling-Salesman Problems with Rectangularly Distributed Distances   (*Gavett* [60])

Number of cities	Rectangular distribution (0, 2)		Number of problems	Average tour lengths			
	Mean	Standard deviation		Random	Optimum	Closest unvisited city	
						Arbitrary origin	All origin
5	1.0	0.576	20	5.0	1.85	2.58 (1.39)*	2.03 (1.10)
10	1.0	0.576	20	10.0	2.61	3.97 (1.52)	3.45 (1.32)
15	1.0	0.576	40	15.0	3.02	5.15 (1.71)	4.38 (1.45)
20	1.0	0.576	20	20.0	3.08	4.78 (1.51)	4.26 (1.38)

A simple example will illustrate the utility of preemption.

Job	1	2
Ready-time, r	0	1
Processing-time, p	4	1

If job 1 is begun at time 0 and is not preempted when job 2 arrives, the mean flow-time is 4.0. If preempt-resume is permitted when job 2 arrives, then the mean flow-time is 3.0.

The situation with intermittent arrivals and a preempt-resume discipline is essentially the same as for simultaneous arrivals. At the moment of an arrival one can review the decision as to what to process, considering only the remaining processing-time of the job currently in process. At each point in time one makes the best selection from among those jobs available, with the comforting knowledge that if a more attractive candidate arrives the decision can be immediately revised without penalty. The obvious generalization of the shortest-processing-time rule is called the *shortest-remaining-processing-time rule.* This minimizes the mean flow-time under these circumstances, and the *other results for the case of simultaneous arrival are similarly applicable.* Note in particular that it is not necessary to have any advance information about job arrivals. The dominating procedures are strictly local, selecting from the jobs on hand without concern for imminent arrivals.

If a preempt-repeat discipline is used in the last example, the mean flow-time is 3.5, which is better than operation without preemption but not as good as the preempt-resume discipline. But analysis is much more difficult for the preempt-repeat case, and no interesting general results have been offered. In particular, the results for the case of simultaneous arrivals that can be extended to preempt-resume do not immediately apply here.

Preempt-repeat also differs from preempt-resume operation in that advance information of job arrival has a definite effect on the schedule. If one knows that another job with characteristics that will cause preemption will arrive before the processing of a current candidate can be completed, one is not likely to begin the futile processing. If there is no job that can be completed before the next arrival, or none that can withstand its preemption, then one would probably hold the machine idle until the next arrival. In the previous example this results in mean flow-time of 3.5. This device, in effect, takes the place of preempt-repeat operation when advance information is available. Again, no general optimal procedures have been given for these circumstances.

4–3 REQUIRED PRECEDENCE AMONG JOBS

In all the preceding sections we assumed that every possible sequence of the jobs was capable of execution, so that whichever best served a given measure of performance could be selected. Now consider a situation in which certain orderings are prohibited, either by technological constraints or by some externally imposed policy.

4–3.1 Required Strings of Jobs

Consider a situation in which some of the decisions of sequence have already been made and in which one has the task of completing the schedule without altering what has been decided; this might occur when setup-times are highly dependent on sequence. One would then group jobs with similar setups, sequence within these groups for minimum changeover time, and arrange the groups to minimize some measure, such as mean flow-time.

In general, assume that the original n jobs have been grouped into k strings, with $n_1, n_2, \ldots, n_k$ as the number of jobs in the various strings. Assume that the membership of each string is fixed, that the order of jobs within each string is fixed, and that once started an entire string must be processed to completion. It remains to select one of the $k!$ ways in which the strings could be sequenced. We shall use the following notation:

p_{ij},   the processing-time of the jth job in the ith string,

F_{ij},   the flow-time of the jth job in the ith string,

$p'_i = \sum_{j=1}^{n_i} p_{ij}$,   the total processing-time for the ith string,

and

$F'_i = F_{i,n_i}$,   the flow-time of the ith string, equal to the flow-time of the last job in that string.

Now if the objective is to minimize the mean of the string flow-times,

$$\overline{F}' = \frac{\sum_{i=1}^{k} F'_i}{k},$$

it is obvious that one should consider each string as a job and apply the shortest-processing-time rule to the p'_i. That is, one should sequence the strings so that

$$p'_{[1]} \leqq p'_{[2]} \leqq \cdots \leqq p'_{[k]}.$$

If the objective is to minimize the mean of the string flow-times weighted by the number of jobs in each string,

$$\overline{F}'_n = \frac{\sum_{i=1}^{k} n_i F'_i}{n},$$

then it follows directly from the results of Section 3–7 that one should sequence the strings so that

$$\frac{p'_{[1]}}{n_{[1]}} \leqq \frac{p'_{[2]}}{n_{[2]}} \leqq \cdots \leqq \frac{p'_{[k]}}{n_{[k]}}.$$

It is also true, although perhaps not immediately obvious, that this ordering of the strings will minimize the mean of the original job flow-times:

$$\overline{F} = \frac{\sum_{i=1}^{n} F_i}{n} = \frac{\sum_{i=1}^{k} \sum_{j=1}^{n_i} F_{ij}}{n}.$$

Theorem 4–1. When one schedules an $n/1//\overline{F}$ problem, without setup, in which there are k disjoint subsets of jobs called strings, within which a job order that may not be preempted is specified, the mean job flow-time is minimized by order-ing the strings so that the ratio of the string-processing-time to the number of jobs in the string forms a nondecreasing sequence. That is, the strings must be so ordered that

$$\frac{p'_{[1]}}{n_{[1]}} \leq \frac{p'_{[2]}}{n_{[2]}} \leq \cdots \leq \frac{p'_{[k]}}{n_{[k]}}.$$

Proof. Let h_{ij} be the time between the completion of the jth job in the ith string and the completion of the string itself; then $h_{i,n_i} = 0$ and

$$h_{ij} = \sum_{k=j+1}^{n_i} p_{ik} \qquad (j = 1, 2, \ldots, n_i - 1).$$

Observe that for every job, h_{ij} is a constant, given in the statement of the problem. It depends on the processing-times and the order of jobs *within* a string, but both these are given:

$$F_{ij} = F'_i - h_{ij},$$

$$\overline{F} = \frac{\sum_{i=1}^{k} \sum_{j=1}^{n_i} F_{ij}}{n} = \frac{\sum_{i=1}^{k} n_i F'_i}{n} - \frac{\sum_{i=1}^{k} \sum_{j=1}^{n_i} h_{ij}}{n}.$$

Note that the second term on the right is a constant unaffected by the ordering of the strings, and that the first term on the right is known (by Theorem 3–10) to be mini-mized by the given ordering.

4–3.2 General Precedence Constraints

More generally, one might wish to specify a weaker ordering of a set of jobs to require certain precedences but not require strings to be processed without interruption. This sort of ordering can be specified by use of the *precedes* and *directly precedes* relations and the precedence graph described in Section 1–2. (Precedence in 1–2 applies to the individual operations of a single job. Since in the present situation each job consists of a single operation, one can readily view the n jobs as n operations of a single over-all job.)

Suppose that there are three jobs to be processed: a, b, and c. With no constraints (as in Chapter 3) each of the six possible sequences could be considered:

$$(1) \ abc \qquad (2) \ acb \qquad (3) \ bac$$
$$(4) \ bca \qquad (5) \ cab \qquad (6) \ cba.$$

If b and c are considered a string in the sense of Section 4–3.1, then there are only two possible sequences:

$$(1) \ abc \qquad \text{and} \qquad (2) \ bca.$$

However, if one simply requires that b precede c without requiring them to be adjacent in sequence, then there are three possibilities:

$$(1) \ abc, \qquad (2) \ bac, \qquad (3) \ bca.$$

Consider the case in which the n jobs are divided into k *chains*. This occurs when precedence constraints give each job at most one predecessor and at most one successor. For example:

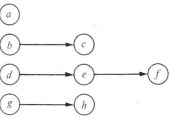

A procedure to minimize mean flow-time subject to this type of constraint is given by the following theorem:

Theorem 4–2. When one schedules an $n/1//\overline{F}$ problem, without setup, in which *directly-precedes* relationships are given between certain pairs of jobs such that a given job has at most one predecessor and at most one successor—resulting in k disjoint subsets of jobs called chains within which job order is specified, but which may be preempted between jobs—the mean job flow-time is minimized by the following procedure:

1) For each job j in chain i compute

$$x_{ij} = \frac{\sum_{h=1}^{j} p_{ih}}{j}.$$

2) For each chain i compute

$$y_i(h_i) = \min(x_{i1}, x_{i2}, \ldots, x_{in_i}),$$

where h_i is the second subscript of the minimum value on the right.

3) Select chain I such that

$$y_I(h_I) \leqq y_i(h_i) \qquad \text{for all } i,$$

and put the first h_I jobs from chain I at the beginning of the sequence on the machine.

4) Neglecting these first h_I jobs on chain I, recompute y_I.

5) Repeat (3) and (4) until all the jobs have been added to the sequence.

Proof. This procedure will produce and order $K \geqq k$ subchains. The manner in which the procedure works, not revealing the second subchain of a given chain to the sequencing mechanism until the first has been assigned a position in sequence, assures that the precedence relationships in the original chains will be observed.

Now if one assumes that these subchains are strings (in the sense of Section 4–3.1) and cannot be interrupted, then by Theorem 4–1, the procedure gives an optimal ordering. What must be shown, then, is that this is a valid assumption, that there is no advantage to be gained by interrupting any of these subchains, so that they are, in effect, strings.

Consider a subchain b, containing n_b jobs. Let

$$y_b = \frac{\sum_{j=1}^{n_b} p_{bj}}{n_b}.$$

The jobs in b are sequenced: $(b, 1), (b, 2), \ldots, (b, n_b)$. Any interruption, or preemption, of b will mean the division of b into two subchains b' and b'':

$$b' = (b, 1), (b, 2), \ldots, (b, i),$$
$$b'' = (b, i + 1), \ldots, (b, n_b).$$

Each of these new subchains will have a ratio of processing-time to number of jobs:

$$y_{b'} = \frac{\sum_{j=1}^{i} p_{bj}}{i}, \qquad y_{b''} = \frac{\sum_{j=i+1}^{n_b} p_{bj}}{n_b - i}.$$

Each of the new subchains will find its place in the sequence (by Theorem 4–1) according to the value of $y_{b'}$ and $y_{b''}$.

What will be shown is that this partition of b is ineffective, that subchain b' will directly precede subchain b'' so that b will be intact and, in effect, be treated as a string.

The key step is to show that

$$y_{b'} > y_b > y_{b''}. \tag{1}$$

The first part of (1) is true by construction: this was the method by which the length of the subchain b was originally determined:

$$y_{b'} = \frac{\sum_{j=1}^{i} p_{bj}}{i} > y_b = \frac{\sum_{j=1}^{i} p_{bj} + \sum_{j=i+1}^{n_b} p_{bj}}{i + (n_b - i)}.$$

Multiplying both sides of the inequality by $i(i + (n_b - i))$, we have

$$\left(i + (n_b - i)\right) \sum_{j=1}^{i} p_{bj} > i\left(\sum_{j=1}^{i} p_{bj} + \sum_{j=i+1}^{n_b} p_{bj}\right).$$

Subtracting

$$i \sum_{j=1}^{i} p_{bj} \quad \text{and adding} \quad (n_b - i) \sum_{j=i+1}^{n_b} p_{bj}$$

to both sides yields

$$(n_b - i) \sum_{i=1}^{n_b} p_{bj} > n_b \sum_{j=i+1}^{n_b} p_{bj}, \qquad y_b = \frac{\sum_{j=1}^{n_b} p_{bj}}{n_b} > \frac{\sum_{j=i+1}^{n_b} p_{bj}}{n_b - i} = y_{b''}.$$

This demonstrates that the partitioning of any subchain created by this procedure will result in two subchains that are out of order insofar as the value of y is concerned, but which must remain so because of precedence constraints. The question is whether b' could possibly be moved ahead in the sequence or b'' moved back to good advantage.

Suppose that in the original sequence subchain b were preceded by subchain a and followed by subchain c so that

$$y_a \leqq y_b \leqq y_c.$$

It is clear that subchain b' cannot move in front of a, since

$$y_a \leqq y_b < y_{b'};$$

nor can it move farther ahead in sequence, since every subchain in front of a has a value of y which is less than or equal to y_a. So b's only chance of advancement is to break into the middle of a, partitioning it into a' and a''; but

$$y_{a''} < y_a \leqq y_b < y_{b'},$$

so that b' cannot move ahead of a''. A similar argument will show that b'' cannot move into or behind subchain c.

Slightly more general precedence constraints can be handled with this procedure. For example, in the precedence graph,

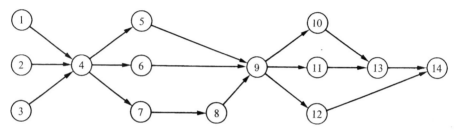

jobs 4 and 9 are break points at which the problem can be partitioned. Jobs 1, 2, and 3 and jobs 5, 6, 7, and 8 are effectively independent problems, each of which can be handled by the procedure of Theorem 4–2. (Jobs 1, 2, and 3 actually constitute a simple $n/1$ problem and all of Chapter 3 applies.) However, this decomposition does not permit treatment of an arbitrary precedence graph, as illustrated by jobs 10, 11, 12, 13, and 14.

4–4 PARALLEL MACHINES

Resuming the assumptions that the jobs are independent, that they arrive simultaneously, and that setup is sequence independent, suppose that there were more than one machine to perform the processing. The capabilities of m machines in this situation might be described by giving an n-by-m matrix of processing-times:

| | Machines | | | | |
	1	2	3	...	m
1	p_{11}	p_{12}			p_{1m}
2	p_{21}				
Jobs 3					
⋮					
n	p_{n1}				p_{nm}

Here p_{ij} gives the time to perform the single operation of job i on machine j, assuming that machine j performed the entire operation. The simplest case occurs when the machines are identical, indicated by having all the elements of a given row of the matrix equal (in which case the second subscript on p can be omitted). In general one can consider machines that are different by having different values in a particular row of the matrix, and admit the possibility that a particular machine might be unable to perform a particular task by making the corresponding p_{ij} prohibitively large. It is, of course, necessary only to consider matrices that cannot be partitioned into independent submatrices, in a manner analogous to that of the previous section, since the machines and jobs of an independent submatrix would be treated as a separate problem.

The key question in this situation is whether or not individual jobs may be divided among two or more machines. If this is not possible, then the problem is one of partitioning the n jobs into m distinct subsets and determining sequence within each subset. If some division of an individual job is allowed, then better schedules are possible, but the determination of the schedule is usually more difficult. The idea of dividing a particular job among two or more machines is not as impractical as it might initially appear, since in many situations a job actually consists of a "lot" of identical pieces, each of which is to receive exactly the same processing. What we have called an operation is then actually the repetition of some smaller element of work on each of these pieces. Clearly such a task could be divided, at the expense of additional setup, and two or more machines could work on the task simultaneously if multiple tooling were available. This practice of using multiple setups is, in fact, reasonably common in some types of manufacturing.

The basic virtue of the division of jobs among machines is easily shown by a simple example. Suppose that the processing-time matrix for a 2/2 problem is the following:

$$
\begin{array}{cc}
 & \text{Machine} \\
 & \begin{array}{cc} 1 & 2 \end{array}
\end{array}
$$

$$
\text{Job} \quad
\begin{array}{c} 1 \\ 2 \end{array}
\begin{array}{|cc|}
\hline
2 & 2 \\
2 & 2 \\
\hline
\end{array}
$$

If one job is assigned to each machine, as in Schedule A (below), then $\bar{F} = 2$. If both machines are used to process each of the jobs (ignoring setup considerations) as in Schedule B, then $\bar{F} = 1.5$.

Schedule A		Schedule B		
Machine 1	Job 1	Machine 1	Job 1	Job 2
Machine 2	Job 2	Machine 2	Job 1	Job 2
	1 2		1	2

More generally, suppose that there are m identical machines, and m jobs to be performed, each with processing-time p. There is a total of mp work to be done, and if this is divided equally among the m machines, they will all finish simultaneously after p time units, regardless of how the jobs are assigned to individual machines. However, this assignment does affect the times at which the jobs finish. At one extreme, if a single job is assigned to each machine, then the jobs also finish simultaneously at time p and $\overline{F} = p$. On the other hand, if each job is divided among all m machines, then the first job will finish at time p/m, the second at time $2p/m$, the third at $3p/m$, until finally the last job is finished at time p. Every job but the last one is finished earlier by this strategy than it would be under separate assignment to an individual machine. The average flow-time is given by

$$\overline{F} = \frac{p}{m}\left(\frac{1}{m} + \frac{2}{m} + \frac{3}{m} + \cdots + \frac{m}{m}\right) = \frac{m+1}{2m}\,p.$$

This would indicate that the advantage for simultaneous processing ranges from a minimum of 25% (two machines) to a limit of 50% for many machines. These specific figures, of course, apply only to this highly artificial case, but the advantage is universal. Ignoring the penalties of multiple setup and tooling, any schedule can be improved by taking advantage of parallel operations on identical machines.

In effect, one could view m machines working simultaneously on a single job as a single machine with m times the power of the basic machine. This permits the observation that, from a scheduling point of view, it is better to provide required capacity with a single machine than with an equivalent number of separate machines. There are often economies-of-scale and economies-of-operation that reinforce this effect, but considerations of reliability work in the opposite direction.

The sequencing of jobs in the situation where each is to be processed simultaneously on m machines corresponds directly to the $n/1$ cases of Chapter 3. One simply treats the m parallel machines as if they were a single machine. If the m machines are identical, a pseudo-processing-time is defined for each job,

$$p_i' = \frac{p_i}{m},$$

and these p' are used whenever the scheduling procedure calls for a processing-time. In some cases, when the decision is based entirely on processing-time, one can use the original processing-times directly, since dividing each p by the same constant would have no effect on the relative values. Even if the machines are not identical, this same approach may be employed by defining a pseudo-processing-time as

$$p_i' = \frac{1}{\sum_{j=1}^{m}(1/p_{ij})}.$$

In either case, in order to sequence to minimize the mean flow-time, mean completion-time, mean lateness, etc., one would use the shortest-processing-time rule with respect to the p'. That is, one would sequence so that

$$p_{[1]}' \leqq p_{[2]}' \leqq \cdots \leqq p_{[n]}'.$$

There are many equally important examples of parallel machines when jobs may not be divided among machines. Tellers' windows at a bank, checkout counters at a supermarket, reservation desks at an air terminal, and toll booths on a highway could all be classed in this category. Manufacturing problems in which multiple setup and tooling is either not possible or not economical must also be treated in this way. While the manufacturing job-shop is in aggregate a more complex structure, as we shall see in Chapters 6 and 11, often in practice scheduling is done on a local basis by a foreman whose responsibility consists of a set of m similar or identical machines and who is concerned with one operation per job.

In this case a scheduling procedure must assign a job to both a particular machine and to a position in sequence on that machine. Let $j[k]$ be the job which is in the kth position in sequence on the jth machine and n_j be the number of jobs processed on the jth machine:

$$n = \sum_{j=1}^{m} n_j.$$

The flow-times are given by

$$F_{j[1]} = p_{j[1]}, \quad F_{j[2]} = p_{j[1]} + p_{j[2]}, \quad F_{j[k]} = \sum_{i=1}^{k} p_{j[i]}.$$

The mean flow-time for all n jobs is

$$\bar{F} = \frac{\sum_{j=1}^{m} \sum_{k=1}^{n_j} (n_j - k + 1)p_{j[k]}}{n}.$$

The numerator consists of a sum of n terms, each of which is the product of one of the processing-times and an integer coefficient. Each of the n processing-times appears exactly once and the coefficients are $n_1, n_1 - 1, \ldots, 2, 1, n_2, n_2 - 1, \ldots, 2, 1, \ldots, n_m, n_m - 1, \ldots, 2, 1$. It is known that such a sum of products is minimized by arranging the terms so that the processing-times form a nondecreasing sequence and the coefficients form a nonincreasing sequence. There are m coefficients of 1; these should be associated with the m largest of the p_i. This means that the m jobs with the longest processing-times will appear at the end of the m machine sequences. It is also apparent that it is immaterial which of these m jobs goes on which machine, regardless of what the earlier sequence was on any machine. Continuing in the same manner, there are m coefficients of 2 (unless perhaps $n < 2m$) and these should be assigned to the m jobs of those remaining which have the longest processing-times. Continue until each processing-time has been assigned a coefficient, and therefore a position in sequence.

Equivalently one could renumber the n jobs in order of increasing processing-time:

$$p_1 \leqq p_2 \leqq p_3 \leqq \cdots \leqq p_n.$$

Then one could number the machines and simply assign jobs in rotation:

Job	1	2	...	m	$m+1$	$m+2$	...	$2m$	$2m+1$	...
Machine	1	2	...	m	1	2	...	m	1	...

In operation, the rule would simply be that each time a machine finished a job, it would be assigned, from among those jobs waiting, the job with the shortest processing-time.

This variation of the shortest-processing-time rule minimizes the mean flow-time, mean waiting-time, and mean lateness. It seems intuitively reasonable, although it may be a bit hard to accept the fact that one can interchange jobs in equivalent positions in sequence on different machines without having any effect on the mean flow-time. For example, if there are six jobs with processing-times 1, 2, 3, 4, 5, and 6 (numbered in that order) to be processed on two machines, each of the following schedules has exactly the same mean flow-time:

	Schedule			
	A	B	C	D
Machine 1	1, 3, 5	2, 3, 5	1, 4, 5	1, 3, 6
Machine 2	2, 4, 6	1, 4, 6	2, 3, 6	2, 4, 5

However, this shortest-processing-time rule does not necessarily minimize the maximum flow-time. For example, if there are four jobs with processing-times 1, 2, 3, and 10 (numbered in that order) to be processed on two machines, the two shortest-processing-time schedules, A and B, have maximum flow-times of 12 and 11, respectively. Schedule C is not a shortest-processing-time schedule; it has a maximum flow-time of 10, but a greater mean flow-time.

	A	B	C
Machine 1	1, 3	2, 3	1, 2, 3
Machine 2	2, 4	1, 4	4

Also, unlike the single machine case, the shortest-processing-time rule does not necessarily minimize the mean number of jobs in the shop. The relationship between flow-time and inventory, for the case of simultaneous arrivals, is given in Section 2–4 as

$$\overline{N}(0, F_{\max}) = \frac{n\overline{F}}{F_{\max}}.$$

The denominator on the right is not constant in this situation, and it is not known what rule minimizes this ratio of mean to maximum.

There are also many cases in which the facilities are similar in function but not identical in capability. For example, a screw-machine department may have a dozen machines that differ in number of spindles, number of tool stations, power, and other characteristics. Each machine could be capable of performing the work on a given job, but a different amount of time would be required on each machine. Also the machines in a given department will be typically acquired over a period of years, so that although they may be of nominally similar type they are capable of different production rates, reflecting progress in machine-tool design over the years.

When a job is to be processed by only one machine and the processing-times are entirely arbitrary, no general results are known. But if the processing-times vary so that individual machines are consistently fast or consistently slow for all the jobs, then a procedure to minimize mean flow-time can be given. This situation can be characterized by assuming a "machine factor," h_j, for each machine. The time to process a particular operation on machine j is expressed as $h_j \cdot p_i$. Flow-times are given by

$$F_{j[k]} = h_j \sum_{i=1}^{k} p_{j[i]}.$$

The mean flow-time for all n jobs is

$$\bar{F} = \frac{\sum_{j=1}^{m} h_j \sum_{k=1}^{n_j} (n_j - k + 1) p_{j[k]}}{n}.$$

As in the previous case, the numerator is a sum of n terms, each of which is the product of one of the processing-times and one of the coefficients:

$$h_1 n_1, h_1(n - 1), \ldots, 2h_1, h_1, h_2 n_2, h_2(n_2 - 1),$$
$$\ldots, 2h_2, h_2, \ldots, h_m n_m, h_m(n_m - 1), \ldots, 2h_m, h_m.$$

A schedule may be constructed by matching the job with the longest processing-time to the smallest of these coefficients, the job with the second-longest processing-time to the second-smallest of the coefficients, etc.

The identical machine case is, of course, just a special version in which all the h_j are equal to one.

Eastman, Even, and Isaacs [46] consider a more general aspect of this problem in which there is a weighting coefficient u_i associated with each job and the criterion is to minimize the weighted mean flow-time:

$$\bar{F}_u = \frac{\sum_{i=1}^{n} u_i F_i}{n}.$$

The one-machine version of this was considered in Section 3–7, in which it was shown that the optimal sequence has

$$\frac{p_{[1]}}{u_{[1]}} \leq \frac{p_{[2]}}{u_{[2]}} \leq \cdots \leq \frac{p_{[n]}}{u_{[n]}}.$$

For the m identical-machine case, in which each job must be assigned to an individual machine, no optimal procedure has been offered, but Eastman, Even, and Isaacs show that the following is a lower bound for the weighted mean flow-time,

$$\bar{F}_u(m) \geq \frac{m + n}{m(n + 1)} \bar{F}_u(1),$$

where $\bar{F}_u(k)$ is the weighted mean flow-time with k identical machines. Moreover, they show that, in general, no greater lower bound is possible by exhibiting a set of jobs that actually attain this bound. However, one should note that for the special case in which all the u_i are equal and for which a solution and not just a bound is known, the value obtained is in general greater than this lower bound.

CHAPTER 5

FLOW-SHOP SCHEDULING

Another special case to be considered before discussing the general job-shop problem is that in which there is a natural ordering of the machines in a shop. In some cases there are two or more machines, and at least some of the jobs have a sequence of operations to be performed before they are completed and can leave the shop. In such cases the collection of machines is said to constitute a *flow-shop* if the machines are numbered in such a way that, for every job to be considered, operation K is performed on a higher-numbered machine than operation J, if $J > K$.

An obvious example of such a shop is an assembly line, where the workers or work stations represent the machines. Any group of machines served by a unidirectional, noncyclic conveyor would be considered a flow-shop, and a strict or pure example would be a case in which all materials handling was accomplished by the conveyor. Of course, any real example is spoiled to some extent by the necessity for "rework" operations. It is not required that every job have an operation on each machine in the shop, nor must all work enter the shop on a single machine, or leave from a single machine. The only requirement is that all movement between machines within the shop be in a uniform direction.

As might be expected, such shops have special characteristics that affect scheduling decisions significantly. In general, they are rather more easily scheduled than a job-shop and somewhat more is known about this class of shops. However, it will be clear, by the end of the chapter, that this is only a relative advantage and that the problem of scheduling even a flow-shop of practical dimensions is unsolved.

5-1 PERMUTATION SCHEDULES

Much of the simplicity of the single-machine models of Chapter 3 can be attributed to the fact that it is sufficient in those cases to consider only permutation schedules, which are completely described by a particular permutation of the job identification numbers. The difficulty of the general job-shop problem of Chapter 6 and the paucity of results for that case is a consequence of the fact that a much larger class of schedules must be considered. The flow-shop is an intermediate situation in which for certain special cases permutation schedules are sufficient, and in general there is some preservation of order of jobs on initial and terminal machines. The situation is described by Theorems 5–1 and 5–2.

Theorem 5–1. When one is scheduling an $n/m/F$ problem with all jobs simultaneously available, to minimize any regular measure of performance one need consider only schedules in which the same job order is prescribed on the first two machines.

Proof. If a schedule does not have the same job ordering on the first two machines, then somewhere in the schedule for the first machine there must be a job *I* that directly precedes a job *J*, where *I* follows *J* (possibly with intervening jobs) on the second machine.

First machine	. . .	*I*	*J*	. . .		
Second machine			. . .	*J*	. . .	*I*	. . .

Obviously the positions of these two jobs can be reversed on the first machine without requiring an increase in the starting-time of any job on the second machine, and therefore on any subsequent machine. This exchange could not cause an increase in the completion-time of any job, and hence not an increase in any regular measure of performance. The exchange may in fact permit a decrease in the starting-time of some jobs on the second machine and consequently an improvement in the measure of performance, but nonincrease is sufficient and all that can be promised.

This result does not exclude the possibility that there may be good schedules that have different orderings on the first two machines. It simply means that it is sufficient to consider only schedules that do have the same ordering on the first two machines, since for any other schedule there is a corresponding and easily obtained schedule of this type that is equivalently good and possibly better.

By specializing the measure of performance, one can obtain a slightly stronger result.

Theorem 5–2. When one is scheduling an $n/m/F/F_{\max}$ problem with all jobs simultaneously available, one need consider only schedules in which the same job order is prescribed on machines 1 and 2, and the same job order is prescribed on machines $m - 1$ and m.

Proof. The first part, specifying common ordering on machines 1 and 2, is simply an application of Theorem 5–1, since maximum flow-time is a regular measure of performance.

Suppose that the ordering on the last two machines is different. Then somewhere on the schedule for machine m there must be a job *I* that directly follows a job *J*, where *I* precedes *J* (possibly with intervening jobs) on machine $m - 1$.

Machine $m - 1$	*I*	. . .	*J*	. . .	
Machine m				*J*	*I*

Obviously the positions of these two jobs can be reversed on machine m without increasing the maximum flow-time of these particular jobs and without changing the flow-time of any other job. There is, of course, the possibility that job *I* could begin on machine m before job *J* so that some reduction in maximum flow-time could be achieved by inverting the order.

A simple example will illustrate that Theorem 5–2 could not be strengthened to hold for any regular measure of performance. Suppose that two jobs are to be scheduled on a three-machine flow-shop to minimize the mean flow-time:

	Processing-times		
Machine	1	2	3
Job I	4	1	1
Job J	1	4	4

There are two schedules that have the same order on machines 2 and 3.

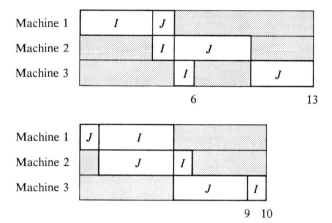

In each case the mean flow-time is 9.5. There is a schedule with different orderings on machines 2 and 3 with mean flow-time of 9.0.

The result is that for *any* regular measure of performance, and in particular for mean flow-time, the two-machine flow-shop is the only case in which it is sufficient to consider only permutation schedules. For *some* measures of performance, in particular for maximum flow-time, permutation schedules are sufficient for a three-machine flow-shop. But another example will indicate that even for maximum flow-time, in flow-shops of four or more machines a more general type of schedule must be considered.

	Processing-times			
Machine	1	2	3	4
Job I	4	1	1	4
Job J	1	4	4	1

There are only two order-preserving schedules, both of which have a maximum flow-time of 14.

14

14

Now consider a schedule which has the same order on machines 1 and 2 and the same on machines 3 and 4, but in which the order is reversed between machines 2 and 3.

12

The maximum flow-time is 12, which is less than that achieved by either order-preserving schedule.

5–2 MINIMIZING MAXIMUM FLOW-TIME IN A TWO-MACHINE FLOW-SHOP ($n/2/F/F_{max}$)

Probably the most frequently cited paper in the field of scheduling is Johnson's solution to the two-machine flow-shop problem [103]. He gives an algorithm for sequencing n jobs, all simultaneously available, in a two-machine flow-shop so as to minimize the maximum flow-time. This paper is important, not only for its own content, but also for the influence it has had on subsequent work. In particular, it is likely that the general acceptance of minimizing the maximum flow-time as a criterion for the general job-shop problem can be attributed to Johnson's result. The actual procedure is rather obvious and intuitive, so that it is quite probable that if any such shops existed and if there were actual interest in minimizing the maximum flow-time the proper ordering would have been used. Johnson's accomplishment is not so much the proposition of an algorithm as it is the offering of a proof that the obvious algorithm is in fact optimal.

In presenting this procedure and proof, it is convenient to temporarily adopt Johnson's simpler notation. Let

$A_i = p_{i,1}$ be the processing-time (including setup, if any) of the first operation of the ith job,

$B_i = p_{i,2}$ be the processing-time (including setup, if any) of the second operation of the ith job, and

F_i be the time at which the ith job is completed (consistent with the current notation).

Each job consists of a pair (A_i, B_i), where A_i is the work to be performed on the first machine of the shop and B_i is the work to be performed on the second machine. This ordering on the machines is the same for each of the n jobs, although it is allowable for some of the A_i's and B_i's to be zero, since some of the jobs may have only a single operation. The normal constraints of the simple shop are assumed:

1) each machine can work on only one job at a time, and
2) each job can be in process on only one machine at a time—for each i, A_i must be completed before B_i can begin.

The problem, then, is: Given the $2n$ values, $A_1, A_2, \ldots, A_n, B_1, B_2, \ldots, B_n$, find an ordering of these jobs on each of the two machines so that neither the precedence (routing) nor the occupancy constraints are violated and so that the maximum of the F_i is made as small as possible.

It follows from either Theorem 5–1 or 5–2 that one need consider only schedules which have the same ordering on each of the two machines. It should also be obvious that, like the problems of Chapter 3, neither preemption nor the insertion of idle-time need be considered, so that a solution is simply a matter of specifying an optimal permutation of the job identification numbers.

An examination of obvious lower bounds on the maximum flow-time will suggest the nature of Johnson's algorithm. Again using the notation of Chapter 3 of a square-bracketed subscript to denote position in sequence, $A_{[2]}$ indicates the time on machine 1 of the job which is second in sequence. The schedule could be drawn as follows.

| Machine 1 | $A_{[1]}$ | $A_{[2]}$ | | $A_{[3]}$ | $\cdots$ | | $A_{[n-1]}$ | | $A_{[n]}$ | |
| Machine 2 | | $B_{[1]}$ | | $B_{[2]}$ | $\cdots$ | | | $B_{[n-1]}$ | | $B_{[n]}$ |

It is clear that the last job cannot be completed earlier than the time required to process each job on machine 1 plus the time needed to perform the second operation of the last job, since $B_{[n]}$ cannot overlap $A_{[n]}$. Thus:

$$F_{\max} \geq \sum_{i=1}^{n} A_{[i]} + B_{[n]}.$$

Similarly, the last job cannot be completed in less time than it takes to process each job on machine 2 plus the time caused by the delay before machine 2 can begin, since $B_{[1]}$ cannot overlap $A_{[1]}$. Thus

$$F_{\max} \geqq A_{[1]} + \sum_{i=1}^{n} B_{[i]}.$$

Now note that the sum of the A's and the sum of the B's are direct consequences of the given information and entirely unaffected by the ordering of the jobs, so that to reduce these bounds one can only influence $B_{[n]}$ and $A_{[1]}$ by the choice of sequence. One would therefore choose the smallest of the set of $2n$ values of the A's and B's. If this value happened to be an A_i one would put that job first in sequence so as to make $A_{[1]}$ as small as possible. If it happened to be a B_i one would put that job last in sequence so as to make $B_{[n]}$ as small as possible. With the position of that one job determined, one can repeat essentially the same argument for the set of $n - 1$ jobs that remain. Of course, these bounds are not generally attainable so that one cannot be sure that this construction would yield a good schedule. Neither is it clear what would happen at the ends as one applied the arguments repeatedly to reduced job sets, but the method is nevertheless heuristically appealing. This bounding argument is precisely the basis for Johnson's procedure and he offers a proof of its optimality.

It should also be apparent that one can restrict attention to schedules in which the work on machine 1 is compact to the left; i.e., there is no idle-time on machine 1 until $A_{[n]}$ has been completed. The assignments of the $A_{[i]}$ values of any arbitrary schedule, then, can be shifted to the left to achieve this compactness without increasing the maximum flow-time, and one can make the following definition:

Let $X_{[i]}$ be the idle-time on machine 2 immediately preceding $B_{[i]}$. A typical schedule might look like the following:

Machine 1	$A_{[1]}$	$A_{[2]}$	$A_{[3]}$	$A_{[4]}$		$A_{[5]}$	
Machine 2	$X_{[1]}$	$B_{[1]}$	$X_{[2]}$	$B_{[2]}$	$B_{[3]}$	$X_{[4]}$	$B_{[4]}$	$B_{[5]}$

The values of the $X_{[i]}$ can be given in terms of the A's and B's by the following relationships:

$$X_{[1]} = A_{[1]},$$
$$X_{[2]} = \max(A_{[1]} + A_{[2]} - B_{[1]} - X_{[1]}, 0),$$
$$X_{[3]} = \max(A_{[1]} + A_{[2]} + A_{[3]} - B_{[1]} - B_{[2]} - X_{[1]} - X_{[2]}, 0).$$

In general, we may write

$$X_{[J]} = \max\left(\sum_{i=1}^{J} A_{[i]} - \sum_{i=1}^{J-1} B_{[i]} - \sum_{i=1}^{J-1} X_{[i]}, 0\right).$$

The partial sums of the X's can be obtained from these expressions:

$$X_{[1]} = A_{[1]},$$
$$X_{[1]} + X_{[2]} = \max(A_{[1]} + A_{[2]} - B_{[1]}, A_{[1]}),$$
$$X_{[1]} + X_{[2]} + X_{[3]} = \max(A_{[1]} + A_{[2]} + A_{[3]} - B_{[1]} - B_{[2]}, X_{[1]} + X_{[2]})$$
$$= \max\left(\sum_{i=1}^{3} A_{[i]} - \sum_{i=1}^{2} B_{[i]}, \sum_{i=1}^{2} A_{[i]} - B_{[1]}, A_{[1]}\right).$$

In general,

$$\sum_{i=1}^{J} X_{[i]} = \max \left(\sum_{i=1}^{J} A_{[i]} - \sum_{i=1}^{J-1} B_{[i]}, \sum_{i=1}^{J-1} A_{[i]} - \sum_{i=1}^{J-2} B_{[i]}, \dots, \sum_{i=1}^{2} A_{[i]} - B_{[1]}, A_{[1]} \right).$$

If Y_J is defined as

$$Y_J = \sum_{i=1}^{J} A_{[i]} - \sum_{i=1}^{J-1} B_{[i]},$$

then

$$\sum_{i=1}^{J} X_{[i]} = \max(Y_1, Y_2, \dots, Y_J).$$

Now if $F_{\max}(\mathbf{S})$ is used to denote the maximum flow-time of a particular schedule $\mathbf{S}$, then

$$F_{\max}(\mathbf{S}) = \sum_{i=1}^{n} B_{[i]} + \sum_{i=1}^{n} X_{[i]} = \sum_{i=1}^{n} B_{[i]} + \max(Y_1, Y_2, \dots, Y_n).$$

Since the sum of the B_i's is independent of sequence, the maximum flow-time depends entirely on the sum of the intervals of idle-time on the second machine, and this is equivalent to the maximum of the Y_i's. Our task of finding a schedule $\mathbf{S}^*$ such that $F_{\max}(\mathbf{S}^*) \le F_{\max}(\mathbf{S})$ for any $\mathbf{S}$ reduces to the task of finding an ordering so that the maximum of the n values of Y_i is minimized.

Theorem 5–3. (Johnson 1954, [103]).   In an $n/2/F/F_{\max}$ problem with all jobs simultaneously available, job j should precede job $j + 1$ if

$$\min(A_j, B_{j+1}) < \min(A_{j+1}, B_j).$$

The working procedure described earlier follows directly from this inequality. The resulting optimal schedule is unique if there is an inequality of this form between each pair of jobs, but there will be a set of alternative optimal schedules if there is equality for some pairs of jobs. The proof is in two parts. The first shows that sequencing according to this inequality minimizes the maximum of the Y_i's; the second illustrates that the relation is transitive so that a complete schedule is in fact specified by this pairwise ordering.

Proof. Part 1—Ordering Minimizes Maximum Y_i. Suppose that we are concerned with filling positions J and $J + 1$ in the sequence—either job j or job $j + 1$ will be put in position J, the other in position $J + 1$. The inequality of the hypothesis yields

$$\max(-B_j, -A_{j+1}) < \max(-B_{j+1}, -A_j). \tag{1}$$

Now consider the quantity

$$\sum_{i=1}^{J+1} A_{[i]} - \sum_{i=1}^{J-1} B_{[i]}, \tag{2}$$

and note that it does not depend on which of the two jobs fills position J. Now we add (2) to each of the terms of (1):

$$\sum_{i=1}^{J+1} A_{[i]} - \sum_{i=1}^{J-1} B_{[i]} - B_j = Y_{J+1},$$

which denotes the value of Y_{J+1} *if* job j is in position J;

$$\sum_{i=1}^{J+1} A_{[i]} - A_{j+1} - \sum_{i=1}^{J-1} B_{[i]} = Y_J,$$

which denotes the value of Y_J *if* job j is in position J;

$$\sum_{i=1}^{J+1} A_{[i]} - \sum_{i=1}^{J-1} B_{[i]} - B_{j+1} = Y'_{J+1},$$

which denotes the value of Y_{J+1} *if* job $j + 1$ is in position J;

$$\sum_{i=1}^{J+1} A_{[i]} - A_j - \sum_{i=1}^{J-1} B_{[i]} = Y'_J,$$

which denotes the value of Y_J *if* job $j + 1$ is in position J.
    The result is

$$\max(Y_J, Y_{J+1}) < \max(Y'_J, Y'_{J+1}),$$

which says simply that if the inequality of the hypothesis holds then the greater of these two values of Y_i is reduced by putting job j in position J (and job $j + 1$ in position $J + 1$). None of the other $n - 2$ values of Y_i is affected by this choice of job for position J, so the maximum of all the Y_i is either unchanged or reduced by putting job j in position J.

Part 2—Transitivity of the Relationship. We must show that for three jobs, i, j, and k, if

$$\min(A_i, B_j) \leqq \min(A_j, B_i) \quad \text{and} \quad \min(A_j, B_k) \leqq \min(A_k, B_j),$$

then

$$\min(A_i, B_k) \leqq \min(A_k, B_i),$$

except possibly for cases in which the relationship is an equality between both i and j and j and k.

CASE 1. $A_i \leqq B_j$, A_j, B_i and $A_j \leqq B_k$, A_k, B_j. Then $A_i \leqq A_j \leqq A_k$ and $A_i \leqq B_i$, so $A_i \leqq \min(A_k, B_i)$.

CASE 2. $B_j \leqq A_i$, A_j, B_i and $B_k \leqq A_j$, A_k, B_j. Then $B_k \leqq B_j \leqq B_i$ and $B_k \leqq A_k$, so $B_k \leqq \min(A_k, B_i)$.

CASE 3. $A_i \leqq B_j$, A_j, B_i and $B_k \leqq A_j$, A_k, B_j. Then $A_i \leqq B_i$ and $B_k \leqq A_k$, so $\min(A_i, B_k) \leqq \min(A_k, B_i)$.

CASE 4. $B_j \leqq A_i$, A_j, B_i and $A_j \leqq B_k$, A_k, B_j. Then $A_j = B_j$ and the relationship is an equality between both i and j and j and k.

There is an interesting geometrical interpretation that may help to illustrate the character of the problem, as well as the virtue of Johnson's algorithm. Consider a plot with machine 1 activity as abscissa and machine 2 activity as ordinate. A schedule may be represented on these axes by a line, **S**, with $2n$ straight segments (possibly some of length zero)—horizontal segments with length corresponding to $A_1, A_2,$ $\ldots, A_n$ and vertical segments with length corresponding to $B_1, B_2, \ldots, B_n$. The segments are taken in the order $A_{[1]}, B_{[1]}, A_{[2]}, B_{[2]}, \ldots, A_{[n]}, B_{[n]}$ as shown in Fig. 5–1. Each of the $n!$ lines that could be constructed begins at $(0, 0)$ and ends at $\{\sum_{i=1}^{n} A_i, \sum_{i=1}^{n} B_i\}$. The maximum flow-time of a particular schedule, **S**, may be determined from the graph by constructing a line V:

1) From $(0, 0)$ to

$$\left(\sum_{i=1}^{n} A_i, \sum_{i=1}^{n} B_i \right) ;$$

2) That lies everywhere on or below line **S**;

3) That is composed of horizontal, vertical, and 45° segments, using 45° segments whenever possible.

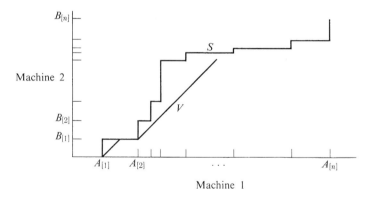

Fig. 5–1. Graph of a two-machine flow-shop schedule.

Such a line is shown in Fig. 5–1. The horizontal segments represent intervals when machine 1 is working and machine 2 is idle; the vertical segments represent intervals when 2 is working and 1 is idle; and the 45° segments represent intervals when both machines are working. The maximum flow-time of a schedule **S** is simply $\sum_{i=1}^{n} A_i$ plus the sum of the vertical segments of the corresponding line V (or equivalently $\sum_{i=1}^{n} B_i$ plus the sum of the horizontal segments of V). Obviously, to minimize the maximum flow-time one would seek an ordering of the jobs so that line **S** would permit the corresponding line V to use as much 45° segment as possible. This requires a line **S** which is essentially concave and which stays "up out of the way of V" as much as possible. It is clear that Johnson's algorithm produces this concavity, starting from $(0, 0)$ in as northerly a direction as possible, and from $(\sum_{i=1}^{n} A_i, \sum_{i=1}^{n} B_i)$ in as westerly a direction as possible.

Example: The following simple example will illustrate the execution of Johnson's procedure and compare the result with several other orderings.

Job number i	First operation A_i	Second operation B_i	Order of selection	Position in sequence
a	6	3	3	4
b	0	2	1	1
c	5	4	4	3
d	8	6	5	2
e	2	1	2	5

A Gantt chart for the Johnson sequence would look like the following:

Although this is not the only sequence producing the optimal maximum flow-time of 23, many other sequences do not work as well. For example, sequencing in order of increasing processing-time on the first operation yields a maximum flow-time of 27.

Simply taking the jobs in the order listed in the table gives a maximum flow-time of 26.

5–3 MINIMIZING MEAN FLOW-TIME
IN A TWO-MACHINE FLOW-SHOP $(n/2/F/\overline{F})$

Even for the two-machine flow-shop the minimization of mean flow-time is a very difficult problem. Although Theorem 5–1 is still applicable, so that it is necessary to consider only schedules in which the sequence of jobs is the same on each machine, no constructive algorithm comparable to Johnson's is known. Johnson's procedure is not optimal with respect to this criterion and, in general, it is not even very good.

For example, in the previous five-job illustration the average flow-time for the Johnson sequence is 15, while for the shortest-processing-time sequence the average is 11.8.

It is instructive to examine the special case with only two jobs, before considering the difficulties of the general n job case. There are only two schedules that must be considered.

Schedule **S**   Machine 1 | A_1 | A_2 | | Schedule **S'**   Machine 1 | A_2 | A_1 | |

Machine 2 | | B_1 | B_2 | Machine 2 | | B_2 | B_1 |

Under schedule **S** the sum of the flow-times is

$$2\bar{F} = 2A_1 + B_1 + B_2 + \max(A_2, B_1).$$

Under schedule **S'** the sum of the flow-times is

$$\bar{F}' = 2A_2 + B_1 + B_2 + \max(A_1, B_2).$$

Clearly job 1 should precede job 2 (that is, schedule **S** should be chosen) if $\bar{F} < \bar{F}'$, which requires that

$$2A_2 - 2A_1 + \max(A_1, B_2) - \max(A_2, B_1) > 0.* \tag{1}$$

One can, of course, evaluate this expression directly for any pair of jobs, but it is interesting to look for a simpler condition on the processing-times that will determine the optimal sequence. One is guided by the knowledge that the shortest-processing-time sequence minimizes the mean flow-time in the comparable single-machine case (see Section 3–2). However, there are a number of ways in which this procedure might be generalized to the two-machine case. It is obviously not sufficient to consider only the first operations of the jobs to determine the best sequence, for although $A_2 > A_1$, statement (1) can be made either true or false by appropriate specification of the values of B_1 and B_2. Similarly it is not sufficient that $B_2 > B_1$, so that the shortest-processing-time concept cannot be applied to either the first or the second operations alone. One might consider the total processing-time of each job, and place job 1 first if

$$(A_2 + B_2) > (A_1 + B_1).$$

A simple counterexample shows that this condition does not ensure the truth of statement (1):

Let $B_1 = 0$; $A_2 = B_2$; $A_1 = A_2 + B_2 - \epsilon$,   where $\epsilon > 0$.

However, if each of the operations of job 1 is shorter than the corresponding operation of job 2, then statement (1) is true and job 1 should precede job 2. The statement that

$$A_2 \geqq A_1 \quad \text{and} \quad B_2 \geqq B_1$$

---

* It is interesting to note that the last two terms of the left side suggest the exact opposite of the Johnson procedure: select the maximum of the A's and B's; if this is an A put that job first in sequence; if a B put that job last in sequence.

implies that

$$2A_2 - 2A_1 + \max(A_1, B_2) - \max(A_2, B_1) \geqq 0.$$

This can be seen by examining the four possibilities for the terms $\max(A_1, B_2)$ and $\max(A_2, B_1)$.

CASE 1. $A_1 > B_2$ and $A_2 > B_1$.
$$2A_2 - 2A_1 + A_1 - A_2 = A_2 - A_1 \geqq 0.$$

CASE 2. $A_1 > B_2$ and $B_1 > A_2$.
Since $A_2 \geqq A_1$, this implies that $B_1 > B_2$, which is contrary to assumption, so this case is impossible.

CASE 3. $B_2 > A_1$ and $A_2 > B_1$.
$$2A_2 - 2A_1 + B_2 - A_2 = (A_2 - A_1) + (B_2 - A_1) \geqq 0.$$

CASE 4. $B_2 > A_1$ and $B_1 > A_2$.
$$2A_2 - 2A_1 + B_2 - B_1 \geqq 0.$$

It is also obvious that while this condition is *sufficient*, it is *not necessary*; there are cases in which job 1 does not dominate job 2 by being shorter on both operations, yet nevertheless job 1 should precede job 2 in sequence. For example,

$$A_1 = 1, \qquad B_1 = 3, \qquad A_2 = 4, \qquad B_2 = 2.$$

A general expression for the mean flow-time of n jobs can be obtained by an argument that follows Johnson's directly:

1) It is sufficient to consider only schedules with the same sequence on each machine.
2) It is sufficient to consider only schedules which are compact on the first machine.
3) Let $X_{[i]}$ be the idle-time that precedes operation $B_{[i]}$ on the second machine.
4) Let

$$Y_J = \sum_{i=1}^{J} A_{[i]} - \sum_{i=1}^{J-1} B_{[i]}.$$

Then the flow-times of individual jobs can be given by

$$F_{[1]} = A_{[1]} + B_{[1]} = X_{[1]} + B_{[1]},$$
$$F_{[2]} = X_{[1]} + B_{[1]} + X_{[2]} + B_{[2]},$$
$$F_{[J]} = \sum_{i=1}^{J} B_{[i]} + \sum_{i=1}^{J} X_{[i]} = \sum_{i=1}^{J} B_{[i]} + \max(Y_1, Y_2, \ldots, Y_J).$$

The mean flow-time is given by

$$\bar{F} = \frac{1}{n}\left[\sum_{J=1}^{n} \sum_{i=1}^{J} B_{[i]} + \sum_{J=1}^{n} \max(Y_1, Y_2, \ldots, Y_J) \right].$$

Using this expression, one can show that for adjacent jobs I and J in a schedule, job I should precede job J if

$$A_J \geqq A_I \qquad \text{and} \qquad B_J \geqq B_I,$$

but this does not constitute a solution since it does not produce a complete ordering of the jobs.

Ignall and Schrage have applied a branch-and-bound technique (see, for example, Section 4–1.1) to this problem [87]. They have observed that it is an appropriately tree-structured problem. Each node represents a sequence of r of the n jobs, with $n - r$ of the jobs remaining to be ordered. The required lower bound on mean flow-time for schedules emanating from a given node is roughly the following:

$$F_r + \max(F^a_{n-r}, F^b_{n-r}),$$

where F_r is the sum of the flow-times of the r jobs assigned in the partial schedule represented by the node, F^a_{n-r} is the sum of the flow-times for the remaining $n - r$ jobs, calculated under the assumption that the A's dominate the B's ($A_i > B_i$, for all i of the $n - r$) and that the jobs are sequenced in order of increasing A, F^b_{n-r} is the sum of the flow-times for the remaining $n - r$ jobs, calculated under the assumption that the B's dominate the A's ($A_i < B_i$, for all i of the $n - r$) and that the jobs are sequenced in order of increasing B.

Table 5–1

Computational Experience with a Branch-and-Bound Procedure
for the $n/2/F/\bar{F}$ Problem

(Ignall and Schrage [87])

Number of job sets	Number n of jobs/set	Number of nodes created					
		Minimum possible	Maximum possible	Observed			
				Minimum*		Median	Maximum
50	5	15	206	15	(26)	15	38
50	6	21	1,237	21	(16)	29	108
50	7	28	8,660	28	(5)	61	190
50	8	36	69,281	36	(3)	123	777
50	9	45	623,530	45	(5)	321	1258

Number of job sets	Number n of jobs/set	Maximum number of nodes on list at one time				
		Minimum possible	Maximum possible	Observed		
				Minimum	Median	Maximum
50	5	11	120	11	11	21
50	6	16	720	16	20	56
50	7	22	5,040	22	43	98
50	8	29	40,320	29	80	293
50	9	37	362,880	37	161	579

* Parenthesized numbers indicate number of times minimum was observed.

They test the procedure on computer-generated job sets with processing-times drawn from a uniform distribution. A portion of their results is shown in Table 5-1. In each case an optimal solution was obtained and the table reports the number of nodes required. In general, the computation *time* is proportional to the number of nodes created and the storage *space* requirements are proportional to the maximum number of nodes that must be stored at any one time. The program was written in the Cornell List Processor language and executed on a CDC 1604 [35]. Execution time ranged from 0.75 second for the five-job problems to about 4 minutes for the hardest of the nine-job problems. They note that the difficulty of the problem is essentially doubled each time n is increased by one, and they argue that this is a consequence of the fact that an optimal sequence for $n - 1$ jobs does not always provide a good starting place for the n job problem. For example, consider

$$A_1 = 2, \quad A_2 = 10, \quad A_3 = 1,$$
$$B_1 = 11, \quad B_2 = 3, \quad B_3 = 8.$$

For a two-job problem, with jobs 1 and 2, the optimal sequence is 1, then 2; for a three-job problem, the optimal sequence is 3-2-1, reversing the order of jobs 1 and 2.

If 2^n is a discouraging rate for the computational task to increase, it is at least a substantial improvement over $n!$, which would be required for exhaustive enumeration.

5-4 THE THREE-MACHINE FLOW-SHOP $(n/3/F/F_{max})$

When a schedule is constructed to minimize the maximum flow-time, Theorem 5-2 states that it is still sufficient to consider only permutation schedules in the three-machine flow-shop. It is somewhat surprising, therefore, that in spite of this similarity to the two-machine case a comparable constructive algorithm cannot be found.

There has been considerable attention given to the problem and a number of special variations have been solved. In his original paper [103] Johnson considers the special three-machine cases in which either min $A_i \geq$ max B_i or min $D_i \geq$ max B_i, where D_i is the processing-time of the ith job on the third machine. In these cases the second machine is completely dominated by either the first machine or the third, and a sequence which minimizes the maximum flow-time is obtained by putting job I ahead of job J if

$$\min(A_I + B_I, D_J + B_J) < \min(A_J + B_J, D_I + B_I).$$

The working procedure is the same as for the two-machine case, with A_i replaced by $A_i + B_i$ and B_i replaced by $B_i + D_i$.

Wagner and Giglio applied this procedure to general three-machine flow-shop problems where the processing-times were randomly generated and did *not* satisfy the conditions min $A_i \geq$ max B_i or min $D_i \geq$ max B_i [66]. There were 20 problems, each of 6 jobs, which was few enough to allow the optimal solution to be obtained by enumeration for comparison. In 9 out of the 20 cases an optimal solution was actually obtained by the procedure, and in 8 of the remaining cases the solution obtained could be made optimal by a simple interchange of two adjacent jobs. The average maximum flow-time yielded by the procedure was 131.7, whereas the average optimal time for the 20 cases was 127.9. Apparently the procedure represents a useful approximation even for cases in which the conditions that ensure optimality are not satisfied, and yields an excellent starting point from which to attempt further modification.

Ignall and Schrage have also applied a branch-and-bound technique to the three-machine problem [87]. Their results, for 150 randomly generated job sets, are shown in Table 5–2. Computation times ranged from less than a second to more than $2\frac{1}{2}$ minutes on the CDC 1604. They observe that the three-machine maximum-flow-time problem is easier for the branch-and-bound procedure than the two-machine mean-flow-time problem, in that a higher proportion of the job sets were solved with the minimum possible number of nodes. They give the operation-times for the particular ten-job problem, which proves hardest for the algorithm.

i	1	2	3	4	5	6	7	8	9	10
A_i	1	5	7	8	3	7	9	8	6	3
B_i	2	9	6	9	2	10	7	9	1	1
D_i	9	7	8	9	3	4	7	4	3	1

There is an optimal ordering of the jobs in the order they are numbered, with maximum flow-time of 66. The reader might find it interesting to conceal the job numbers, shuffle the jobs, and try to find an optimal ordering.

Jackson has considered a very special problem which is actually a variant of the three-machine problem [91]. He considers a case in which the n jobs have a common machine for their first operation and a common machine for their last (third) operation, but in which the second operation of each job is performed on a different machine. There are thus $n + 2$ machines, n of them corresponding to a second machine in a flow-shop which can process many jobs simultaneously. Using A_i, B_i, and D_i as before, Jackson shows that an ordering minimizes the maximum flow-time

Table 5–2

Computational Experience with a Branch-and-Bound Procedure
for the $n/3/F/F_{\max}$ Problem

(*Ignall and Schrage* [87])

		Number of nodes created					
				Observed			
Number of job sets	Number n of jobs/set	Minimum possible	Maximum possible	Minimum†		Mean*	Maximum
50	8	36	69,281	36	(33)	284	1455
50	9	45	623,530	45	(39)	661	1687
50	10	55	6,235,301	55	(38)	711	2570

* Mean excluding problems that achieved the minimum; 284 is mean number for 17 job sets, etc.
† Parenthesized numbers indicate number of times minimum was observed.

if job I precedes job J when

$$\min(A_I + B_I, B_J + D_J) < \min(A_J + B_J, B_I + D_I).$$

This is again Johnson's three-machine condition so that Jackson has exhibited another special case for which this ordering is optimal. An interesting interpretation of this case is to consider the B_i an arbitrary delay which must be observed between the first and second operations in a two-machine flow-shop—say for inspection, materials handling, cooling, drying, etc. Moreover, the B_i can be permitted to take on negative values covering the situation in which lap-scheduling is allowed in a two-machine flow-shop; that is, portions of a job may be moved so that work can begin on the second operation before the final cycles of the first operation have been completed. Mitten [140] and Johnson [104] also consider this special problem. It is worth noting that the peculiar properties of the second machine in this case (its ability to process many jobs at once) make it unnecessary for an optimal schedule to have the same job order on the first and third machines.

Wagner, with Story [189] and Giglio [66], has used integer programming to formulate and solve the $n/3/F/F_{\max}$ problem. The program is defined as follows.* Let:

$z_{ij} = \begin{cases} 1 & \text{if job } i \text{ is scheduled in order position } j \\ 0 & \text{otherwise,} \end{cases}$

$Z(j) = [z_{1j}, z_{2j}, \ldots, z_{nj}]$, a column vector,

$X_j^k = $ idle-time on machine k before the start of the job in position j,

$Y_j^k = $ idle-time for the job in position j between the end of the operation on machine k and the beginning on machine $k + 1$,

$A = $ row vector of (integer) processing times for jobs $1, 2, \ldots, n$ on machine 1,

$B = $ row vector of (integer) processing times for jobs $1, 2, \ldots, n$ on machine 2,

$D = $ row vector of (integer) processing times for jobs $1, 2, \ldots, n$ on machine 3.

Note that there are n^2 different z_{ij} variables, n of which are to be set equal to one, the rest to zero. For example, if there are four jobs, the permutation 3–1–4–2 would be expressed by $z_{31} = z_{12} = z_{43} = z_{24} = 1$; all other $z_{ij} = 0$.
The constraints are

$$\sum_i z_{ij} = 1, \qquad \sum_j z_{ij} = 1,$$

$$X_{j+1}^2 + BZ(j + 1) + Y_{j+1}^2 - Y_j^2 - DZ(j) - X_{j+1}^3 = 0, \qquad j = 1, 2, \ldots, n - 1,$$

$$AZ(j + 1) + Y_{j+1}^1 - Y_j^1 - BZ(j) - X_{j+1}^2 = 0, \qquad j = 1, 2, \ldots, n - 1.$$

Minimizing the maximum flow-time is equivalent to minimizing the idle-time on machine 3, which is used as the objective function:

$$\text{Minimize} \sum_{j=1}^{n} X_j^3.$$

---

* A more detailed description of the integer-programming model for scheduling problems will be given in Section 6-4.

Wagner states that "previous experience with integer programming problems has demonstrated that the convergence properties of the algorithm are highly sensitive to the form in which the problem is stated initially" ([66], page 310). He considers six variations of the problem statement, adding various additional constraints and initial bounds on the solution, testing each against six sample problems with $n = 6$. He reports the number of iterations required for solution,* with the routine set to terminate at 10,000 whether or not a solution has been obtained. These results are shown in Table 5–3.

Table 5–3

Number of Iterations to Solution for
Integer Programming Approach to Six Different
$6/3/F/F_{max}$ Problems

(Giglio and Wagner [66])

Program variation	Sample problem					
	1	2	3	4	5	6
A	745	42	1973	188	1,555	10,000*
B	404	44	1647	218	10,000*	9,568
C	1256	50	710	123	1,025	2,467
D	85	50	710	63	858	2,811
E	85	50	208	58	245	304
F	220	146	903	–	581	–

* Terminated without solution.

These data do not appear very encouraging. In many cases the number of iterations is of the same order of magnitude as the number of possible permutations (720) and an iteration involves more computation than the generation of a permutation. However, they do confirm that the exact form in which the problem is stated has an important effect on the efficiency of the algorithm and further work is being directed at the development of more efficient constraints and bounds. At least at the moment the branch-and-bound technique appears to have a substantial advantage over integer programming as a practical computational procedure for this problem.

Dudek and Teuton [45] have proposed an algorithm to minimize the maximum flow-time in a three-machine flow-shop, which is also applicable to larger shops if one arbitrarily limits consideration to permutation schedules. The algorithm suggests a method for selecting the job to place first in sequence, the job to select from the remaining $n - 1$ to place second, etc. Expanding on the notation used in Section 5–2, let $X_{[i]}^j$ be the idle-time on machine j directly preceding the ith job in sequence. Inductive formulas for the X's in terms of the processing-times and the X's for lower-

---

* Gomory's integer programming algorithm was used (SHARE PKIP91). This uses a dual solution so that no primal feasible solution is available until the algorithm terminates.

numbered machines are given. Let

$\sum\limits_{i}^{j}$ denote $\sum_{k=1}^{i} X_{[k]}^{j}$,   the sum of all of the idle intervals on machine j up to the ith position in the sequence, and

$\sum\limits_{i}^{j} ([1], [2], \ldots, [i])$ represent the sum with explicit identification of the particular jobs in the i positions of the sequence.

Now select a job r for first position in sequence such that

$$\sum\limits_{2}^{j} (r, s) \leqq \sum\limits_{2}^{j} (s, r) \quad \text{for all } s \neq r \text{ and for } j = 2, 3, \ldots, m.$$

If such a job does not exist, then a procedure is given for selecting a subset of jobs and proceeding for each member of this subset as if it had been the unique selection. With the job in first position now determined, select a job s for second position such that

$$\sum\limits_{3}^{j} (r, s, t) \leqq \sum\limits_{3}^{j} (r, t, s) \quad \text{for all } t \neq s, t \neq r \text{ and for } j = 2, 3, \ldots, m.$$

One continues in this manner until the entire sequence has been determined. There is a considerable amount of computation required, but undeniably less than complete enumeration if an appreciable number of jobs is involved. This is apparently a very effective procedure, for when Dudek and Teuton compared it to complete enumeration in 168 problems with $m = 3$ or 5 and $n = 3, 4, 5, 6$ or 7, in every case the algorithm obtained at least one optimal sequence. Unfortunately they conclude from this test that their procedure is a theoretically general one that will guarantee at least one optimal sequence for an arbitrary problem. Karush [105] effectively demonstrated the fallacy of this claim by the exhibition of a counterexample.

	Processing-times		
	A_i	B_i	D_i
Job I	3	22	2
Job J	22	20	20
Job K	20	14	18

The algorithm would select job I for first position in sequence, since

$$\sum\limits_{2}^{2} (I, J) = 3 < \sum\limits_{2}^{2} (J, I) = 22, \quad \sum\limits_{2}^{3} (I, J) = 43 < \sum\limits_{2}^{3} (J, I) = 44,$$

$$\sum\limits_{2}^{2} (I, K) = 3 < \sum\limits_{2}^{2} (K, I) = 20, \quad \sum\limits_{2}^{3} (I, K) = 37 < \sum\limits_{2}^{3} (K, I) = 38.$$

However, the maximum flow-times for all six possible schedules are

$$F_{\max}(I, J, K) = 83, \quad F_{\max}(I, K, J) = 85, \quad F_{\max}(J, K, I) = 82,$$
$$F_{\max}(J, I, K) = 96, \quad F_{\max}(K, I, J) = 96, \quad F_{\max}(K, J, I) = 86,$$

clearly indicating that job I does not belong in the first position. Pairwise comparison of adjacent jobs in sequence is often a useful technique, but it is simply not strong enough to guarantee an optimal sequence in this problem.

5–5 SEQUENCING IN LARGE FLOW-SHOPS

A special case that is highly artificial but has an interesting graphical solution occurs when there are only two jobs to be scheduled through a flow-shop of any number of machines. This is a special case of a procedure given by Akers [3]. Suppose that the machines are numbered $1, 2, \ldots, m$ in the order that they are to process the jobs, so that the processing-times are given by

$$p_{1,1}, p_{1,2}, p_{1,3}, \ldots, p_{1,m} \qquad \text{for job 1,}$$
$$p_{2,1}, p_{2,2}, p_{2,3}, \ldots, p_{2,m} \qquad \text{for job 2.}$$

These times can be marked off on axes for the two machines, as shown in Fig. 5–2.

A schedule can be represented by any line:

1) from $(0, 0)$ to
$$\left(\sum_{j=1}^{m} p_{1,j}, \sum_{j=1}^{m} p_{2,j} \right),$$

2) that is composed of horizontal (work on job 1 only), vertical (work on job 2 only), and 45° (work on both jobs) line segments,

3) that does not enter the interior of any of the shaded regions (which would imply one machine working on both jobs simultaneously).

Line S in Fig. 5–2 is such a "schedule line."

There are clearly 2^m different schedules, since the two jobs may be taken in either order on each machine, and there are 2^m corresponding schedule lines for the graph. These lines form a type of graph called a *tree*. That is, one starts at $(0, 0)$ and explores both branches (above and below) each time that a line encounters an obstructing shaded region. The encounter and branching take place in one of the three ways shown in Fig. 5–3.

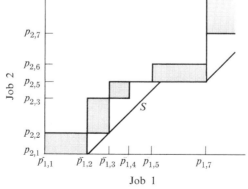

Fig. 5–2. Graph of a two-job flow-shop schedule.

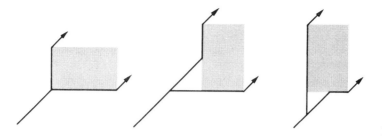

Figure 5-3

The maximum flow-time obtained by a schedule is related to the sum of the lengths of the vertical segments of the corresponding schedule line; it is equal to this sum plus the processing-times on the first machine. (Alternatively, it is equal to the sum of the horizontal segments plus the processing-times on the second machine.) The optimal schedule is given by that schedule line that has the least length in vertical segments; if the line passes under the shaded region for machine J, then job 1 precedes job 2 on machine J.

Thus one could at least conceptually draw each of the 2^m schedule lines* and select one with a minimum total vertical component. However, what makes the procedure useful is that far fewer than 2^m lines actually need be considered, since some of the potential branches do not occur—when a line does not encounter a particular shaded region—and the larger shaded regions on the graph cause many of the schedule lines to run together. The result is that, for a problem of, for example, twelve machines in which there would potentially be about a thousand schedules (2^{10}) to consider, unless the processing-times are pathologically contrived the optimal schedule can be found by comparison of a dozen or so lines.

However, the real point of describing this special case is not so much to offer a solution procedure as to examine several insights that it can provide. A graph such as that shown in Fig. 5-2 rather clearly indicates the character of a schedule in a flow-shop, and the virtue of a good schedule. The desirability of adhering to a diagonal path makes it obvious why inversions of job order (lines crossing the connected shaded region) can be required for larger flow-shops. The graph also illustrates the essential difficulty of the problem. It is clear that even in this very simple two-job problem there is no way of determining the proper order of two jobs on a particular machine from the characteristics of the particular operations and the schedule up to that point; that is, one cannot decide whether to go above or below a particular shaded region without examining the lines for the balance of the schedule. For example, from point X in Fig. 5-4, with four machines to go, if one takes the apparently attractive path under block $m-3$, one is trapped in a region in which all alternative lines are worse than one that could be attained by going over block $m-3$.

Finally, a comparison of Fig. 5-2 with Fig. 5-1 illustrates an interesting duality about the problem, a duality which will become even more important in Chapter 6.

---

* One can of course take advantage of the fact that the ordering of jobs will be the same on the first pair of machines, and on the last pair to ignore lines that divide the first pair of shaded regions, and the last pair. There are therefore 2^{m-2} potential lines.

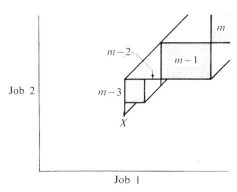

Fig. 5-4. Graph of the final stages of a two-job flow-shop schedule.

In each case the processing-times of the operations are marked off on the axes, but in Fig. 5-1 the axes represent machines, while in Fig. 5-2 they represent jobs. The point is that the *operations* are the central entities of a scheduling problem and there are two precedence orderings of these operations. One ordering is given: the *routing* is an ordering of the operations onto jobs. The other ordering is sought: the *schedule* is an ordering of the operations onto machines.

Except for this special two-job case and the simple results of Theorems 5-1 and 5-2 concerning the ordering on the first and last pairs of machines, analytical work on flow-shop scheduling ends abruptly with the three-machine shop. At the same time the "computational" procedures, enumeration, integer programming, and branch and bound, move out of the realm of practical computability. A problem with n jobs and m machines has $(n!)^{m-2}$ possible schedules. Even with $m = n = 6$, dimensions which could hardly be considered in any sense practical, there are 2.7×10^{11} schedules and the problem is not computable. So beyond the tiny core of problems that have been solved by constructive algorithms and the slightly larger set of problems that are computable, there is nothing available except sampling results and various heuristic procedures—methods that hopefully offer high probability of finding a relatively good schedule, but little assurance of this for a specific problem and no hope of optimality.

Heller [78] reports an investigation of a ten-machine flow-shop in which he (a) studies the form of the distribution of maximum flow-time of different schedules. (In his study maximum flow-time is called schedule-time); (b) seeks a method of "biasing" the schedule generation procedure to emphasize the lower tail of this distribution.

He generated a set of 100 jobs, with integer processing-times apparently uniformly distributed between zero and nine. He then generated schedules at random by taking 10 independent permutations of the integers 1–100 and computing the resulting maximum flow-time. The values for 3000 such schedules are shown in Fig. 5-5. He concludes that "the numerical experiments show that the distribution of schedule-times is normal; theoretical analysis indicates that the schedule-times are asymptotically normally distributed for schedules with large numbers of jobs" ([78], page 178). The argument continues that this normality can be used to determine decision-

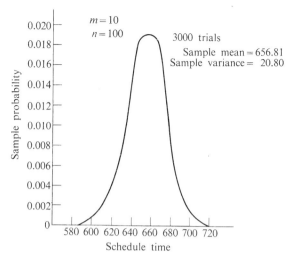

Fig. 5–5. Randomly generated schedules for a $100/10/F/F_{max}$ problem (Heller [78]).

theoretical rules to terminate sampling when the cost of continued sampling exceeds the expected gain from further sampling. There is some question about the asymptotic normality of the distribution, but aside from that there are at least two obstacles that lie in the path of a practical procedure:

a) If the distribution of schedule-times is approximately and/or asymptotically normal, the departures from normality will be most pronounced in the tails of the distribution, which is precisely the area of interest. No one is concerned with estimating the mean of the schedule-time distribution for a particular problem.

b) It is inconceivable that anyone would be sampling from this distribution, for there are obviously more efficient subpopulations readily available.

The most interesting aspect of this study is the consideration of the question of common job ordering on the 10 machines. Considering only the first 20 of the set of 100 jobs, Heller generated 9037 schedules with an independent permutation of the integers 1–20 for the ordering on each machine. The distribution of schedule-times for these schedules is shown on the right in Fig. 5–6. He then produced 12,000 more schedules for the same set of 20 jobs, with a single permutation being used to determine the schedule—the same job order is observed on each machine. The distribution of schedule-times for these order-preserving schedules is shown on the left in Fig. 5–6. It is interesting to note that the samples are entirely nonoverlapping; the maximum schedule-time observed for an order-preserving schedule is much smaller than the minimum schedule-time observed for an independent-order schedule. The reason for this wide disparity is easily seen by examining an independent-order schedule in detail. Consider the job that happens to be in first position in the permutation that determines the schedule on machine J. This particular job is equally likely to be in any one of the 20 positions of the permutation for machine $J - 1$. Machine J *cannot do any work on any job* until this particular job has been completed on machine $J - 1$. There can

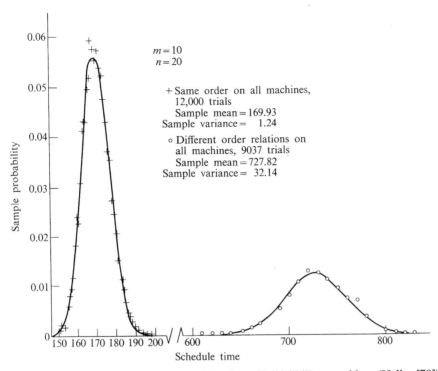

Fig. 5–6. Randomly generated schedules for a $20/10/F/F_{max}$ problem (Heller [78]).

be anywhere from 1 to 19 other jobs waiting, completed on $J - 1$ but unable to proceed on J. On the average, each machine will not start until its predecessor has completed half its work. Schedules of this kind are so patently inadmissible that it seems pointless to consider them or the form of the distribution of their schedule-times, or to claim particular virtue for a procedure that does better than this. Heller concludes that "... if we were looking for the minimum schedule, we would sample from those schedules which have the same ordering on all machines" ([78], page 184). This is undoubtedly an appropriate starting place—one which any reasonable sched-uler would have selected on intuitive grounds—but one should not stop there. There are better schedules to be found. Nugent [157] reports an investigation of what he calls "probabilistic priority rules," applying these to exactly the same 20-job problem Heller concerned himself with. Although these rules were developed principally for the general job-shop problem, they yielded schedules with lower maximum flow-times than any of Heller's sample of order-preserving schedules. On detailed exami-nation of the best schedules, Nugent discovered that they did not strictly preserve order over the 10 machines, but rather specified occasional and local reversals of order of neighboring jobs on consecutive machines. It would appear that procedures which permit judicious departures from common ordering would sample from an even more attractive subpopulation than the order-preserving schedules and that there is room for further work in identifying conditions for departure.

CHAPTER 6

THE GENERAL n/m JOB-SHOP PROBLEM

The general job-shop problem is a fascinating challenge. Although it is easy to state, and to visualize what is required, it is extremely difficult to make any progress whatever toward a solution. Many proficient people have considered the problem, and all have come away essentially empty-handed. Since this frustration is not reported in the literature, the problem continues to attract investigators, who just cannot believe that a problem so simply structured can be so difficult, until they have tried it.

6–1 A GRAPHICAL DESCRIPTION OF THE PROBLEM

It is very easy to visualize the requirements of the job-shop problem as a sort of graphical jigsaw puzzle, in which one rearranges blocks representing the given set of operations. Each job i is given as a set of g_i blocks, one for each of its g_i operations. Each operation has three identifiers, i, j, and k: i, the job number to which the operation belongs; j, the sequence number of the operation—1, 2, $\ldots$, g_i; and k, the number of the machine required to perform the operation.

The length of each block is proportional to the processing-time required to perform the operation. As the information is given, these operation-blocks are arranged in rows according to their parent jobs and in numbered order. For example, a 5/3 problem might be given as shown in Fig. 6–1.

Arrangement by job implies that the first identifier of each operation in a row is the same. The second identifier forms an increasing sequence in each row. Minimally, what is required is to rearrange these 16 operation-blocks into rows by machine, without changing the identifiers or the length of any blocks so that the last identifier of each operation in a row is the same. This can easily be done, as shown in Fig. 6–2, and is useful in that it describes the total amount of work facing each machine, but it

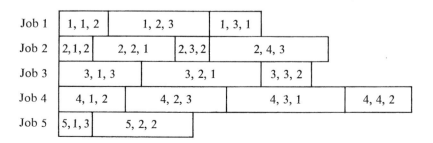

Fig. 6–1. A 5/3 problem.

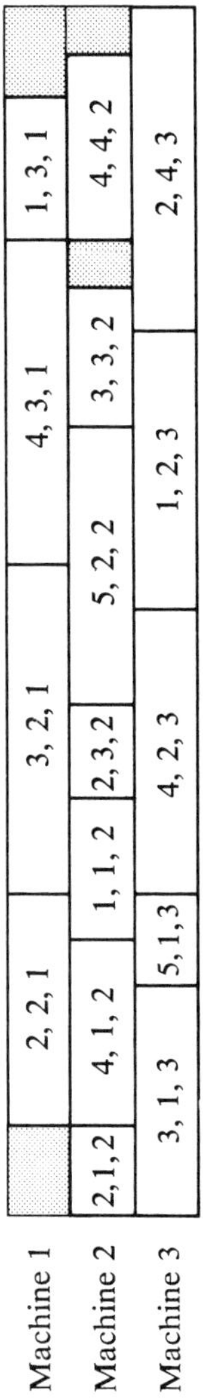

Fig. 6-2. A 5/3 work-load analysis.

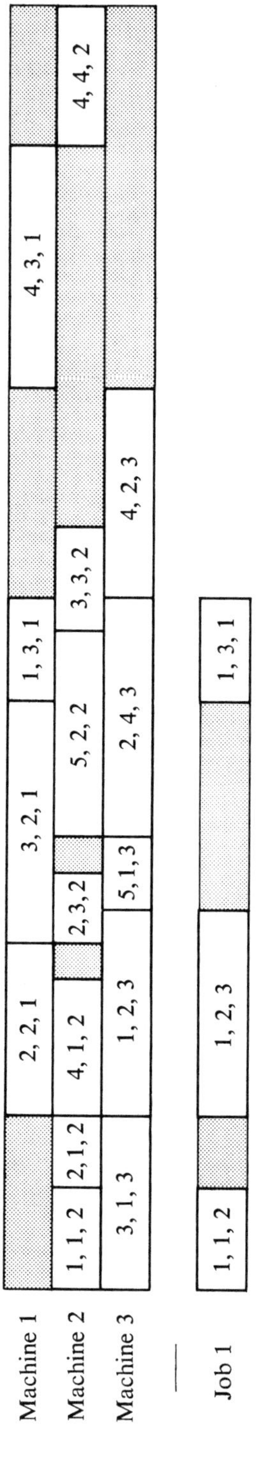

Machine 1: 2, 1, 2 | 2, 2, 1 | 3, 2, 1 | 4, 3, 1 | 1, 3, 1
Machine 2: 4, 1, 2 | 1, 1, 2 | 2, 3, 2 | 5, 2, 2 | 3, 3, 2 | 4, 4, 2
Machine 3: 3, 1, 3 | 5, 1, 3 | 4, 2, 3 | 1, 2, 3 | 2, 4, 3

Fig. 6-3. A schedule for the 5/3 problem of Fig. 6-1.

Machine 1: 1, 1, 2 | 2, 2, 1 | 3, 2, 1 | 1, 3, 1 | 4, 3, 1
Machine 2: 2, 1, 2 | 4, 1, 2 | 2, 3, 2 | 5, 2, 2 | 3, 3, 2 | 4, 4, 2
Machine 3: 3, 1, 3 | 1, 2, 3 | 5, 1, 3 | 2, 4, 3 | 4, 2, 3

Job 1: 1, 1, 2 | 1, 2, 3 | 1, 3, 1

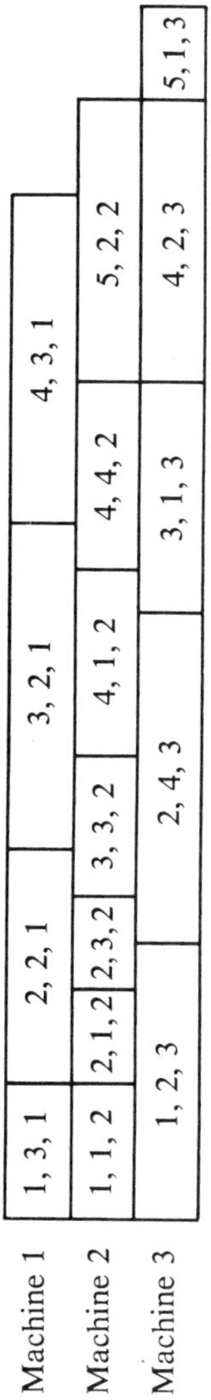

Fig. 6-4. An alternative schedule for the 5/3 problem of Fig. 6-1.

Machine 1: 1, 3, 1 | 2, 2, 1 | 3, 2, 1 | 4, 3, 1
Machine 2: 1, 1, 2 | 2, 1, 2 | 2, 3, 2 | 3, 3, 2 | 4, 1, 2 | 4, 4, 2 | 5, 2, 2
Machine 3: 1, 2, 3 | 2, 4, 3 | 3, 1, 3 | 4, 2, 3 | 5, 1, 3

does not constitute a schedule for the work, since the operations cannot be done in the order indicated and at the times implied. The key to the problem is that one must construct an arrangement by machine such that if one were to project the operations from this arrangement back onto an arrangement by job, the operations would be in the original order and would not overlap. An admissible machine arrangement is shown in Fig. 6–3, including the projection of the operations of Job 1. The other four could be similarly checked. The graph, or Gantt chart, of Fig. 6–3 represents a feasible schedule for this 5/3 problem, for the operations could be performed by the machines in the order and at the times indicated without violating any of the constraints of the simple job-shop process. (For example, one job on a machine at a time, no overlap of the operations of a single job, and operations in the required order.) While this schedule is possible, it is neither unique nor particularly good. There are many schedules for this problem and many that are better than Fig. 6–3 for any reasonable measure of performance. For example, Fig. 6–4 shows another schedule for this same problem that is quite obviously better than that of Fig. 6–3. Clearly there is no schedule with a smaller maximum flow-time than that in Fig. 6–4, but for other measures of performance better schedules may exist.

6–2 THE TWO-MACHINE JOB-SHOP PROBLEM

The only job-shop problem for which a solution can really be said to be known is the two-machine problem $(n/2/G/F_{\max})$ with the special restriction that each job can have, at most, two operations. Jackson [91] has shown that this is a direct generalization of Johnson's two-machine flow-shop problem (Section 5–2) and that his algorithm can be applied with only minor modification. This means that a problem with an arbitrarily large number of jobs can be solved with a very modest amount of computation.

The procedure starts by partitioning the n jobs to be scheduled into four sets as follows:

$\{A\}$ is the set of jobs which have only one operation, which is to be performed on machine 1.

$\{B\}$ is the set of jobs which have only one operation, which is to be performed on machine 2.

$\{AB\}$ is the set of jobs which have two operations, the first to be performed on machine 1, and the second on machine 2.

$\{BA\}$ is the set of jobs which have two operations, the first to be performed on machine 2, and the second on machine 1.

Initially, sequence the jobs of $\{AB\}$ by Johnson's procedure, just as if they were the only work to be done. Similarly, determine an ordering for the jobs of $\{BA\}$ by Johnson's rule, independent of the other jobs. The ordering of the jobs within $\{A\}$ and $\{B\}$ has no effect on the maximum flow-time whatever, so select any arbitrary ordering of these jobs. Jackson points out that an optimal schedule is obtained simply by combining the sets of jobs in the following way, without changing the order within each set.

On Machine 1: Jobs in $\{AB\}$ before jobs in $\{A\}$ before jobs in $\{BA\}$.

On Machine 2: Jobs in $\{BA\}$ before jobs in $\{B\}$ before jobs in $\{AB\}$.

The optimality of the arrangement is easily seen. The goodness of a schedule depends essentially on the amount of idle-time between operations, since the operation-times are a fixed obligation. Now, if there were no jobs in $\{BA\}$, the jobs of $\{A\}$, $\{B\}$, and $\{AB\}$ would be ordered so that the schedule is optimal; in this optimal schedule, the idle-time on machine 2 is a minimum. Now consider the work that must be done by machine 2 on jobs in $\{BA\}$. This work can be compactly scheduled on machine 2 with no idle-time between operations. By placing it first on that machine, the work in $\{BA\}$ can move on to machine 1, and idle-time within the machine 2 work on $\{AB\}$ may well be reduced by the deferment. There is simply no reason to consider schedules in which an operation on machine 2 on a job in $\{AB\}$ occurs before the end of all $\{B\}$ and $\{BA\}$ work on that machine. Suppose one were to attempt to fit a job K of $\{BA\}$ into the idle-interval X_J that occurs between jobs I and J in the initial scheduling of $\{AB\}$. Even if the time required for K is less than X_J, it might just as well be done before job I. At worst it will cause I to be delayed, reducing X_J, but it will not affect job J and hence not affect the maximum flow-time. A similar and symmetric argument can be used to justify the ordering on machine 1.

6–3 THE TWO-JOB JOB-SHOP PROBLEM

The graphical procedure for the two-job flow-shop problem given in Section 5–5 can also be applied to the two-job job-shop problem $(2/m/G/F_{\max})$. This can provide a solution to a problem of reasonable dimension, but the procedure is neither as efficient nor as interesting as that for the two-machine case and the problem itself seems even more unrealistic. The approach was first suggested by Akers and Friedman [4] and has been stated more completely and formally by Hardgrave and Nemhauser [72].

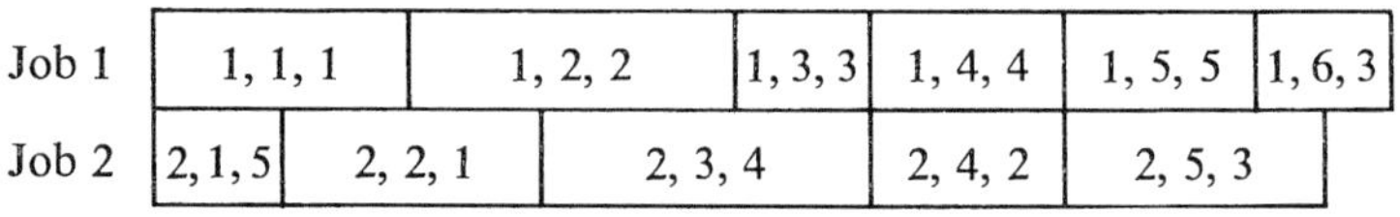

Job 1	1, 1, 1		1, 2, 2	1, 3, 3	1, 4, 4	1, 5, 5	1, 6, 3
Job 2	2, 1, 5	2, 2, 1	2, 3, 4		2, 4, 2	2, 5, 3	

Fig. 6–5. A 2/5 job-shop problem.

The procedure is exactly as given for the flow-shop case and the discussion of Section 5–5 applies. As an example, the two jobs shown in Fig. 6–5 are plotted on a schedule graph in Fig. 6–6. The graph for the job-shop case is slightly different, in that the shaded areas may be anywhere on the graph, depending on the routing of the two jobs, and are not connected in the same way as the flow-shop example. A job's routing may be completely arbitrary; there is no restriction that each machine be required only once. It is interesting that this increase in generality actually makes the problem easier, since some of these shaded areas will presumably be "out of the way" of the schedule line and fewer branches should be required.

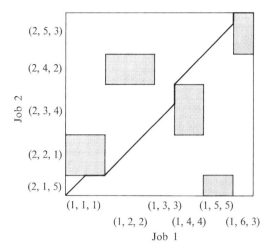

Fig. 6–6. Graph of a two-job job-shop schedule.

This graphical procedure could, of course, be generalized to higher dimensions, for three or more jobs, but even for three jobs it is rather unwieldy.

6–4 INTEGER PROGRAMMING
FORMULATION OF THE JOB-SHOP PROBLEM

The general job-shop problem can be modeled as an integer programming problem, as can the flow-shop problem (Section 5–4) and the single-machine sequencing problems of Chapter 3. In order of chronological appearance as well as increasing simplicity, the models are due to Bowman [21], Wagner [197], and Manne [126]. The description here will follow Manne's model. For simplicity of exposition and notation we assume that each job requires processing by each machine once and only once; this restriction can be circumvented, but with some notational complexity. Other sophistications, such as lap-scheduling, are also possible but are omitted here so as not to obscure the simplicity of the formulation of the model. Let:

p_{ik} = the processing-time of job i on machine k,

r_{ijk} = 1 if the jth operation of job i requires machine k,

= 0 otherwise, and

T_{ik} = the starting-time of job i on machine k.

From the requirement that only one job may be in process on a machine at any instant of time, we have for two jobs, I and J, either

$$T_{Ik} - T_{Jk} \geqq p_{Jk} \quad \text{or} \quad T_{Jk} - T_{Ik} \geqq p_{Ik},$$

and clearly not both.

Simply stated, either job J precedes job I or else job I precedes job J. Such either-or restrictions cannot be handled by ordinary linear programming and require the introduction of integer variables.

Let $Y_{IJk} = 1$ if job I precedes job J (not necessarily directly) on machine k; let $Y_{IJk} = 0$ otherwise.

There is no need to define both Y_{IJk} and Y_{JIk}; one will suffice. The two either-or constraints above may now be written as two independent restraints, both of which must hold. They are

$$(M + p_{Jk})Y_{IJk} + (T_{Ik} - T_{Jk}) \geqq p_{Jk},$$

$$(M + p_{Ik})(1 - Y_{IJk}) + (T_{Jk} - T_{Ik}) \geqq p_{Ik}.$$

The M is a constant and is chosen sufficiently large so that only one of the above constraints is binding for $Y_{IJk} = 0$ or 1. (For example, set $M = \sum_i \sum_k p_{ik}$.)

The operation precedence-constraints are handled by noting that $\sum_k r_{ijk} T_{ik}$ is the starting-time of the jth operation of job i. For all but the last operation of a job, one must have

$$\sum_k r_{ijk}(T_{ik} + p_{ik}) \leqq \sum_k r_{i,j+1,k} T_{ik}.$$

For m machines and n jobs the system of variables and restraint equations is:

Variables	*Number*
$T_{ik} \geqq 0,$ $Y_{IJk} = 0$ or 1	$mn,$ $m\dfrac{n(n-1)}{2}$

Equations	*Number*
$\sum_k r_{ijk}(T_{ik} + p_{ik}) \leqq \sum_k r_{i,j+1,k} T_{ik}$	$(m - 1)n$
$(M + p_{Jk})Y_{IJk} + (T_{Ik} - T_{Jk}) \geqq p_{Jk}$	$m\dfrac{n(n-1)}{2}$
$(M + p_{Ik})(1 - Y_{IJk}) + (T_{Jk} - T_{Ik}) \geqq p_{Ik}$	$m\dfrac{n(n-1)}{2}$

For even small-size problems this is a formidable system of inequalities. With 4 machines and 10 jobs there are 220 variables and 390 restraint equations.

A variety of objective functions may be handled. For min $\bar{F}$, one takes the equivalent objective of minimizing the sum of the start-times of the last operation of each job,

$$\min \sum_i \sum_k r_{imk} T_{ik}.$$

For min $F_{\max}$, one adds restrictions of the following form:

$$\sum_k r_{imk}(T_{ik} + p_{ik}) \leqq F_{\max}, \quad \text{for each } i,$$

where $F_{\max}$ is a variable and the objective is to minimize it.

For minimizing mean tardiness the equations, $T_i - E_i = F_i - d_i, i = 1, \ldots, n$, are added to the restraints. The objective is then min $\sum T_i$.

Only one research effort has employed integer programming for the solution of job-shop scheduling problems. This study, by Wagner with Story [189] and with Giglio [66], was for the flow-shop structure and was summarized in Section 5–4.

Apparently the size of the resultant integer programming problem, the time-consuming and often erratic behavior of existing integer programming computer codes, and the limited availability of such codes have discouraged other investigators from employing this approach. For the modest experimental investigations of heuristic procedures which are described in the following sections, an optimal solution would have been highly desirable for the purpose of absolute rather than relative comparison; yet the cost of obtaining an optimum solution via integer linear programming was deemed sufficiently large that it was not obtained.

6-5 TYPES OF SCHEDULES

Most of the rest of this chapter is concerned with approximate solutions to the job-shop problem, procedures that are intended to produce a relatively good solution with high probability, but which have little chance of obtaining a truly optimal solution. To facilitate discussions of these procedures it is useful to first discuss characteristics of schedules that are associated with goodness, and to identify certain types of schedules that might be expected to have relatively good values of the measure of performance.

It is clear that there are infinitely many schedules for any job-shop problem, since idle-time can be inserted into any given schedule in infinitely many ways. For example, in Fig. 6-4 an amount of idle-time equal to any nonnegative real number can be inserted between operations (1, 2, 3) and (2, 4, 3) and the result is still a schedule—an arrangement of the operations that satisfies all the constraints of the simple job-shop process.

It is also clear that most such schedules are patently uninteresting for any reasonable measure of performance and only schedules that are reasonably compact are worth considering. Consider the infinite set of all schedules that possess identically the same ordering of the operations on each machine. These schedules differ only in the amount of idle-time that has been inserted between the operations. Within this set there is a unique schedule called a *semiactive schedule* [157] which *dominates* all other schedules in the set for any regular measure of performance. All other schedules in the set are said to be inadmissible. The semiactive schedule for a particular ordering-set can easily be obtained from any arbitrary schedule in the set by repeated execution of an operation called a *limited-left-shift*. This is simply to move each operation as far to the left as possible, on a graph like that of Fig. 6-4, until it is blocked either by the preceding operation on that machine or the preceding operation on that job, but the ordering of the operations is not to be altered. In a semiactive schedule no operation can be moved by a limited-left-shift. It should be obvious that for any regular measure of performance it is sufficient to consider only semiactive schedules, since all the good schedules will be of this type. When one encounters any schedule not of this type one can easily obtain the corresponding dominating schedule.

It is also obvious that there is now only a finite number of schedules to be considered, since there is only one semiactive schedule for each ordering of the operations, and that there are only finitely many orderings. Unfortunately the number is still very large so that solution by exhaustive enumeration and comparative evaluation is not feasible even for relatively small problems and large computers.

Job 1	1, 1, 1	1, 2, 2
Job 2	2, 1, 2	2, 2, 1

Fig. 6–7. A 2/2 problem.

It is even difficult to give a good upper bound on the number of semiactive schedules. It is usually much less than $G!$ where G is the total number of operations on the n jobs. For one machine ($m = 1$) this is exact, but it becomes a very poor upper bound as m increases. If $h_1, h_2, \ldots, h_m$ represent the number of operations to be performed on each of the m machines, then $(h_1!)(h_2!)(\ldots)(h_m!)$ is an upper bound. This bound is attained only if each job consists of a single operation so that there are no routing precedence constraints. For example, consider the simple case of two jobs and two machines shown in Fig. 6–7. There are two operations to be performed on each of the two machines, so that there are $2!2! = 4$ different arrangements of operations, but there are only three semiactive schedules, which are shown in Fig. 6–8. The fourth arrangement just does not represent a schedule, since it would necessarily violate the routing precedence requirements of both jobs.

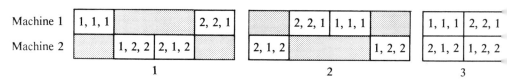

Fig. 6–8. Three semiactive schedules for the 2/2 problem of Fig. 6–7.

The expression $(n!)^m$ is often cited for the number of semiactive schedules in an n/m problem [61, 63] but provides in general neither a very good estimate nor an upper bound. It is presumably based on the special symmetric problem in which each job has one and only one operation on each machine. In this case $h_i = n$ for each machine and this is a convenient form of the previous bound. It would not be attained since a large proportion of the operation arrangements would not be valid schedules. On the other hand, in the general case in which it is not required that each job have one operation on each machine, the number of operations on a given machine could be greater than n, and the number of semiactive schedules far greater than $(n!)^m$. In any event it is hardly worth very much effort to find such a bound, or estimate, since the only practical use would be to demonstrate the enormous size of the problem and to discourage attempts at enumeration. It should be sufficient to note that for a 6/5 problem, $(6!)^5$ is approximately 1.93×10^{14}, which is more than the number of microseconds in six years. It hardly seems likely that adequate computer time would ever be made available to solve a problem of this size by exhaustive enumeration, or that computer capacity will ever exist that could solve problems of realistic dimension by this approach.

Some further slight improvement is possible by restricting consideration to *active schedules*, schedules on which it is not possible to perform a *left-shift* on any operation [64]. A left-shift of an operation is any decrease in the time at which the operation starts that does not require an increase in the starting-time of any other operation. A limited-left-shift is a left-shift, but not conversely, for the limited-left-shift preserves the order of operations on a machine. The left-shift permits an operation to "jump over" another operation into an interval of idle-time if that interval is large enough to accommodate the shifted operation. Figure 6-8 provides a good illustration of this difference. Schedules 1, 2, and 3 are all semiactive, since in no case can an operation be subjected to a limited-left-shift. However, Schedule 1 is not active since a left-shift can be performed on operation (2, 1, 2); it can be moved into the initial idle interval on machine 2 without deferring the start of operation (1, 2, 2). Once this is done, operation (2, 2, 1) can also be left-shifted and the result is Schedule 3. Similarly, left-shifting in Schedule 2 also results in Schedule 3. In this simple example, 3 is the only active schedule that exists and it is obviously optimal for any regular measure of performance.

In general, although the set of active schedules is usually a proportionately small and proper subset of the set of semiactive schedules, there are still an impossibly large number. Also, it is sufficient to consider only active schedules since the optimum schedule(s) is active, and for any nonactive schedule some dominating active schedule is readily obtained by left-shifting.

A further subclassification of schedules depends on detailed examination of the intervals of idle-time. Let us say that an operation is *"scheduleable" on machine j at time t* if t is greater than or equal to the completion-times of all other operations of the job that must precede the given operation. Perhaps more simply, an operation is "scheduleable" after all preceding operations on the job have been assigned a place in the schedule and the last of these has been completed. If for every instant of idle-time t and for every machine j in an active schedule, there is no operation that is "scheduleable" on j at t, then the schedule is said to be a *nondelay schedule** [157]; simply stated, there is no instance in which a job is delayed when the machine that is to process the next operation is available and idle.

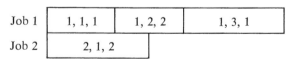

Fig. 6-9. A 2/2 problem.

Nondelay schedules are by definition a subset of the active schedules, but not a dominating subset in the same sense that the active schedules dominate the semiactive. It is not true that in every problem there is an optimal schedule among the nondelay schedules. For example, a 2/2 problem is shown in Fig. 6-9, and the only two active schedules for this problem are shown in Fig. 6-10. Schedule 1 is not a nondelay

---

* This is what Jackson has called an "availability schedule" [88].

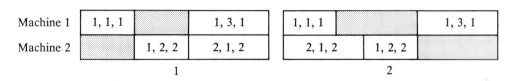

Fig. 6–10. The two active schedules for the 2/2 problem of Fig. 6–9.

schedule, since machine 2 is initially idle while operation (2, 1, 2) is available for processing, yet this is clearly superior to the nondelay Schedule 2 for at least some regular measures of performance (maximum flow-time and others). Moreover a nondelay schedule is not as easily constructed from an arbitrary active schedule as an active schedule is from a semiactive one. Yet the nondelay schedules are a very important class, for while they are not readily derived from active schedules, they are equally easy to generate, and there is strong empirical evidence that while they may lack an optimal schedule, on the whole the nondelay schedules are better than the remainder of the active schedules. This suggests that when one lacks a procedure for constructing an optimal schedule directly and must resort to a heuristic or sampling approach, it may be more profitable to address the nondelay schedules than the active schedules even though one may, in doing so, forfeit the infinitesimal probability that an optimal schedule may be obtained.

6–6 GENERATION OF SCHEDULES

Many of the heuristic, sampling, and computational approaches to the job-shop problem require the generation of a number of schedules for a particular problem. The types of schedules defined in the previous section are useful in describing these approaches, but it is also useful to define different types of procedures, for there are many distinctly different ways of generating a set of active schedules for a problem.

Schedule-generation procedures have a basic similarity in that each operates on the set of $G = \sum_{i=1}^{n} g_i$ operations, selecting operations one at a time and assigning a starting-time to each. The order in which the operations are selected and the manner in which the starting-time is determined characterize a schedule-generation procedure.

The most important distinction is between *single-pass* and *adjusting* procedures. Under a single-pass procedure, once a starting-time is assigned to a particular operation, this time is permanent and may not be changed to accommodate a later assignment. However involved the decision rule may be that selects the operation to be assigned and its starting-time, there is just one pass through the set of operations and precisely G starting-time decisions are made. Under an adjusting procedure, each starting-time assignment is tentative and subject to repeated modification until the entire schedule has been completed. With very rare exceptions the generation procedures that have been programmed for computers have been of the single-pass type. There is, of course, no reason why adjusting procedures cannot be implemented on a computer, but the explicit statement of reasonable rules for adjustment is a very difficult task. Allegedly, human schedulers operating on some form of Gantt chart can employ various heuristic adjusting procedures which presumably give them an advantage over machine competition. This hypothesis has been proposed but never really tested, mainly due to the drawbacks in adjusting a manually constructed Gantt chart. With the advent of graphical input-output devices for computers, it is now conceivable that a meaningful experiment of this nature may be performed.

One might initially suppose that the restrictions placed on single-pass procedures would make some schedules inaccessible to this type of generation, so that adjusting procedures would be theoretically preferable. Strictly speaking, this is not the case, since for any schedule at all there is a corresponding single-pass procedure capable of producing it. It follows that for any active schedule there is a single-pass procedure. However, while this implies that the class of single-pass procedures is sufficient, it does not necessarily imply that it is desirable to restrict attention to this class. An optimal single-pass generation procedure assuredly exists for a particular problem, but it may have neither rhyme nor reason, and no similarity to an optimal procedure for any other problem. It may be easier to search among adjusting procedures than to pursue a patternless single-pass procedure.

A special type of single-pass procedure that has been particularly important in research and in actual industrial practice is the class of *dispatching* procedures. These are single-pass procedures in which the starting-times for any given machine are determined in such an order that they form a strictly nondecreasing sequence of numbers. This means that the decisions are made in the same order as they will be implemented, and that the scheduling process can be spread out in time, making each decision immediately before it is to be implemented. The class of dispatching procedures is sufficient in the same sense as the class of single-pass procedures; for any given active schedule, and therefore any optimal schedule, there is a corresponding dispatching procedure that would produce that schedule. However, the same objection also applies, in that this dispatching procedure may not be systematic or understandable, so that it may be worth while and necessary to consider procedures outside this theoretically sufficient class.

An important concept in many schedule-generation procedures is the *set of scheduleable operations*. At any moment this set, denoted $\{S_{so}\}$, is a subset of the G operations consisting of all those operations which have had their predecessors scheduled; for example, an operation is in $\{S_{so}\}$ if a starting-time has been assigned to each of the preceding operations of that job. This means that for an n/m problem $\{S_{so}\}$ initially consists of exactly n operations, the first operation on each job. When one of this initial set is selected and scheduled, it is replaced by the second operation on that job, if there is a second operation. Whenever the last operation of a job is selected and scheduled, there is no replacement in $\{S_{so}\}$ and the size of the set is reduced by one. The schedule is completed whenever $\{S_{so}\}$ becomes empty. This concept is important for both single-pass and most adjusting procedures. For a single-pass procedure, once an operation has been removed from $\{S_{so}\}$ it can never re-enter, but with an adjusting procedure whole sequences very frequently need to be rescheduled, and will re-enter $\{S_{so}\}$ one at a time.

There are two common ways of partitioning $\{S_{so}\}$. The first is by job, partitioning $\{S_{so}\}$ into n subsets, each consisting of at most one operation,* which become empty as the last operation of the particular job is scheduled. This partition is used by *job-at-a-time schedule-generation procedures* in which all the operations of a particular job are scheduled consecutively; the g_i starting-time decisions for this particular job are made consecutively, without intervening decisions on any operation of any other job. There are, of course, exactly $n!$ orderings of the n jobs so that, considering procedures which produce only active schedules, there are precisely $n!$ job-at-a-time single-pass schedule-generation procedures. Each of these procedures produces a unique (active) schedule, but these schedules are not always distinct; several procedures can produce identically the same schedule, for example, when certain jobs do not compete for the same machines.

The second partitioning of $\{S_{so}\}$ is by machine, into m subsets, each denoted $\{S_{so}^k\}$ ($k = 1, 2, \ldots, m$) and consisting of the operations "scheduleable" on each particular machine at the given moment. Each subset consists of operations waiting before that machine, and of operations all of whose direct predecessors are in process on other machines. Initially there are n operations distributed into the m subsets $\{S_{so}^k\}$. Particular subsets can alternate between empty and nonempty status until eventually all the subsets are empty and the schedule is completed.

There are two particularly important methods of selecting the subset $\{S_{so}^k\}$, from which the next assignment is to be made, which provide the basis for the class of dispatching procedures. Consider the set of operations which are "in process," denoted $\{S_{ip}\}$. An operation is placed in $\{S_{ip}\}$ as soon as it is removed from $\{S_{so}\}$. The set $\{S_{ip}\}$ is partitioned by machine into m subsets, and there is exactly one operation in each of these subsets† (in process on each machine). When an operation is placed in one of the subsets, denoted by $\{S_{ip}^k\}$, the previous occupant is removed from the set Let

C_k be the process completion-time of the operation in $\{S_{ip}^k\}$,

s_{jk} be the potential starting-time of operation j in $\{S_{so}^k\}$ (this is the completion-time of the previous operation or zero if this is an initial operation of a job), and

p_{jk} be the processing-time of operation j in $\{S_{so}^k\}$.

Then,

1) $\max(C_k, s_{jk})$ is the soonest possible starting-time for operation j in $\{S_{so}^k\}$, and

2) $\max(C_k, s_{jk}) + p_{jk}$ is the soonest possible completion-time for operation j in $\{S_{so}^k\}$.

---

* This is true only for the simple job-shop process. When assumption (2) of the process (see Section 1–2) requiring a strictly ordered sequence of operations on each job is relaxed, then two or more operations of a job may be simultaneously members of $\{S_{so}\}$.

† To begin the process, $\{S_{ip}^k\}$ consists of an imaginary operation, with a process completion-time of zero.

If a nonempty subset, $\{S_{so}^k\}$, is chosen for selecting the next operation by the criterion

$$\min_{k} \ \min_{j \in \{S_{so}^k\}} \ [\max(C_k, s_{jk})],$$

and if the operation selected achieves this value, then the procedure is of a dispatching nature. Further, only nondelay schedules will result, so that the procedure might be called *nondelay dispatching*.

If a nonempty subset, $\{S_{so}^k\}$, is chosen for selecting the next operation by the criterion

$$\min_{k} \ \min_{j \in \{S_{so}^k\}} \ [\max(C_k, s_{jk}) + p_{jk}],$$

and if the operation selected has an s_{jk} less than this value, then the procedure is also a dispatching one. Further, any active schedule can be generated by this method, so that the procedure might be called *active-schedule dispatching*. Under this procedure an idle period can be scheduled on a machine with the assurance that there is no operation which could be left-shifted into the interval of idle-time.

The difference between the two procedures can be illustrated by using the 2/2 problem of Fig. 6–9. Suppose that

$$\{S_{ip}^1\} = \{(1, 1, 1)\}, \qquad \{S_{ip}^2\} = \{\text{empty}\},$$

$$\{S_{so}^1\} = \{\text{empty}\}, \qquad \{S_{so}^2\} = \{(1, 2, 2), (2, 1, 2)\},$$

$$P_{111} = 2, \qquad\qquad P_{122} = 2,$$

$$P_{212} = 3;$$

then

$$C_1 = 2, \qquad C_2 = 0, \qquad s_{122} = 2, \qquad s_{212} = 0.$$

Under both procedures, machine 2 is chosen for dispatching an operation. Under nondelay-dispatching,

$$\min_{k} \ \min_{j \in \{S_{so}^k\}} \ [\max(C_k, s_{jk})] = 0,$$

and only operation $(2, 1, 2)$ can be started at this time. Schedule 2 of Fig. 6–10 would result. Under active-schedule dispatching,

$$\min_{k} \ \min_{j \in \{S_{so}^k\}} \ [\max(C_k, s_{jk}) + p_{jk}] = 3,$$

and both operations $(2, 1, 2)$ and $(1, 2, 2)$ have potential starting-times less than this value. If $(2, 1, 2)$ is selected, the resulting schedule is Schedule 2 of Fig. 6–10, whereas if $(1, 2, 2)$ is selected, an active schedule with inserted idle-time, Schedule 1 of Fig. 6–10 results.

Any active schedule may be obtained by a generation procedure which selects an operation from $\{S_{so}\}$ and schedules it at the soonest possible time consistent with the precedence constraints and the existing schedule for the machine which the operation requires. This procedure is, however, not a dispatching procedure since an operation can be inserted before another previously scheduled operation.

A computer program that produces a schedule by the use of a dispatching procedure is often called a *job-shop simulation*. The operation of the program appears to simulate the activity of a physical job-shop process over time in the sense that the assignment decisions are made in the same order that they would be implemented in the actual process.

The concept of job or operation *priority* is inherent in many schedule-generation procedures. A priority is simply a numerical attribute of a job or operation on which selection is based. For example, in job-at-a-time schedule generation there must be some mechanism for determining the order in which the jobs are selected and it is convenient to view this in terms of a priority system; jobs are always selected in order of increasing value of a numerical attribute assigned to each job. Then there is really only one job-at-a-time schedule-generation procedure, but there are $n!$ different ways of assigning priority to the jobs.

A priority system must always have sufficient precision to lead to a unique selection so that two competing jobs should never have precisely the same value of priority. This may require that in support of the primary priority-attribute there may have to be secondary attributes assumed in order to resolve ties. For example, in job-at-a-time generation the primary job-priority might be taken equal to the number of operations. Since the priorities would be small integers one would expect frequent ties and the job identification number might be arbitrarily specified as the secondary priority-attribute, so that between two jobs with an equal number of operations, the job with the smaller identification number is selected first.

For even greater generality one can consider priority values to be obtained from a probability distribution by the use of some random mechanism. A different distribution is assumed to exist for each job-operation combination receiving a priority. With such probabilistic priority schedule-generation procedures, the concept of replication becomes important, since for a given problem (a certain set of jobs and machines) one could randomly obtain different priority values and consequently different schedules. Meaningful replication involves either repeated generation of priority values to make statistical statements about schedules for a particular problem and priority mechanism, or repeated generation of job-sets to make statistical statements about schedules for a particular problem class.

Two extreme types of probabilistic priority have been frequently employed. At one extreme the distributions have been degenerate point-distributions with zero variance so that in effect a deterministic priority value is assigned to each job. For example, this would be the case when priority is taken equal to the operation processing-time. In this case replication of the first kind is meaningless, since repeated scheduling of a given problem with a certain priority mechanism always results in identically the same schedule. At the other extreme the distribution for each job can be exactly the same nondegenerate distribution. The distribution used is irrelevant, since only the relative ordering between values has bearing on the schedule. This provides random selection whenever two jobs (or operations) are in competition and represents the case in which there is no rational basis for selection. Replication of both kinds is meaningful for this type of priority schedule-generation.

Dispatching procedures for schedule generation always include a priority mechanism to select among operations competing for machine assignment. In some cases the priority may be implicit in the mechanism of the computer program, but it must be present. Operation selection is actually in two stages for such procedures. The first stage is the selection of a machine, say K, to which an assignment is to be made; the second stage is the selection of one operation from among possibly several operations in $\{S_{so}^k\}$. Once the machine has been determined it is not automatic that the operation with minimum s_{jK} is selected, for to do so implies a particular priority mechanism: first-come, first-served. For example, under nondelay dispatching, any of the operations with $s_{jK} \leqq C_K$ is eligible for selection and a priority mechanism is required for resolution. In the most general sense this would be a probabilistic mechanism, ranging from the degenerate distribution extreme (such as first-come, first-served) to the common distribution extreme in which each eligible operation is equally likely to be selected. The degenerate distribution extreme results in a unique schedule for a particular problem and replication of the first kind is meaningless. Therefore, to be completely descriptive, one might speak of probabilistic priority dispatching, but there really is no other kind of dispatching; there are only special cases in which either the probability or the priority is not evident. Thus the term *dispatching* includes schedule-generation procedures which are based on some probabilistic priority mechanism.

6–7 BRANCH-AND-BOUND APPROACH TO THE JOB-SHOP PROBLEM

The branch-and-bound technique has been described previously in connection with the traveling-salesman problem (see Section 4–1.1) and the flow-shop problem (Sections 5–3 and 5–4). A similar approach to the $n/m/G/F_{\max}$ problem has been given by Brooks and White [22].

In the tree of unsolved problems generated by this approach, each node represents an active schedule for a particular subset of the operations, and a point in time at which there is more than one operation which could be scheduled next on the particular machine chosen (as described in the preceding section). Each branch descending from a node represents the selection of one of the competing operations. Selection of a particular branch is based on determining the minimum value of a lower bound on the maximum flow-time for each possible branch. As usual, the power of the procedure depends heavily on the quality of the lower bounds, particularly those used in the early stages of branching. Three of the more promising possibilities for these bounds are given below. In each case the computations are based on the conditions which would exist in the partial schedule at the node to which the branch would lead.

1) Construct any complete schedule starting from the partial schedule in-hand using one of the simple dispatching procedures discussed in the next section. Take as a bound the maximum flow-time for this complete schedule (this is actually an upper bound).

2) For each job, find the earliest time at which it could possibly start on its next unscheduled operation. Add to this the sum of the processing-times of all the unsched-

uled operations of the job. Take as a lower bound the maximum of these quantities. This bound is good if there are several jobs whose remaining processing-time requirements are large relative to the others.

3) For each machine find the minimum time at which an unscheduled operation could be started. Add to this the sum of the processing-times of the unscheduled operations which require this machine. Take as a lower bound the maximum of these quantities. This bound is reasonably good if there are a few machines whose workload is considerably higher than the others.

Computationally, procedure 1 involves the most work and procedure 2 the least. However, both procedures 2 and 3 are effective only under special circumstances. The difference in the operation of these three bounding procedures is illustrated by the following 3/4 problem.

		Operation							
		1		2		3		4	
		Machine	p	Machine	p	Machine	p	Machine	p
	1	1	6	4	8	3	9	2	4
Job	2	1	1	2	3	3	9	4	6
	3	1	5	3	5	2	3	3	6

Since all the jobs have their initial operation on machine 1, there is a conflict on machine 1 at time 0. The three branches out of the top node of the solution tree correspond to selection of job 1, 2, and 3 at time zero; call these branches B_1, B_2, and B_3. The branch chosen by each of the bounding procedures will be determined.

1) A complete schedule for each branch is required. If the ones chosen are those given in the Gantt charts of Fig. 6–11 (the third identifier of an operation is omitted), the branch with the least lower bound is B_3. The next branch and bound would be based on the partial schedule which has job 3 started at time 0 on machine 1.

2) If job 1 is chosen, then each job could potentially start at time 6 and the remaining processing-times would be 21, 19, and 19, so that the lower bound for B_1 is 27. For B_2 it is 28 $\big(1 + \max(27, 18, 19)\big)$ and for B_3 it is 32 $\big(5 + \max(27, 19, 14)\big)$. Branch B_1 would be followed.

3) If job 1 is chosen, then the earliest possible times that each machine could star. on the rest of the operations are 6, 7, 10, 6 (machines 1 through 4, respectively). The sum of the processing-times of the unscheduled operations on each of the four machines would be 6, 10, 29, and 14. The lower bound associated with B_1 is then max(6 + 6, 7 + 10, 10 + 29, 6 + 14) = 39. For B_2 the lower bound is 33 and for B_3 it is 34, so that B_2 would be followed.

Brooks and White conclude that the procedure is computationally prohibitive for problems of practical dimension, but it could nevertheless be an attractive alternative to integer programming for solution of small problems. Furthermore their results suggest some interesting dispatching procedures, corresponding to following a particular path in the branch-and-bound solution tree, without backtracking.

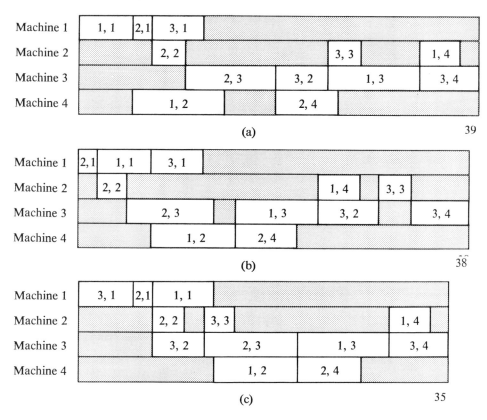

Fig. 6–11. (a) A complete schedule for B_1; bound of 39. (b) A complete schedule for B_2; bound of 38. (c) A complete schedule for B_3; bound of 35.

6–8 EXAMPLES OF SCHEDULE GENERATION

There have been many investigations in which a digital computer has been used to generate schedules for a variety of different n/m problems. Rather than making a complete report of this work, we have selected several of the more extensive studies with which we have been directly associated to illustrate the approach and the results that may be obtained. In aggregate we would conclude that the results are only mildly interesting and that the approach has not been tremendously successful. However, there is a basis for some insight into the problem and some comparisons that would be useful if, in fact, anyone ever encounters an actual n/m problem. Moreover the approach is so obvious that each investigator must discover what can be done in this way before resorting to more sophisticated methods, so that a report of what has been tried may help to forestall endless repetition of this work.

The properties of active and nondelay schedules were compared in a study by Bakhru and Rao [13]. They used a 10/6 problem with *permutation routings*; each job had exactly m operations, one on each machine, in a randomly determined order. A sample of 11 such problems was generated, and for each of these problems a

Table 6–1

Random Sampling of 10/6 Schedules (Permutation Routing)

Mean flow-time

Problem number	Random sample of 50 active schedules		Random sample of 50 nondelay schedules	
	Minimum of 50	Maximum of 50	Minimum of 50	Maximum of 50
1	61.9	74.9	59.1	69.4
2	61.5	75.4	61.3	70.3
3	62.8	81.2	69.2	73.8
4	51.7	63.6	49.0	59.8
5	56.0	74.7	54.1	67.9
6	67.7	82.8	65.2	75.6
7	51.8	65.4	51.0	62.1
8	57.4	72.3	55.0	67.5
9	60.0	75.5	58.9	69.6
10	52.6	62.4	51.3	61.0
11	61.4	77.3	61.0	68.6

Maximum flow-time

Problem number	Random sample of 50 active schedules		Random sample of 50 nondelay schedules		
	Minimum of 50	Maximum of 50	Minimum of 50	Maximum of 50	Lower bound
1	76	106	71	103	71
2	83	115	78	99	67
3	86	119	82	106	73
4	68	99	64	91	64
5	74	109	70	93	56
6	95	122	85	116	71
7	69	112	66	83	59
8	76	108	73	95	67
9	82	119	75	113	65
10	67	100	62	88	54
11	84	116	86	105	78

random sample of 50 active schedules and a random sample of 50 nondelay schedules were generated. Both the sample minimum and maximum of the mean flow-time from each sample of 50 is displayed in Table 6–1. Considering mean flow-time, for each of the 11 problems, the maximum nondelay schedule was better than the maximum active schedule. For 10 of the problems the minimum nondelay schedule was better than the minimum active schedule. The same statements can be made using maximum flow-time as a measure of performance (the single reversal is not the same problem under the different measures).

A lower bound on maximum flow-time can be easily obtained by summing the operation times on each job and the operation times on each machine. The maximum of these $n + m$ sums is a lower bound, but not necessarily an attainable one. These values are shown in the second part of the table. In at least two of the 11 problems the minimum of the 50 nondelay schedules is an optimal schedule since its maximum flow-time equals the lower bound. Optimal schedules may have been obtained for other of the problems (i.e., the bounds given may not be attainable). Averaging over the 11 problems the maximum flow-time of the best nondelay schedule is 12% greater than the lower bound; the best active schedule is 19% greater than the lower bound.

These results suggest that if one were to seek a good schedule simply by selecting the best of a randomly generated set of schedules it would be more efficient to sample from the population of nondelay schedules. An example in Section 6–5 demonstrated that for an arbitrary problem it is not true that the set of active schedules is dominated by its nondelay subset; for some problems, there *may not* be an optimal nondelay schedule, but it is obvious from the evidence of Table 6–1 that for some problems there is. However, the existence and attainment of an optimal schedule is almost irrelevant since a willingness to settle for high probability of obtaining a good schedule is almost implied by the resort to a sampling procedure. In general, it would seem that the population of nondelay schedules offers better prospects in this regard than the larger set of active schedules.

Further evidence on this question is available from a study by Jeremiah, Lalchandani, and Schrage [102]. In this case, the experimenters, rather than generating schedules at random, selected operations from the set of "scheduleable" operations according to some attribute of the operations and/or the jobs associated with the operations. Once an operation was selected, the starting-time was determined so as to produce an active schedule. The procedures for selecting operations from $\{S_{so}\}$ that were considered were the following:

RANDOM Select operation at random. (This is the same as the procedure used to generate active schedules in [13].)

MOPNR Select operation for the job with the largest number of operations remaining to be processed.

MWKR-P Select operation for the job that has the most processing-time on operations subsequent to the "scheduleable" operation.

MWKR/P Select operation for the job that has the greatest ratio between work remaining to be done and the processing-time of the "scheduleable" operation.

SPT Select operation which has the shortest processing-time.

MWKR Select operation for the job that has the most work remaining.

LWKR Select operation for the job that has the least work remaining.

The schedules produced by these procedures were compared to the nondelay schedules produced by the corresponding dispatching procedures. A machine is first

Table 6–2 Performance Summary: Mean Flow-Time

Results \ Problem type	10/4/Rdm/$\overline{F}$	20/4/Rdm/$\overline{F}$	20/6/Rdm/$\overline{F}$	10/4/FixR/$\overline{F}$ No. sch. = 1	10/4/FixR/$\overline{F}$ No. sch. = 50	20/6/FixR/$\overline{F}$	10/4/Perm/$\overline{F}$
Replications	14	10	10	10	10	20	20
Best Nondelay better than best Active	14	–	–	–	–	–	20
SPT or RANDOM rule best for Active	14	10	10	1	0	18	20
LWKR rule best for Active	–	–	–	9	10	–	0
SPT rule best for Active	3	7	5	1	0	17	19
LWKR rule best for Nondelay	7	–	–	–	–	–	7
SPT rule best for Nondelay	5	–	–	–	–	–	13

Table 6–3 Performance Summary: Maximum Flow-Time

Results \ Problem type	10/4/Rdm/$F_{\max}$	20/4/Rdm/$F_{\max}$	20/6/Rdm/$F_{\max}$	10/4/FixR/$F_{\max}$ No. sch. = 1	10/4/FixR/$F_{\max}$ No. sch. = 50	20/6/FixR/$F_{\max}$	10/4/Perm/$F_{\max}$
Replications	14	10	10	10	10	20	20
Best Nondelay better than best Active	7 (2 ties)	–	–	–	–	–	18 (4 ties)
MWKR best for Active	14 (2 ties)	–	–	–	–	–	7 (2 ties)
MWKR or SPT best for Nondelay	14	–	–	–	–	–	14
MWKR, MOPNR, MWKR-P, or MWKR/P best for Active	14	10	10	9 (1 tie)	9 (1 tie)	20	20
MWKR best for Active	10 (7 ties)	6 (2 ties)	4	9 (1 tie)	9 (1 tie)	6 (1 tie)	6

selected for the next assignment, and then an operation selected from among those "scheduleable" on that machine according to one of the following rules:

RANDOM   Select operation at random.

SPT   Select operation which has the shortest processing-time.

MWKR   Select operation for the job which has the most work remaining.

LWKR   Select operation for the job that has the least work remaining.

LPT   Select operation which has the longest processing-time.

FCFS   Select operation which first entered $\{S_{so}\}$ (first-come, first-served).

Although the procedures on these two lists appear to be similar, the distinction between selecting from all of $\{S_{so}\}$ and from the subset associated with the earliest available machine is important, and the resulting schedules are quite different.

A portion of the results of this study for mean flow-time, maximum flow-time, and mean machine finish-time are given in Tables B–1 through B–3 of Appendix B and are summarized in Tables 6–2 and 6–3.

A total of 84 different problems were considered in six different categories. Three different procedures were used to generate the routings for these sample problems. One category consisted of *permutation routing* (denoted Perm in the tables) in which each job has m operations, one on each machine. Three of the categories employed *random routing* (denoted Rdm in the tables), in which each machine is equally likely to be the machine which handles the first operation of a job; after each operation the job is equally likely to move to each of the other machines for its next operation, or to be considered completed. The number of operations on a job is a random variable with a geometric distribution; the expected number of operations is m. It is obviously not necessary for each job to have an operation on each machine, and in many cases a job will have two or more operations on the same machine, although consecutive operations on the same machine are not allowed. The two remaining categories employed what was called *fixed random routing* (denoted FixR in the tables), in which each job has exactly m operations, but they are determined at random without the constraint that each job must have an operation on each machine. This type of routing is considered to be intermediate between the permutation and random extremes.

The summarized results for mean flow-time are given in Table 6–2. This provides additional evidence of the general superiority of nondelay schedules. For all 14 of the $10/4/\text{Rdm}/\bar{F}$ problems, the best nondelay schedule obtained was better than the best active schedule. The contrast is particularly interesting for the SPT and MWKR procedures, which were used to produce both active and nondelay schedules. In every case the nondelay schedules were superior. The same result was obtained for the $10/4/\text{Perm}/\bar{F}$ problems; in all 20 cases, the best nondelay schedule was superior (only the best of the various active schedules obtained is listed in the table) and in direct contrast to the SPT procedures, the nondelay SPT schedule was, in each case, better than the active SPT schedule.

In the case of active schedules three procedures are dominant: SPT, RANDOM, and LWKR. In the 84 problems, there were only two cases in which one of these rules did not yield the best performance.

However, these results are disappointing with respect to the identification of one procedure that consistently produces better schedules than the others, with five different procedures (including random generation) producing the best schedules for at least one of the 84 problems. There is also little basis for saying anything positive about the effect of either the size of the problem or the manner of determining the routing.

The results for maximum flow-time (Table 6–3) are substantially different in that there is less of a choice between active and nondelay schedules, but a clear indication of type of generating procedure. Of the fourteen $10/4/Rdm/F_{max}$ problems there were only five in which the best nondelay schedule was better than the best active schedule obtained. Two resulted in a tie. In fourteen of the twenty $10/4/Perm/F_{max}$ problems the nondelay schedule was better (with four ties).

When one is determining which job to assign next, it is worth one's while to consider the total remaining processing-time, since MWKR often produced the best schedule, and when it did not, the best would be associated with a minor variation that simply represented another way of measuring the same consideration. It seems preferable to favor jobs with much work remaining rather than those with little, and to use this consideration rather than the processing-time of the operations in $\{S_{so}\}$. There seems a clear advantage to this selection as compared to random generation.

For some problems, the average time it took the machines to complete their processing was also obtained; these results are presented in Table B–3, for general interest, even though this criterion has not generally been considered in other models. Curiously, although this is an average over machines, remarks would be much more like those for *maximum* flow-time than *mean* flow-time.

6–9 PROBABILISTIC DISPATCHING

The investigation of [102] also touched on the question of truly probabilistic dispatching, in which some attribute of the operation or job is used to randomize in a biased manner the selection from $\{S_{so}\}$. With some amount of randomness present in the selection, different schedules will be produced if a procedure is repeatedly applied to the same problem. One could produce a set of schedules in this way and then by examination select the best of these. The argument for such a procedure is something like the following. Suppose that one had identified an attribute, X, that was important to the ordering of operations in a schedule so that a dispatching procedure based on this attribute tended to generate better schedules than those selected at random. For problems of a certain class, this could be represented as in Fig. 6–12, in which $E(X)$ represents the expected schedule value for problems of this class if strictly deterministic dispatching based on attribute X is employed, $E(R)$ denotes the expected schedule value if schedules are generated at random, and M represents the expected schedule value using probabilistic dispatching. Now if a dispatching procedure is employed that causes selection to be biased but not completely determined by attribute X, the result might be as pictured by the distribution in Fig. 6–12. If the attribute, X, is actually beneficial, one would anticipate that M would lie to the right of $E(X)$, and presumably to the left of $E(R)$. The point of the procedure is that the variability

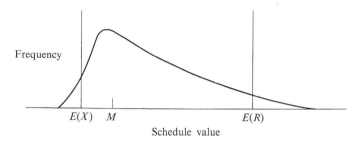

Fig. 6–12. Probabilistic dispatching.

admitted may cause some of the schedules that could be generated to have values better than $E(X)$. Depending on the variance of the distribution and the distance between $E(X)$ and M, a sample of reasonable size might contain one of these good schedules.*

Ten examples of 10/4/FixR problems were examined in this way. Schedules were generated by strict application of SPT, MWKR, and LWKR dispatching procedures and then 50 schedules were generated by a probabilistic variation of each of these procedures. Also, 50 were generated by the use of the RANDOM procedure. These results are given in Tables 6–4 and 6–5. In each of these procedures the members of $\{S_{so}\}$ were ranked according to the particular attribute and then geometrically decreasing probabilities were assigned to the operations in ranked order. For example, under probabilistic SPT, the operation with the shortest processing-time was *most likely* to be selected from $\{S_{so}\}$, but by chance any of the operations *could* be selected.

For mean flow-time, in every single case the minimum of the set of 50 was better than the single schedule produced by the corresponding deterministic procedure. The same was true for maximum flow-time, except for two cases which produced equiva-

Table 6–4

Probabilistic Generation of Active Schedules: Mean Flow-Time ($10/4/\mathrm{FixR}/\overline{F}$)

Problem number	Repli-cation	SPT Pure	SPT Min. of 50	MWKR Pure	MWKR Min. of 50	LWKR Pure	LWKR Min. of 50	RANDOM Min. of 50
46	1	133.2	123.4	146.2	133.3	**123.5**	*109.0*	124.0
47	2	146.5	136.4	187.0	171.8	**140.1**	*131.6*	152.4
48	3	158.0	141.0	170.9	161.3	**150.5**	*130.1*	142.8
49	4	131.7	121.8	143.4	139.8	**128.9**	*113.8*	131.0
50	5	123.2	110.8	145.8	138.6	**110.2**	*107.4*	115.7
51	6	**121.4**	112.6	152.0	139.4	130.7	*110.4*	117.0
52	7	148.0	125.3	152.4	144.9	**126.3**	*120.1*	124.3
53	8	150.7	132.7	204.6	186.5	**134.7**	*125.1*	144.4
54	9	159.8	146.9	179.8	175.7	**151.3**	*141.6*	153.9
55	10	150.1	131.1	181.7	153.7	**129.4**	*122.8*	144.1

* This argument might apply equally well to either active or nondelay schedule-generation.

Table 6–5

Probabilistic Generation of Active Schedules: Maximum Flow-Time $(10/4/\text{FixR}/F_{\max})$

Problem number	Repli- cation	SPT Pure	Best of 50	MWKR Pure	Best of 50	LWKR Pure	Best of 50	RANDOM Best of 50
46	1	206	171	**164**	*160*	222	201	188
47	2	253	221	**219**	*213*	279	237	214
48	3	219	206	**205**	*196*	263	224	235
49	4	247	228	**164**	*164*	273	207	180
50	5	248	196	**195**	195	229	209	*191*
51	6	214	173	**193**	*170*	225	187	*170*
52	7	234	188	**193**	*185*	222	197	191
53	8	373	288	**243**	*237*	333	274	249
54	9	300	227	**210**	*204*	263	231	218
55	10	258	207	**248**	216	239	234	*210*

Table 6–6

Probabilistic Dispatching Procedures: Maximum Flow-Time

Problem number	n	m	Routing	Number of schedules	Type schedule	RANDOM	SPT	MWKR	FCFS
96	20	9	Rdm	1	Nondelay		20.7	28.3	22.5
				100	Nondelay	21.2	19.9	21.2	
97				1	Nondelay		15.0	19.0	16.8
				100	Nondelay	15.2	14.1	15.2	
98				1	Nondelay		24.3	29.5	25.8
				100	Nondelay	24.3	22.2	24.3	
99				1	Nondelay		23.1	30.3	25.0
				100	Nondelay	23.3	22.7	24.3	
100	60	9	Rdm	1	Nondelay		32.5	53.3	32.8
				50	Nondelay	30.8	29.7	30.8	
101				1	Nondelay		30.9	55.5	33.8
				50	Nondelay	33.0	29.7	33.0	
102	100	9	Rdm	1	Nondelay		45.7	93.4	58.0
				40	Nondelay	56.0	44.8	56.0	
103				1	Nondelay		47.8	90.9	54.3
				40	Nondelay		46.9	73.2	
104	6	6	Perm	1	Nondelay		52.7	55.8	54.7
				100	Nondelay	48.2	46.7	48.2	
105	10	10	Perm	1	Nondelay		834	1011	1036
				100	Nondelay	875	791	875	
106	10	7	Perm	1	Nondelay		68.9	73.5	74.5
				100	Nondelay	68.0	66.8	68.0	
				1000	Nondelay		66.4		

lently good schedules. There is also an interesting consistency between the deterministic rule that produced the best schedule and the probabilistic rule that produced the best schedule. Under both measures of performance, with only three exceptions, these were corresponding rules.

Nugent [157] has pursued the matter of probabilistic dispatching to much greater depth. He parametrized the distance between a particular deterministic-dispatching procedure and purely random schedule-generation and tested probabilistic procedures that lie on different points on the scale between these extremes. He was able to vary the amount of randomness that entered into operation selection and attempted to determine experimentally how much to permit in order to obtain a good schedule from a small sample. He conducted these tests both on his own randomly generated problems and on problems that had previously appeared in the literature in reports of other investigations. A portion of his results is given in Tables 6–6 and 6–7. The tests

Table 6–7

Probabilistic Dispatching Procedures: Maximum Flow-Time

Problem number	n	m	Routing	Number of schedules	Type schedule	RANDOM	SPT	MWKR	FCFS
96	20	9	Rdm	1	Nondelay		48.7	41.9	48.5
				100	Nondelay	43.6	41.4	39.6	
97				1	Nondelay		39.5	37.7	40.7
				100	Nondelay	37.1	34.8	31.3	
98				1	Nondelay		54.4	49.1	55.3
				100	Nondelay	49.1	48.6	46.1	
99				1	Nondelay		50.1	49.5	54.5
				100	Nondelay	51.1	48.8	44.5	
100	60	9	Rdm	1	Nondelay		108.5	76.4	105.5
				50	Nondelay	94.2	89.9	74.9	
101				1	Nondelay		100.6	82.6	107.7
				50	Nondelay	101.9	93.3		
				40	Nondelay		80.3		
102	100	9	Rdm	1	Nondelay		132.0	106.9	131.2
				40	Nondelay	126.7	116.4		
				20	Nondelay		106.4		
103				1	Nondelay		128.6	103.0	124.6
				40	Nondelay		111.8	107.5	
104	6	6	Perm	1	Nondelay		88	61	67
				100	Nondelay	59	58	58	
				1000	Nondelay		57		
105	10	10	Perm	1	Nondelay		1074	1108	1235
				100	Nondelay	1050	1022	1047	
				300	Nondelay		960		
106	10	7	Perm	1	Nondelay		108	81	86
				100	Nondelay	82	78	78	

were considerably more extensive than would appear from these tables, for where the table shows a sample of 100 schedules produced by a probabilistic SPT dispatching procedure, there were actually 100 schedules for each of several probabilistic SPT procedures, differing in the amount of randomness introduced. The value reported is the best schedule from any of the probabilistic procedures, so that it often represents the best of 500 to 1000 schedules.

These results give a rather clear indication that probabilistic dispatching can produce better schedules for either maximum flow-time or mean flow-time as a criterion than strictly deterministic dispatching; but the improvement appears to be modest, and considerable effort is required. By direct comparison of this procedure with every other procedure that has been published with data adequate for testing, Nugent shows that this procedure can produce better schedules for any n/m finite problem than any other sampling, generation, dispatching, or heuristic procedure that has been offered. It even matches the solution obtained by integer programming in the few cases in which that has been used.

Nevertheless we regard this as something less than a breakthrough on the problem, since the effort required is prodigious. Although it was possible to parametrize the amount of randomness in these probabilistic procedures, there was little consistency in the values of the parameter that yielded good results for different problems, and it appears unlikely that this experience can be summarized in a way that would reduce the effort of a subsequent investigator. Moreover, the whole concept of probabilistic dispatching seems like something of a trick that is applicable only to relatively small problems. As the number of decisions contributing to the value of a schedule increases, one would expect that the relative variability of the process would diminish, and that the chances of obtaining a schedule better than the one produced by the corresponding deterministic dispatching procedure would become disappointingly small, while at the same time the effort required to generate replicate schedules would increase. This diminishing return is evident in the results in Tables 6–6 and 6–7. In problems 102 and 103, the largest problems tested, the best schedule obtained by probabilistic dispatching is hardly any better than the best obtained by deterministic dispatching.

However, there does appear to be something more than variability at work in these procedures and further inquiry would be warranted. Certainly the most surprising and intriguing result in Nugent's work is the manner in which the performance of several of the probabilistic procedures varies as the amount of randomness is increased. For a procedure that can be varied between the extremes of random generation and SPT dispatching, one would think that for a given problem the expected (average over many schedules) mean flow-time (an average over n jobs) would decrease smoothly and monotonically from one extreme to the other, perhaps with a curve something like that in Fig. 6–13. In fact, Nugent discovered a number of procedures that exhibited performance such as shown in Fig. 6–14. These are sampling results and subject to error, but the confidence limits on each of the points were such that there was just no question about the general shape of the curve. The interpretation is most interesting. For this combination of problem and procedure, it is not just variability that permits the probabilistic procedure to produce good schedules, because in expectation these procedures are better than the corresponding deterministic dispatching procedure.

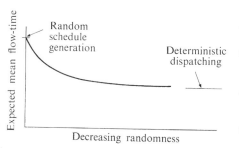

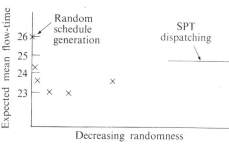

Fig. 6–13. Conjectured performance of probabilistic dispatching procedure.

Fig. 6–14. Sample of actual performance of probabilistic SPT dispatching procedure $(20/9/\text{Rdm}/\bar{F}\text{-Problem No. 98})$.

Nugent also examined the performance of a form of active-schedule dispatching. A machine could be held idle even though there were jobs available for assignment, if some more attractive job would soon be available. The problem was, of course, to find appropriate meanings for "more attractive" and "soon." Nugent was successful in the sense that in each of the 9 problems in which he tried to do this he was able to find a better schedule (for mean flow-time) than was obtained by any of the non-delay procedures. Again the searching effort was considerable and the successful values were not consistent from one problem to another. The improvement in mean flow-time was 3% to 5%, but no general improvement in maximum flow-time was observed. The fact that Nugent's best schedules were not nondelay schedules does not contradict previous indications that the nondelay schedules are a preferred subset of the active schedules; the delay schedules produced were carefully contrived modifications of nondelay schedules and in no sense a random sampling of the complement of the nondelay active schedules.

6–10 HEURISTIC PROCEDURES

Each of the preceding studies has essentially been the exhaustive application of a set of relatively simple and direct single-pass procedures. At the other extreme a human being faced with an n/m job-shop problem would rely on a much more complex mixture of procedures, undoubtedly including adjusting procedures. It would seem that the gap between these two would permit many interesting inquiries. If one cannot at this point reproduce on a computer the complex intellectual process by which a man manipulates a Gantt chart, one should at least be able to contrive procedures an order of magnitude more complex than single-pass dispatching procedures. One paper promising "A Heuristic Approach to Job-Shop Scheduling" [61] on close examination turns out to be a comparison of dispatching procedures which were no more complex than the ones described in the preceding pages. The whole question of adaptive or learning procedures is open to exploration and we think promising, although the results of the single study thus far [52] are admittedly not encouraging. However, that study seems strangely conceived and was subject to the computational limits of a very modest machine (IBM 650), so that we do not consider it adequate basis for discouragement.

One rather limited study that does provide some basis for optimism about heuristic procedures was performed by Crabill [40]. The procedure he examined might be classified as a *job-at-a-time adjusting procedure*. All the operations of a job were scheduled in succession, yet in this scheduling the starting-times of previously scheduled operations of other jobs could be changed. At each operation to be scheduled, the local effect on the maximum flow-time of jobs is determined and the best position for inserting the operation into the existing schedule is established. The operation is inserted into this best position, and its ramifications for previously scheduled operations are determined by computing new operation starting-times and left-shifting operations so as to achieve an active schedule for the set of scheduled operations.

The procedure used by Crabill in selecting the best position to insert an operation of Job I into an existing schedule is to select the position such that a lower-bound estimate of the flow-time of jobs is minimized over the set of jobs already scheduled and the job being scheduled (job I). Suppose that the potential starting-time of job I on machine K is time T and that the jobs scheduled on machine K but uncompleted at time T are jobs $i = 1, 2, \ldots, L$ and that they are scheduled on machine K in that order. There are $L + 1$ possible positions which job I could occupy, before job 1 or right after each of the L scheduled jobs. Let

D_i be the current finish-time of job i on machine K,

C_i be the current completion-time of job i,

C_I be the smallest possible completion-time of job I,

E_{il} be the finish-time of job i $(I, 1, 2, \ldots, L)$ on machine K if job J is inserted into schedule position $l = (0, 1, \ldots, L)$, and

p_I be the processing-time of job I on machine K.

For each schedule position, l, the following quantity is calculated:

$$\max(C_I + E_{Il} - T - p_I, \max_{i=1,\ldots,L} [C_i + E_{il} - D_i]).$$

The position with the smallest value for this quantity is chosen as the schedule position of job I. If job i is finished later than D_i its final completion-time may be potentially increased by $E_{il} - D_i$ (the bound is not precise, for waiting-time at a subsequent operation may nullify this increase or precedence constraints may amplify this increase). The waiting-time of job I if started in position l is $C_I + E_{Il} - T - p_I$. The position with the least increase in the lower-bound completion-time of any job is the position selected.

The procedure requires an order in which jobs are inserted and here an iterative approach was used. The first iteration inserted jobs in decreasing order of total processing-time. Any subsequent iteration inserted jobs in decreasing order of the completion-times acquired in the previous iteration.

Table 6–8 summarizes the limited experimental runs with this procedure. The problems were a subset of those employed in [102] and reported in Appendixes B–1 through B–3, so that direct comparisons may be made. In 13 out of 20 problems, an optimal schedule was produced by the adjusting procedure. This compares very favorably to 7 out of 20 problems obtaining an optimal solution by the selection of the

Table 6-8

Job-At-A-Time Adjusting Procedure: Maximum Flow-Time $(10/4/\text{Perm}/F_{max})$

Problem number	Replication	Number of iterations	Adjusting procedure	Best from Table B-2
76	1	9	161*	161†*
77	2	9	211	223
78	3	9	251*	251†*
79	4	9	166*	203
80	5	9	183	177†
81	6	9	243	241†
82	7	9	190*	193
83	8	9	167*	167†*
84	9	9	225*	225†*
85	10	9	170*	170†*
86	11	9	241*	241†*
87	12	9	223*	223†*
88	13	9	214	206†
89	14	9	137	132†
90	15	9	227	241†
91	16	9	161*	171
92	17	9	196*	202†
93	18	9	211*	216†
94	19	9	225	232†
95	20	9	142*	144†

* Optimal result, by a lower-bound argument.
† Nondelay schedule.

best result from a set of dispatching procedures. Whenever a dispatching procedure achieved an optimal solution, the adjusting procedure did also. In 16 of the 20 problems, the adjusting procedure yielded an equal or better value for maximum flow-time.

A statement made by Gantt four decades ago in support of the use of the Gantt chart still seems like an appropriate note on which to end this chapter: "This graphic layout makes it possible to group orders and distribute them over the available machines in a much more intelligent manner than by the hit-or-miss method of deciding what the next job will be whenever a machine runs out of work" [27].

CHAPTER 7

GENERAL NETWORK PROBLEMS
RELATED TO SCHEDULING

In more general terms a job-shop might be considered to be simply a network. It consists of *nodes* (machine groups) with a certain type of traffic between various nodes. This traffic or flow between nodes consists of discrete particles (jobs), which require a time resource at a node (the processing-time) and which may be stored at a node if insufficient resources are available. Sequencing the processing of the particles which arrive at a node—the topic of Chapters 5 and 6—is but one aspect of the general question of flow in networks. There are a number of closely related questions which are concerned with internode traffic and interference rather than in-node sequencing. The purpose of this chapter is to introduce three of the more important of these problems and to indicate their relationship to the scheduling problem. Many of these problems originally arose in contexts which are completely unrelated to the environment of a manufacturing-oriented production shop but which nevertheless fall within the domain of job-shop problems when job-shop is used as a generic description of a class of network flow situations.

For the description of the problems and of the algorithms for their solution it is convenient to use some of the terminology from the theory of graphs and networks. The notation adopted is that of Ford and Fulkerson [54].

A *graph* consists of (a) a *set* of *nodes*, and (b) a *set* of *arcs* which connect the nodes. The set of nodes will be denoted by $\{N\}$. Nodes in this set will be denoted by the letters s, t, x, y, or z or by the numbers $1, 2, \ldots, n$, where n is the number of nodes.

The set of arcs will be denoted by $\{A\}$. Arcs in this set will be denoted by the letter ν, by the ordered pair (x, y), where x and y are nodes or by the numbers $1, 2, \ldots, m$, where m is the number of arcs. For the problems considered it is meaningful to assume that a graph may not contain arcs of the form (x, x) or more than one arc of the form (x, y). Pictorially a graph is displayed as:

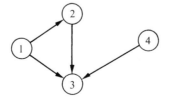

Figure 7–1

The circles represent nodes, the numbers within the circles are node numbers, and a directed line segment is an arc which exists.

132

For an arc (x, y), x is called the *initial node* of the arc and y is called the *terminal node* of the arc. The set of arcs which have x as the initial node is denoted by $\{A(x)\}$, or the set of nodes *after* x. The set of arcs which have x as the terminal node is denoted by $\{B(x)\}$, or the set of nodes *before* x.

Nodes s and t will be called the initial and terminal nodes of the graph, respectively. For the problems considered, it will be meaningful to require that node s be the only node of the graph with $\{B(x)\}$ empty and that node t be the only node of the graph with $\{A(x)\}$ empty.

A *path* is a sequence of distinct arcs, $(v_1, v_2, \ldots, v_k)$, such that two adjacent arcs in the sequence have one node in common. A path is a *directed path* or *route* if there is the further stipulation that for two adjacent arcs in the sequence the terminal node of the first arc corresponds to the initial node of the second arc. A *cycle* is a path in which arcs 1 and k have one node in common, and a *directed cycle* or *closed route* is a directed path which is a cycle. It is often convenient to represent a path by the sequence of nodes which are common to the arcs on the path. It is also possible to define the notions of path, route, cycle, and closed route in terms of a sequence of nodes rather than a sequence of arcs.

In Fig. 7-1, $\{(1, 2), (2, 3), (4, 3)\}$ is a path but not a route, $\{(1, 2), (2, 3)\}$ is a route, $\{(1, 2), (2, 3), (1, 3)\}$ is a cycle but not a closed route, and there are no closed routes.

A graph is called a *connected graph* if there is a path from every node to every other node. In this chapter all the graphs considered will either be connected graphs or will be graphs which can be transformed into connected graphs without loss of generality.

Often there are functions defined on the set of nodes and on the set of arcs. These functions associate with each node or arc a specific numerical value. They are of the form $g(\cdot)$, where g will be chosen to have some mnemonic significance. It is convenient to introduce the functional notation $g(X)$, where $\{X\}$ is a set of nodes, $g(y, X)$, and $g(V)$, where $\{V\}$ is a set of arcs. In such cases,

$$g(X) = \sum_{x \in \{X\}} g(x), \qquad g(y, X) = \sum_{x \in \{X\}} g(y, x), \qquad \text{and } g(V) = \sum_{v \in \{V\}} g(v).$$

A graph may be used to display precedence relationships among the nodes. This notion of a *precedence graph* was introduced in Section 1-2 and used in Chapter 4. In such graphs, an arc describes a precedence relationship between a pair of nodes: arc (x, y) means that node x *directly-precedes* node y, or in the notation used in Section 1-2,

$$x \gg y.$$

A directed path from x to y thus indicates a sequence of directly-precedes relations;

$$x \gg z_1 \gg z_2 \gg \cdots \gg z_k \gg y, \qquad k \geq 0.$$

If there exists a directed path from x to y, x is said to *precede* y, and this is written as follows:

$$x > y.$$

7–1 CRITICAL ROUTE ANALYSIS

Often the description of a job requires the use of a nontrivial precedence graph for the description of precedence among its operations. A linear string of operations does not suffice, for example, in cases in which a job consists of several subassemblies or in which a subassembly must be subjected to intermediate assembly, test, and disassembly of component parts.

The material of Section 4–3 represents one generalization of the work of Chapter 3 which involves precedence among operations. Some of the mechanisms presented in Chapter 6 are capable of handling jobs of this complex nature, yet the results presented did not invoke them; in no cases did the jobs consist of more than a linear sequence of operations. The difficulty of dealing with such jobs in the scheduling environment is that one must consider not only delays caused by the limited resources (machines), but also deal with the delays of operations which cannot start until the precedence constraints are satisfied.

However, there are many situations in which jobs of a complex nature must be scheduled and for which one can assume that there exists an unlimited supply of machines. There is, then, no competition for machines, either across or within jobs. Queuing delays are nonexistent, but there are delays due to precedence restrictions. Jobs can be scheduled and processed independently of each other, so we need only examine the scheduling of the operations of a single job, which is often called a project.

Without queuing delays one may assume that an operation may proceed immediately upon the completion of all its predecessor operations. Clearly, at least one of these predecessor operations determines the earliest time at which the operation can start. An operation could be termed critical if a small increase in its processing-time requirements or a delay in starting it were to delay the processing of its successor operations by an equal amount. A *critical route* would be a route in which all operations represent critical operations. There always exists at least one critical route and there may be several; the purpose of critical route analysis is to identify the critical route or routes. During the processing of the job, attention can be focused on the operations of the critical routes, for any delay in their processing will extend the completion-time of the job. Using critical route analysis, it is also possible to measure the "criticality" of every operation of the job.

Critical route analysis is the basis for the many project planning tools which have been developed. Foremost among these are PERT [124], CPM [106], and GERT [166]. A fundamental discussion of these project analysis techniques can be found in Moder and Phillips [142]. These techniques add many embellishments such as variability in processing-times and cost-time structures, but the underlying procedure is a critical route analysis.

The algorithm for this analysis will be stated formally and then applied to an example problem. A job is depicted as a precedence graph in which each node represents an operation. A function $p(x)$, defined on the operations $x = 1, 2, \ldots, n$, represents the processing-time for operation x. An arc (x, y) means that operation x directly-precedes operation y, so that it must be completed before operation y can start. Let operations s and t denote the initial and terminal operations, respectively (such operations may have to be artificial operations with zero processing-times). The problem is to find a route from s to t such that any increase in $p(x)$ of an operation x on the route will increase the completion-time of operation t. This problem can also be stated slightly differently: find the route or routes from s to t for which the sum of the processing-times of operations on the route is a maximum.*

A related problem is to find the set of operations, $\{C\}$, such that for $x \in \{C\}$ any increase in $p(x)$ will increase the completion-time of operation t. The set $\{C\}$ can then be used to generate all critical routes, if more than one exists.

Let $e(x)$ represent the earliest time at which operation x can start $\left(e(s)\right.$ is assumed to be zero). If $y \in \{B(x)\}$, then the earliest time at which y can be finished is $e(y) + p(y)$. Since operation x cannot be started before all of its predecessors are completed we must have

$$e(x) = \max_{y \in \{B(x)\}} \left(e(y) + p(y)\right).$$

When one starts with operation s one can develop $e(x)$ for every other operation of the job by working "forward" through the graph (assuming that there are no closed routes; for example, no operation must precede itself). The completion-time of the job is $e(t) + p(t)$.

The development of the function $e(x)$ does not in itself provide sufficient information for the identification of critical operations. Although one might suspect that if an operation determines the earliest starting-time of a directly following operation, then it should be a critical operation, but this is a necessary yet not sufficient condition. Consider the example

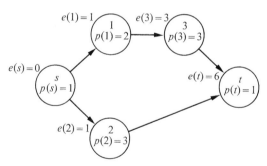

Here $e(2)$ is determined by operation s but $p(2)$ can be increased to 5 without delaying the start of operation t.

The criticality of an operation can be developed by determining the latest time at which it could start without increasing the completion-time of operation t beyond $e(t) + p(t)$. Let $l(x)$ denote this time. An operation would be critical if $l(x) = e(x)$ and, further, $l(x) - e(x)$ is a measure of the degree of criticality. Clearly, $l(t) = e(t)$. Furthermore, since the latest time at which an operation can be started is determined

---

* In this alternative form the problem is a linear programming problem; the solution of its dual problem leads to an algorithm which is very similar to the one given here.

by the latest time at which its directly-following operations can start, we have

$$l(x) + p(x) = \min_{y \in \{A(x)\}} l(y).$$

Starting with operation t, $l(x)$ can be developed for every operation of the job by working backward through the precedence graph.

Algorithm

In addition to the functions already defined on the nodes there is a set of nodes $\{C(x)\}$ for each node x. This set will contain the nodes of $\{B(x)\}$ which would increase $e(x)$ if their processing-times were increased; if $y \in \{C(x)\}$, then $y \in \{B(x)\}$ and $e(y) + p(y) = e(x)$.

There are two main stages of the algorithm, the forward stage and the reverse stage. In each stage and at any phase of the algorithm the set of nodes is partitioned into two sets:

1) $\{L\}$, the set of nodes which has been labeled $(e(x)$ or $l(x)$ has been determined),

2) $\{U\}$, the set of nodes which has not been labeled.

Forward stage

Initially, $e(x) = 0$ for all $x \in \{N\}$, $s \in \{L\}$, and the rest of the nodes are contained in $\{U\}$.

1) Remove from $\{U\}$ and place in $\{L\}$ any node, x, such that for all $y \in \{B(x)\}$, $y \in \{L\}$.

2) Let

$$e(x) = \max_{y \in \{B(x)\}} (e(y) + p(y)).$$

3) Place in $\{C(x)\}$ all nodes $y \in \{B(x)\}$ such that $e(x) = e(y) + p(y)$.

4) If $\{U\}$ is empty, then stop. Otherwise, go to step 1.

Reverse stage

Initially, $l(t) = e(t)$, $t \in \{L\}$ and all other nodes are contained in $\{U\}$.

1) Remove from $\{U\}$ and place in $\{L\}$ any node, x, such that for all $y \in \{A(x)\}$, $y \in \{L\}$.

2) Let

$$l(x) = \min_{y \in \{A(x)\}} (l(y)) - p(x).$$

3) If $l(x) > e(x)$, then empty $\{C(x)\}$.

4) If $\{U\}$ is empty, then stop. Otherwise, go to step 1.

At termination,

a) $l(t) + p(t)$ is the shortest length of time in which the job can be completed,

b) any operation with a nonempty set $\{C(x)\}$ is a critical operation,

c) starting from t it is possible to construct the critical routes from s to t by using the sets $\{C(x)\}$, starting at $\{C(t)\}$.

An example problem and solution are given in Fig. 7–2.

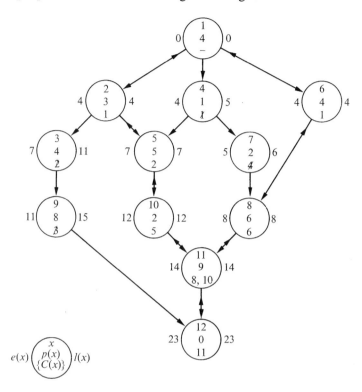

Fig. 7–2. Critical route analysis. The critical operations are 1, 2, 5, 6, 8, 10, 11, 12. There are two critical routes: (a) 1, 2, 5, 10, 11, 12, and (b) 1, 6, 8, 11, 12. Arrows here mean: → precedence; ►− reverse tracking of critical route.

It is possible to modify the forward stage to detect a closed route in the precedence graph. A closed route can be represented as a sequence of nodes $(x_1, x_2, \ldots, x_k)$ where $x_i \in \{B(x_{i+1})\}$, $i = 1, 2, \ldots, k - 1$, and $x_k \in \{B(x_1)\}$. For the nodes of such a route, it is impossible for any one of them to be placed in $\{L\}$. For if x_i is to be selected from $\{U\}$ and placed in $\{L\}$, then it must be true that $x_{i-1}, x_{i-2}, \ldots, x_1$, $x_k, x_{k-1}, \ldots, x_{i+1}$ and finally x_i must be in $\{L\}$, contradicting the assumption that $x_i \in \{U\}$. Thus at step 1 of the forward stage if $\{U\}$ is not empty and no node can be removed from $\{U\}$, then there is a closed route in the graph.

7–2 SHORTEST-ROUTE DETERMINATION

Often there is a function defined on the set of arcs of a graph, which takes on non-negative values and which could be considered to represent the distance or traverse-time from one node to another. In general, it need not be true that the traverse-time for an arc (x, y), say, $p(x, y)$, be equal to the time on the reverse arc, $p(y, x)$, or even that both (x, y) and (y, x) be arcs of the graph. It is then an obvious question to ask for the shortest route between the initial node s and terminal node t of the graph, where the length of the route is the sum of the $p(v)$ for the arcs v in the route.

This problem has received considerable attention (see [164] for a summary), and numerous solution techniques have been developed. The algorithm given below, by Minty [139], has both the closest alliance to the network terminology presented and the most intuitive appeal.

Algorithm

Two functions defined on the set of nodes are used as working variables in the algorithm:

1) $a(x)$ During the course of the algorithm this will be a lower bound on the time required to get from s to x, and

2) $b(x)$ will represent the node before x on the route of smallest total time from s to x.

At any phase of the algorithm the set of nodes, $\{N\}$, is partitioned into two sets:

1) $\{U\}$ a set of unpassed nodes. If $x \in \{U\}$, then the shortest route from s to x has not been determined, and

2) $\{P\}$ a set of passed nodes. If $x \in \{P\}$, then the shortest route from s to x has been determined.

Initially, $a(s) = 0$, $a(x) = \infty$ for $x \neq s$, and $\{U\}$ contain all the nodes.

General steps in algorithm

1) Let x be a node in $\{U\}$ such that

$$a(x) = \min_{y \in \{U\}} a(y).$$

2) If x equals t, then $a(t)$ is the shortest time from s to t and the shortest route is $(s, \ldots, b(b(t)), b(t), t)$. Stop.

3) If x is not t, then remove x from $\{U\}$ and place x in $\{P\}$. For all

$$y \in \{\{A(x)\} \cap \{U\}\},$$

if $a(y) > a(x) + p(x, y)$, then let $a(y) = a(x) + p(x, y)$ and $b(y) = x$. Return to step 1.

To find the shortest route from s to all other nodes, step 2 is omitted and steps 1 and 3 are cycled until $\{U\}$ is empty or until $a(x) = \infty$ at step 1. If $a(x)$ equals ∞ at termination, then there is no route from s to x.

Example

Consider the network of seven nodes given in Fig. 7–3. The node number is given in the circle representing the node; below this is the current value of $a(x)$ for that node, and below that the current value of $b(x)$. The traverse-time is given above or beside

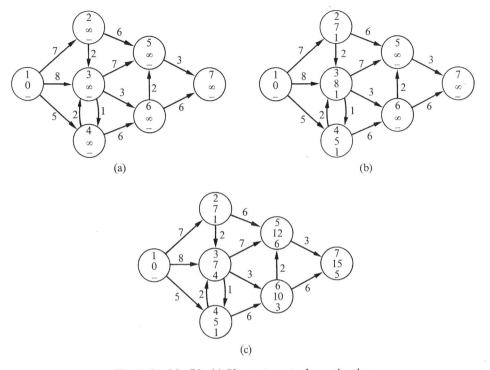

(a)   (b)

(c)

Fig. 7-3. (a), (b), (c) Shortest-route determination.

each arc. Consider node 1 to be the initial node and node 7 the terminal node. Initially the graph corresponds to Fig. 7-3(a). The set $\{U\}$ consists of all seven nodes. The algorithm selects node 1 from $\{U\}$ and places it in $\{P\}$. The result is shown in Fig. 7-3(b). Node 4 is selected next from $\{U\}$. The algorithm is repeated until finally $\{U\}$ is empty and the graph appears as in Fig. 7-3(c). The shortest route from 1 to 7 is 1, 4, 3, 6, 5, 7, with a total traverse-time of 15.

The shortest-route-times from any node to every other node can be determined by applying the algorithm n times, once with each node as the initial node and using the modification mentioned in order to find the shortest route and shortest traverse-time from s to any other node x. In this case, however, there is a matrix procedure which uses min-addition and which involves less total computation [164].

7-3 ASSEMBLY-LINE BALANCING PROBLEMS

With this background it is appropriate to mention assembly-line balancing as a network problem which has received considerable attention during the past two decades and which is often mentioned in conjunction with job-shop scheduling problems. The assembly of a complex product is described in terms of a precedence graph. Each node, x, of the graph represents an assembly operation, requiring a processing-time $p(x)$. Assembly of the product is to be accomplished by moving the products along an assembly line past various subassembly stations. Each of these stations may perform one or more operations subject to precedence constraints; it can perform an operation only if all of its predecessor operations are not performed in a subsequent station. The problem is to partition the set of nodes into g sets, $\{A_1\}, \{A_2\}, \ldots, \{A_g\}$, where the set $\{A_i\}$ contains the operations assigned to station i. To obey the precedence constraints one must have the following:

$$\text{if } x \succ y, x \in \{A_i\}, \text{ and } y \in \{A_j\}, \qquad \text{then } i \leqq j.$$

The total processing-time assigned to station i is $p(A_i)$.

There are two different versions of the assembly-line balancing problem:

1) Minimize g subject to $p(A_i) \leqq$ given constant, all i;

2) $\min\limits_{g} \max\limits_{i} p(A_i)$.

Short of brute-force enumeration schemes, no direct analytic solutions to this combinatorial problem have been given. A branch-and-bound technique (cf. Chapters 4, 5, and 6) can be applied as a form of subtle enumeration. Several heuristic solutions have also been proposed. Ignall [86] has given an excellent review of the status of the problem.

CHAPTER 8

SELECTION DISCIPLINES IN A SINGLE-SERVER QUEUING SYSTEM

Perhaps the strongest and the most unrealistic of the conditions assumed in Chapters 3 through 6 is the simultaneous arrival of all the jobs to be considered. As noted in Section 4–2, this assumption is essential for most of the results that were obtained, but it is nonetheless patently unrealistic. Most real situations that can be identified with the job-shop model are continuous processes. Jobs may arrive periodically and in batches, but in general there is the possibility of carryover from one batch to the next, since all the jobs of one batch may not be completed at the time that the next work is released to the shop.

When we relax this assumption, we can still obtain many strong and interesting results but we must employ an entirely different approach. Although previous arguments have been almost entirely algebraic, a process in which arrivals are intermittent and at times not known in advance must be studied by probabilistic methods. Such a process can be studied as a queuing system with emphasis on the *selection discipline*, i.e., the rule by which one of the jobs waiting in a queue is selected whenever the machine becomes available for reassignment.

Instead of being given a certain set of jobs with known attributes, now the given information that describes a problem is a characteristic of the generation process that produces jobs and values of attributes. In general, one is given the form of a distribution, and the values of the attributes are assumed to be a sequence of independent random variables obtained from this distribution. Similarly, the result of the scheduling process, the flow-times and waiting-times of the jobs, will be determined as a distribution function, rather than a set of specific values.

Attention is centered on mean flow-time as a measure of performance. (Mean completion-time is essentially meaningless in a continuous process.) The relationship of Section 2–4 between mean flow-time and mean number of jobs in the shop still holds for the continuous process, so that a selection discipline that minimizes mean flow-time will also minimize the mean number of jobs in the shop. The general approach will be to derive an expression for the transform of the distribution function of flow-time. In theory this will provide the entire distribution of flow-times, but in practice the transform can seldom be inverted symbolically. However, one can obtain the moments of the distribution by differentiation.

Notation in this chapter (and the two following ones) is a compromise between what has been employed in the preceding chapters and the symbols and conventions that have become fairly standard in the queuing literature. Although previously upper- and lower-case letters have been used to denote derived and given values, respectively, an upper-case letter will now be used to indicate a random variable. Functions with upper-case letters will denote distribution functions; functions with lower-case letters will denote density functions. The Laplace-Stieltjes transform* of the distribution function will be denoted by a Greek letter. The principal symbols are the following:

Quantity	Random variable	Distribution function	Laplace transform	Density function
Processing-time	P	$G(p)$	$\gamma(z)$	$g(p)$
Waiting-time of a job	W	$A(w)$	$\alpha(z)$	$a(w)$
Flow-time of a job	F	$B(f)$	$\beta(z)$	$b(f)$
Length of a busy period	T	$H(t)$	$\eta(z)$	$h(t)$

The average utilization of a single-machine shop has been given bv (Section 2–4):

$$\bar{U} = \lambda\bar{p},$$

where λ is the mean arrival rate and $\bar{p}$ is the mean amount of work per job. In the queuing case, except for certain special conditions of preemption and sequence-dependent setup, the mean amount of work per job is equal to the expected value of the random variable representing processing-time, denoted $E(P)$. The product $\lambda E(P)$ is universally denoted as ρ in the queuing literature, and that notation will be employed in these chapters. Except for the cases of preemption and setup, $\bar{U} = \rho$, so that utilization is a consequence of the given distributions of arrival-time and processing-times and is unaffected by the choice of selection discipline. For the analysis of the following sections to hold, it is a very general condition that these distributions be selected so that $\rho < 1$, in which case the system is said to be *nonsaturated*.

8–1 QUEUES WITH POISSON ARRIVALS

Most of the queuing literature pertains to the special case in which the times between job arrivals are a sequence of independent observations from a fixed exponential distribution. Equivalently, the number of arrivals in a given interval of time is a random variable with a Poisson distribution.† This is called a *Poisson process*, and in the context of queuing theory, *Poisson arrivals* or *random arrivals*. This model has been used partly because it is a plausible representation of many physical processes, but another and more important reason for its use is the tremendous analytical convenience that this process affords.

If V is the time elapsing between two successive arrivals, the exponential inter-arrival-time assumption implies that

$$\text{Prob}\,(V \leq v) = 1 - e^{-\lambda v}, \qquad v \geq 0.$$

---

* The Laplace-Stieltjes transform of the distribution function will be referred to frequently in the text simply as the Laplace transform. A definition of this transform and a summary of its useful properties are given in Appendix A.

† A proof of this equivalence is given in [38].

The mean and variance of interarrival-times are $1/\lambda$ and $1/\lambda^2$, respectively. The Poisson distribution for the number of arrivals in an arbitrary time interval of length t has the form

$$\text{Prob } (N = n) = e^{-\lambda t}\, \frac{(\lambda t)^n}{n!}, \qquad n = 0, 1, \ldots$$

The mean and standard deviation of N are λt and $\sqrt{\lambda t}$.

These dual distributions can be derived from a set of basic postulates which state that, in a small interval of time, the probability of an arrival is proportional to the length of the interval, that the probability of more than one arrival is negligible, and that the occurrence or nonoccurrence of an arrival in a small time interval is independent of any other arrivals, or the time since the last arrival.

The important properties and consequences of the Poisson arrival process will be developed in the balance of the section.

The memoryless property

Suppose that R_a and R_b are the times of two consecutive arrivals in an arbitrary arrival process. Suppose that y time units after R_a, but before R_b, we wish to find the probability of the next arrival occurring within an additional t units of time. We are asking for the conditional probability

$$\text{Prob } (V \le y + t \mid V > y).$$

It is uniquely a property of the Poisson process that this conditional probability is equal to the unconditional probability $\text{Prob } (V \le t)$, and does not depend on the value of y. Thus

$$\text{Prob } (V \le y + t \mid V > y) = \frac{\text{Prob } (y < V \le y + t)}{\text{Prob } (V > y)}$$

$$= \frac{e^{-\lambda y} - e^{-\lambda(y+t)}}{e^{-\lambda y}} = 1 - e^{-\lambda t}.$$

The process is said to be *memoryless* in that, when one is writing a probability statement about the time remaining before the next arrival, one does not have to consider when the last arrival occurred.

Aggregation and branching of Poisson streams

Many arguments in queuing theory are simplified by the fact that the aggregation of several Poisson input streams results in a Poisson stream and that probabilistic selection of jobs from a single Poisson stream into several output paths yields independent Poisson streams.

Consider an arrival stream which is formed by accepting input from k sources. If each input stream is Poisson, with the ith component source having rate λ_i, the combined stream is also Poisson with a rate of $\lambda = \lambda_1 + \lambda_2 + \cdots + \lambda_k$ jobs per unit time.

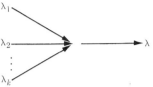

For the case of two input streams, let $N_1(t)$ and $N_2(t)$ be the numbers of events occurring in the two input streams over a time interval of length t. If both these random variables have the Poisson distribution, we wish to show that their sum, denoted by $N(t)$, is Poisson with rate $\lambda_1 + \lambda_2$. The probability that $N(t)$ takes the value n is given by

$$\text{Prob}\big(N(t) = n\big) = \sum_{n_1=0}^{n} \text{Prob}\big(N_1(t) = n_1, N_2(t) = n - n_1\big)$$

$$= \sum_{n_1=0}^{n} \frac{(\lambda_1 t)^{n_1}}{n_1!} e^{-\lambda_1 t} \frac{(\lambda_2 t)^{n-n_1}}{(n - n_1)!} e^{-\lambda_2 t}$$

$$= t^n \frac{e^{-(\lambda_1+\lambda_2)t}}{n!} \sum_{n_1=0}^{n} \binom{n}{n_1} \lambda_1^{n_1} \lambda_2^{n-n_1} = \frac{(\lambda t)^n}{n!} e^{-\lambda t}.$$

By substituting $(\lambda_1 + \lambda_2 + \cdots + \lambda_{k-1})$ for λ_1 and λ_k for λ_2 in this computation, we obtain the general inductive step which proves the assertion for any number of input streams.

For the case in which a Poisson stream of arrivals branches into k output paths, let r_i be the probability that a job takes output path i.

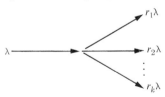

If the input rate is λ and the output paths are chosen independently, the ith output stream is Poisson with rate $r_i \lambda$.

Let $N(t)$ be the number of input jobs in t time units, and let $N_i(t)$ be the number of jobs taking the ith output path over the same interval. The conditional probability of a particular distribution of jobs over the output streams, when the number of input jobs is known, is given by the multinomial distribution as

$$\text{Prob}\big(N_1(t) = n_1, \ldots, N_k(t) = n_k \mid N(t) = n\big) = \frac{n!}{n_1! n_2! \cdots n_k!} r_1^{n_1} r_2^{n_2} \cdots r_k^{n_k}.$$

The conditioning on the value of $N(t)$ is removed by multiplying by the probability that $N(t) = n$ to yield

$$\text{Prob}\left(N_1(t) = n_1, \ldots, N_k(t) = n_k\right) = \frac{n!}{n_1! \cdots n_k!} r_1^{n_1} \cdots r_k^{n_k} e^{-\lambda t} \frac{(\lambda t)^n}{n!}$$

$$= \prod_{i=1}^{k} \frac{(r_i \lambda t)^{n_i}}{n_i!} e^{-r_i \lambda t}.$$

That the output streams are independent Poisson processes is proved by the fact that the joint probability above factors into k Poisson probabilities.

The random property

Poisson arrivals are often called random arrivals because of the following property: Given that a job arrives in an interval of length t, the moment of its arrival has a uniform distribution over the interval

$$\text{Prob}\left(y \leq \text{arrival moment} \leq y + dy \mid \text{arrival in } t\right) = \frac{dy}{t}, \qquad \text{for } 0 \leq y \leq t. \qquad (1)$$

To prove this assertion, one first finds the joint probability that there are n arrivals in the interval and one out of the n arrival instants falls in the subinterval, $(y, y + dy)$. Consider the interval to be divided into three segments having lengths y, dy, and $(t - y - dy)$ as shown below:

Arrivals		k		1		$n - 1 - k$	
Time	0		y		$y + dy$		t

If the number of arrivals falling in the first segment is k, $(k = 0, 1, \ldots, n - 1)$, the number occurring in the last segment must be $(n - 1 - k)$, in order to have one in $(y, y + dy)$ and preserve the total of n. The probability that arrivals in $(0, t)$ occur in this way is the product of the three appropriate Poisson probabilities:

$$\left(\frac{(\lambda y)^k}{k!} e^{-\lambda y}\right) \left(\lambda\, dy e^{-\lambda dy}\right) \left(\frac{[\lambda(t - y - dy)]^{n-1-k}}{(n - 1 - k)!} e^{-\lambda(t-y-dy)}\right)$$

$$= \frac{\lambda^n e^{-\lambda t}}{(n - 1)!} \binom{n - 1}{k} y^k (t - y - dy)^{n-1-k}\, dy.$$

The probability of n arrivals in $(0, t)$, one of which takes place in $(y, y + dy)$, is obtained by summing this expression over the possible values of k with the aid of the binomial theorem:

$$\frac{\lambda^n e^{-\lambda t}}{(n - 1)!}\, dy \sum_{k=0}^{n-1} \binom{n - 1}{k} y^k (t - y - dy)^{n-1-k} = \frac{(\lambda(t - dy))^{n-1}}{(n - 1)!} e^{-\lambda t} \lambda\, dy.$$

Ignoring factors of the order $(dy)^2$ and smaller, this becomes

$$\text{Prob}\left(n \text{ arrivals in } (0, t) \text{ with one arrival in } (y, y + dy)\right) = \frac{(\lambda t)^{n-1}}{(n - 1)!} e^{-\lambda t} \lambda\, dy.$$

This joint probability is converted to a conditional probability by dividing by the probability that there are n arrivals over the whole interval to yield

Prob $\big($one out of n arrival instants in $(y, y + dy) \mid n$ arrivals in $(0, t)\big)$

$$= \frac{(\lambda t)^{n-1}}{(n-1)!} e^{-\lambda t} \lambda \, dy \bigg/ \frac{(\lambda t)^n}{n!} e^{-\lambda t} = n \frac{dy}{t}.$$

It is equally likely that any one of the n jobs is responsible for causing an arrival instant to fall into the dy subinterval. Therefore the conditional probability that a job arrives in $(y, y + dy)$, given that it is one of n arriving in $(0, t)$ for $(0 \le y \le t)$, is dy/t. This conditional probability is independent of the conditioning on n, so it is the unconditional probability.

The random modification

Another useful property is based on the solution to the following problem: Given a job that arrives while another job is being processed, what is the distribution of the remaining processing-time for the job at the moment of arrival? This quantity is the *random modification* of a processing-time.

Denote by X the total processing-time of the job which the new arrival finds in process, and let Y be the amount of processing-time remaining for that job at the instant of arrival. A sequence of processing-times (with idle periods deleted) and the relationship between X and Y is shown below:

It is shown in reference [7] that if $G(p)$ is the processing-time distribution function and $E(P)$ is the mean processing-time, then

$$\text{Prob} \, (p \le X \le p + dp) = \frac{p \, dG(p)}{E(P)}. \tag{2}$$

This result has an intuitive interpretation in that the desired probability should be proportional to the long-run portion of time devoted to processing-times of length p. This, in turn, should be proportional to the product of the processing-time p and the frequency, $dG(p)$, with which such intervals occur. The denominator, $E(P)$, is a normalizing factor required to make the integral of this probability equal to one, since $\int_0^\infty p \, dG(p) = E(P)$.

To determine the distribution of Y, note that, by the random property, the conditional distribution of Y, given that $X = p$, is the uniform distribution. That is,

$$\text{Prob} \, (y \le Y \le y + dy \mid X = p) = \frac{dy}{p} \quad \text{for } 0 \le y \le p.$$

Multiplication by (2) gives the joint probability

Prob $(y \le Y \le y + dy, p \le X \le p + dp)$

$$= \frac{dG(p)}{E(P)} dy, \quad 0 \le y \le p, 0 \le p \le \infty. \tag{3}$$

The marginal distribution of Y is obtained by integrating this expression over the allowable values of p. This yields the density function of Y as

$$\text{Prob } (y \leq Y \leq y + dy) = \int_{p=y}^{\infty} \frac{dG(p)}{E(P)}\, dy = \frac{1 - G(y)}{E(P)}\, dy \qquad \text{for } 0 \leq y \leq \infty.$$

Letting $\gamma_Y(z)$ be the Laplace transforms associated with the distribution of the random modification, we obtain

$$\gamma_Y(z) = \int_0^{\infty} e^{-zy} \frac{1 - G(y)}{E(P)}\, dy = \frac{1 - \gamma(z)}{zE(P)}, \tag{4}$$

from which we obtain the kth moment of Y as

$$E(Y^k) = \frac{E(P^{k+1})}{(k + 1)E(P)}. \tag{4a}$$

Cases in which the underlying random variable is something other than processing-time are also important. The above formulas may be used in such instances by replacing P, $G(p)$, and $\gamma(z)$ by the symbol for random variable, distribution function, and Laplace transform of the random variable concerned.

Poisson-exponential queues

Many analyses of a steady-state queuing system have depended on the special properties of the Poisson arrival process and exponentially distributed processing-times to construct a system of differential-difference equations relating system states in time. For example, Morse [143] employs this technique throughout his book.

As an example of this type of analysis, consider the simple queuing situation with Poisson arrivals, in which the processing-times are exponentially distributed with mean μ^{-1}. If the system is not idle, and if the selection discipline does not depend on processing-times, the probability of a job completion in a dt interval is $\mu\, dt$. System state is described by the number of jobs in the system. Let $p_n(t)$ be the probability that there are n jobs in the system at time t; the set of equations satisfied by these state probabilities is

$$\frac{dp_0(t)}{dt} = \mu p_1(t) - \lambda p_0(t)$$

and

$$\frac{dp_n(t)}{dt} = \lambda p_{n-1}(t) + \mu p_{n+1}(t) - (\lambda + \mu)p_n(t) \qquad \text{for } n = 1, 2, \ldots$$

It can be shown that, for $\rho = \lambda/\mu < 1$, the system will reach equilibrium, and the set of difference equations describing the equilibrium-state probabilities is obtained by setting the time derivatives to zero.

The steady-state solutions are then easily found to be

$$p_n = \lim_{t \to \infty} p_n(t) = \rho^n(1 - \rho). \tag{5}$$

Much of the early literature on selection disciplines (until the end of the 1950's) employed this type of approach. However, the resulting systems of equations were often prohibitively difficult. The following sections use an alternative approach that is simpler and more powerful.

8–2 SYSTEM STATES

A queuing system can be analyzed by defining a set of states which the system may occupy. For our purpose these states will often be much grosser than those needed for a detailed description of the system at a point in time. For example, state 0 might correspond to an idle machine and state 1 to a busy machine. The state obviously does not completely describe the system, since it could be in state 1 with only one job being processed, or it could have a dozen jobs waiting in queue.

When this device is used, the states will be defined so that the state selection process is Markovian; the probability of a transition from state j to state k will depend only on j and k and not on the path by which the system reached state j. Results from the theory of semi-Markov processes (see, for example, [39]) can then be used to make probability statements about finding the system in some particular state at some moment in time. Most important is a statement about the steady-state probability of being in some state in terms of the amount of time that the system occupies the state in the short run.

For some state, j, a representation of its start and recurrence are illustrated in Fig. 8–1. Measured from the time of entrance into state j, the system remains in state j for a time M_j. It then moves out of state j, passes through a sequence of other states, and returns to state j again L_j time units after its previous entrance to that state.

For the interval illustrated in Fig. 8–1, the proportion of time that the system is in state j is M_j/L_j. For several such consecutive intervals, with time in state j of M_{jk} and times between entrance to state j of L_{jk} (for the kth interval), the proportion would be $\sum_k M_{jk}/\sum_k L_{jk}$. The asymptotic probability is given by Smith [188]. If

$$\pi_j = \text{steady-state probability of state } j, \quad m_j = E(M_j), \quad l_j = E(L_j),$$

then

$$\pi_j = \lim_{t \to \infty} \text{Prob (state } j \text{ at time } t) = \frac{m_j}{l_j}. \tag{6}$$

This result, with the properties of a Poisson process for arrivals, is the basis of the analysis in this chapter. States can be defined so that statements can be made about the flow-time of a job conditioned on the particular state that the system occupies at the moment of its arrival. It is a property of the Poisson process of arrivals that the probability that an arriving job finds the system in state j at the moment of its arrival is equal to the steady-state probability of being in state j, π_j. The states defined will be Markovian due to the Poisson process of arrivals, so that Eq. (6) may be used to determine these steady-state probabilities. The mean values, m_j and l_j, will always be easily determined.

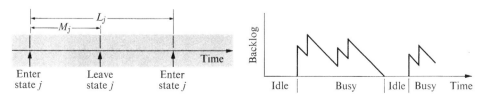

Figure 8–1   Figure 8–2

8-3 THE BUSY PERIOD OF A QUEUE WITH POISSON ARRIVALS

In a queuing system with Poisson arrivals the series of idle periods and busy periods form two alternating and independent sequences of identically distributed random variables. The memoryless property of the exponential distribution implies this independence and also that the idle periods have exponentially distributed lengths. The distribution of the busy-period lengths may depend on the selection discipline, but only if the discipline entails extra processing as, for example, with some of the preemptive rules. This fact will be shown presently. The distribution of the busy-period lengths will be important for the development of flow-time distributions under various selection procedures.

The state of a queuing system can be measured by the backlog of work, measured in units of time, consisting of the sum of the processing-times of the jobs in queue plus the remaining processing-time of the job on the machine. A typical plot of backlog is given in Fig. 8-2. An upward jump equal in magnitude to the processing-time of a job occurs at the time of each arrival. When the backlog is nonzero it decreases at rate -1 so long as there is no additional work imposed due to preemption, or no period of inserted-idleness. When the backlog becomes zero it remains so until the next arrival occurs. The order in which the jobs are selected from queue cannot affect this graph; it is completely determined by the moments of arrival and the processing-times of arriving jobs.

Since the busy periods correspond to the unbroken periods when the backlog is greater than zero, this simple argument shows that all selection disciplines not involving inserted idleness or extra processing share a common distribution of busy-period lengths.

Therefore we are free to choose a convenient discipline for analysis even though it might be uninteresting and artificial from a practical point of view. Consider one such discipline, which allows us to view a busy period as consisting of the processing of the initial job that starts the busy period, plus a series of random variables which are themselves the lengths of busy periods. To implement this special discipline two separate waiting lines are established: the main queue and the initial queue. The n jobs, possibly zero, which arrive during the processing of the initial job go into the initial queue and all other jobs arriving during the course of the busy period join the main queue. A job is always selected from the main queue, if possible; otherwise one is selected from the initial queue. The busy period ends when both queues are empty. Within either queue jobs are selected on a first-come, first-served basis. Figure 8-3 is a plot for a busy period in which three jobs arrive during the processing of the initial job. The upper plot shows the number of jobs in the initial queue while the lower plot shows the number of jobs in the main queue. The crosses on the time axes indicate job completion-times.

Each job from the initial queue goes into processing only when the main queue is empty. As it is being processed, other jobs may arrive at the main queue and these jobs, as well as some future arrivals, are processed until the main queue again becomes empty. Thus the time period between the selection of two successive jobs from the initial queue has all of the characteristics of a busy period which is started by an initial job. Further, these pseudo busy periods have lengths which are independently

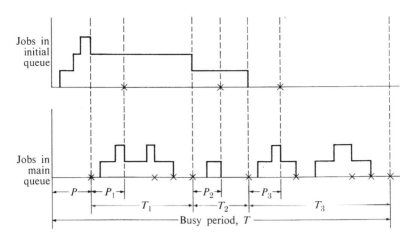

Figure 8–3

and identically distributed and which in fact have the same distribution as the length of the real busy period.

These observations enable one to derive the Laplace transform associated with the distribution of the busy-period length. Let

P = processing-time of the initial job,

N = number of jobs arriving during the processing of the initial job,

T = busy-period length,

T_j = length of pseudo busy period started by the jth job, which arrived during the processing of the initial job. (By convention, $T_0 = 0$.)

From the independence and distribution properties of the T_j's we can write

$$E(e^{-zT} \mid P = p, N = n, T_0 = 0, T_1 = t_1, \ldots, T_n = t_n) = \exp\left(-z\left(p + \sum_{j=0}^{n} t_j\right)\right)$$

or

$$E(e^{-zT} \mid P = p, N = n) = e^{-zp}(\eta(z))^n.$$

Since the distribution of N is Poisson with mean λp,

$$E(e^{-zT} \mid P = p) = e^{-zp}e^{-\lambda p} \sum_{n=0}^{\infty} \frac{(\lambda p \eta(z))^n}{n!} = e^{-p(z+\lambda-\lambda\eta(z))}.$$

The final result is obtained by integrating with respect to the processing-time distribution:

$$\eta(z) = E(e^{-zT}) = \int_{p=0}^{\infty} e^{-p(z+\lambda-\lambda\eta(z))} \, dG(p)$$

or

$$\eta(z) = \gamma(z + \lambda - \lambda\eta(z)). \tag{7}$$

Unfortunately $\eta(z)$ appears on both sides of this functional equation and it is usually impossible to obtain an explicit solution. Nevertheless, moments of busy-

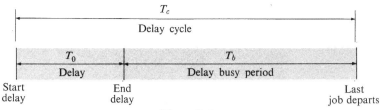

Figure 8–4

period length can be derived from Eq. (7), and these will be useful in the study of flow-time. The first two moments of T, as determined from $-\eta'(0)$ and $\eta''(0)$, are

$$E(T) = \frac{E(P)}{1 - \rho}, \tag{7a}$$

$$E(T^2) = \frac{E(P^2)}{(1 - \rho)^3}. \tag{7b}$$

Frequently it will be useful to consider a more general kind of busy period which is initiated by the performance of some task other than the processing of an initially arriving job. The initiating task will be called a *delay*, while the time spent processing jobs is a *delay busy period*. The delay and the delay busy period taken together are a *delay cycle*. The relationship between these three time intervals is illustrated in Fig. 8–4. The essential features of this situation are that the delay commences when there are no ordinary jobs in the system, and the delay cycle ends when a job is completed and there are no jobs in the system. The delay busy period may be viewed as a series of pseudo busy periods of the type used above.

Associated with the lengths of the delay, the delay busy period, and the delay cycle are the random variables T_0, T_b, and T_c, the distribution functions H_0, H_b, and H_c, and the Laplace transforms $\eta_0(z)$, $\eta_b(z)$, and $\eta_c(z)$. The notation used for the delay is made analogous to that for busy periods rather than processing-times because the concept of delay cycles will provide models useful in the study of priority queues, where the interesting types of delays are often busy periods.

We shall always be given $\eta_0(z)$. The derivations for $\eta_b(z)$ and $\eta_c(z)$ in terms of $\eta_0(z)$ and $\eta(z)$ proceed in the same manner as the derivation of Eq. (7). The Laplace transforms and first two moments of the distributions of T_b and T_c are:

$$\eta_b(z) = \eta_0(\lambda - \lambda\eta(z)), \tag{8}$$

$$E(T_b) = \frac{\rho E(T_0)}{1 - \rho}, \tag{8a}$$

$$E(T_b^2) = \frac{\lambda E(P^2)}{(1 - \rho)^3} E(T_0) + \frac{\rho^2}{(1 - \rho)^2} E(T_0^2), \tag{8b}$$

$$\eta_c(z) = \eta_0(z + \lambda - \lambda\eta(z)), \tag{9}$$

$$E(T_c) = \frac{E(T_0)}{1 - \rho}, \tag{9a}$$

$$E(T_c^2) = \frac{\lambda E(P^2)}{(1 - \rho)^3} E(T_0) + \frac{E(T_0^2)}{(1 - \rho)^2}. \tag{9b}$$

8–4 THE DISTRIBUTION OF FLOW-TIME
UNDER THE FIRST-COME, FIRST-SERVED DISCIPLINE

In practice, the first-come, first-served (FCFS) discipline is often a convenient and natural selection discipline. In theory, it provides a norm for the comparison of disciplines and is embedded as a secondary rule in many other disciplines. The Laplace transform associated with the distribution of flow-time under FCFS will be derived in terms of the Laplace transform of the processing-time distribution. The analysis is easily extended to situations which involve delay cycles, and these in turn can be used to analyze priority-class disciplines.

The flow-time of a job which arrives when the machine is idle is simply its processing-time. Thus with probability $1 - \rho, F = P$ and $\beta(z) = \gamma(z)$. With probability ρ, a job arrives during a busy period and has to wait. For such jobs we use the notation $E(e^{-zF} \mid \text{busy})$ for the Laplace transform of their flow-time distribution. To derive this expression, a busy period is viewed as a sequence of intervals which are dependent random variables. The Laplace transform of flow-time conditional on a job arriving in one of these intervals is then determined and is combined with the nature of the dependence between the intervals to obtain the desired result.

Let the first interval of the busy period be the processing-time of the initial job and denote its length by T_0. The next interval, of length T_1, is the time required to process all jobs which arrive during interval number zero. When we continue in this manner, T_j is the sum of the processing-times of all jobs that arrive in interval number $j - 1$. The sequence of intervals is depicted in Fig. 8–5, where N_j represents the number of arrivals in the jth interval.

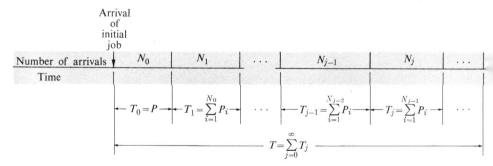

Figure 8–5

Let $\eta_j(z)$ and $H_j(t)$ be the Laplace transform and distribution function of T_j. It is possible to express $\eta_j(z)$ in terms of the Laplace transform of T_{j-1} by a derivation similar to that used for busy-period length in Section 8–3. From the convolution property of the Laplace transform, we have

$$E(e^{-zT_j} \mid T_{j-1} = t, N_j = n) = (\gamma(z))^n,$$

so that

$$\eta_j(z) = E(e^{-zT_j}) = \int_{t=0}^{\infty} e^{-\lambda t} \sum_{n=0}^{\infty} \frac{(\lambda t \gamma(z))^n}{n!} \, dH_{j-1}(t) = \eta_{j-1}(\lambda - \lambda\gamma(z)). \quad (10)$$

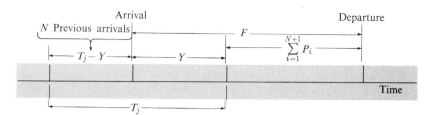

Figure 8–6

The busy period will be represented as an infinite series of intervals, with the ith interval having length T_j. This infinite representation is valid under nonsaturation conditions, since with probability 1 there is a finite j for which T_j is zero. Nonsaturation also implies that $\lim_{j \to \infty} \eta_j(z) = 1$.

Ignoring the idle periods, define the system to be in state j at a particular instant if a jth interval of a busy period is in progress. Due to the assumption of Poisson arrivals, the probability that a job arriving during the busy period arrives when the system is in state j is equal to the steady-state probability that the system is in state j. By Eq. (6), and referring to Fig. 8–5, this probability, denoted by π_j, is given by

$$\pi_j = \text{Prob (state } j \mid \text{busy)} = E(T_j)/E(T). \tag{11}$$

Let $E(e^{-zF} \mid j)$ be the conditional Laplace transform associated with the flow-time of a job which arrives during a jth interval. The flow-time of a job is affected by the interval number through the length of the interval; after it arrives it must first wait until the interval is over and then it must wait until processing is completed on all jobs which arrived before it in the interval. The situation is shown in Fig. 8–6.

By the convolution property of the Laplace transform we have

$$E(e^{-zF} \mid T_j = t, Y = y, N = n) = e^{-zy}\big(\gamma(z)\big)^{n+1}.$$

Eliminating the condition on N yields

$$E(e^{-zF} \mid T_j = t, Y = y) = e^{-zy}\gamma(z)e^{-\lambda(t-y)} \sum_{n=0}^{\infty} \frac{\big(\lambda(t-y)\gamma(z)\big)^n}{n!}$$

$$= e^{-zy}\gamma(z)e^{-\lambda(t-y)}e^{\lambda(t-y)\gamma(z)}.$$

The random variable Y plays the role of a random modification for the jth interval. We may use Eq. (3) of Section 8–1, replacing P by T_j and $G(p)$ by $H_j(t)$, to obtain Prob $(y \leq Y \leq y + dy, t \leq T_j \leq t + dt)$

$$= \frac{dH_j(t)}{E(T_j)} \, dy, \qquad 0 \leq y \leq t, 0 \leqq t \leqq \infty.$$

The conditioning on T_j and Y may be removed to obtain

$$E(e^{-zF} \mid j) = \gamma(z) \int_{t=0}^{\infty} \int_{y=0}^{t} \big(e^{-t(\lambda - \lambda\gamma(z))}e^{-y(\lambda\gamma(z) - \lambda + z)}\big) \frac{dH_j(t)}{E(T_j)} \, dy.$$

$$= \frac{\gamma(z)}{E(T_j)(\lambda\gamma(z) - \lambda + z)} \int_{t=0}^{\infty} \big(e^{-t(\lambda - \lambda\gamma(z))} - e^{-zt}\big) \, dH_j(t).$$

The second half of the integral is $\eta_j(z)$. When we replace j with $j + 1$ in Eq. (10), we recognize the first half to be $\eta_{j+1}(z)$ so that

$$E(e^{-zF} \mid j) = \frac{\gamma(z)\big(\eta_{j+1}(z) - \eta_j(z)\big)}{E(T_j)\big(\lambda\gamma(z) - \lambda + z\big)}. \tag{12}$$

Using Eqs. (11) and (12), we have

$$E(e^{-zF} \mid \text{busy}) = \sum_{j=0}^{\infty} \pi_j E(e^{-zF} \mid j)$$

$$= \frac{\gamma(z)}{E(T)\big(\lambda\gamma(z) - \lambda + z\big)} \sum_{j=0}^{\infty} \big(\eta_{j+1}(z) - \eta_j(z)\big).$$

The summation above telescopes into $\big(1 - \eta_0(z)\big)$, which in this case is $\big(1 - \gamma(z)\big)$, since interval zero consists of a single processing-time. Also, substituting in Eq. (7a) for $E(T)$, we have

$$E(e^{-zF} \mid \text{busy}) = \frac{(1 - \rho)\gamma(z)\big(1 - \gamma(z)\big)}{E(P)\big(\lambda\gamma(z) - \lambda + z\big)}. \tag{13}$$

As the final step, we account for jobs which arrive during an idle period to obtain

$$\beta(z) = (1 - \rho)\gamma(z) + \rho E(e^{-zF} \mid \text{busy}) = \frac{(1 - \rho)z\gamma(z)}{\lambda\gamma(z) - \lambda + z}. \tag{14}$$

The first and second moments of flow-time are obtained from $-\beta'(0)$ and $\beta''(0)$. (The derivative of $\beta(z)$ evaluated at zero is an indeterminate of the form $0/0$ and L'Hôpital's rule must be applied twice.) The results are

$$E(F) = E(P) + \frac{\lambda E(P^2)}{2(1 - \rho)}, \tag{14a}$$

$$E(F^2) = E(P^2) + \frac{\rho E(P^2)}{1 - \rho} + \frac{\lambda E(P^3)}{3(1 - \rho)} + \frac{\big(\lambda E(P^2)\big)^2}{2(1 - \rho)^2}. \tag{14b}$$

Corresponding results for waiting-time follow easily since flow-time is the sum of the two independent random variables: waiting-time and processing-time. From $\beta(z) = \alpha(z)\gamma(z)$ we have

$$\alpha(z) = E(e^{-zW}) = \frac{(1 - \rho)z}{\lambda\gamma(z) - \lambda + z}. \tag{15}$$

The first and second moments of waiting-time are

$$E(W) = \frac{\lambda E(P^2)}{2(1 - \rho)}, \tag{15a}$$

$$E(W^2) = \frac{\lambda E(P^3)}{3(1 - \rho)} + \frac{\big(\lambda E(P^2)\big)^2}{2(1 - \rho)^2}. \tag{15b}$$

These Laplace transforms may be easily inverted when processing-times are exponentially distributed. If

$$G(p) = 1 - e^{-\mu p}, \qquad p \geq 0,$$

then

$$\gamma(z) = \frac{\mu}{\mu + z}, \qquad \beta(z) = \frac{\mu - \lambda}{\mu - \lambda + z}, \qquad \alpha(z) = 1 - \rho + \frac{\rho(\mu - \lambda)}{\mu - \lambda + z}.$$

The form of $\beta(z)$ corresponds to the Laplace transform associated with an exponential distribution, indicating that the flow-time has the distribution function

$$B(f) = 1 - e^{-(\mu - \lambda)f}, \qquad f \geq 0. \tag{16}$$

The distribution of waiting-time is mixed. There is a probability mass of $1 - \rho$ at zero waiting-time and an exponential form over nonzero values of waiting-time:

$$\begin{aligned} A(0) &= (1 - \rho), \\ A(w) &= (1 - \rho) + \rho(1 - e^{-(\mu - \lambda)w}), \qquad w \geq 0. \end{aligned} \tag{17}$$

In Section 8–3, a delay circle was defined as a combination of an initial delay followed by a busy period in which jobs are processed. The waiting-time and flow-time for jobs that are processed during a delay cycle will be needed in the discussion of priority disciplines. Their Laplace transforms and moments are obtained in a manner analogous to the method for FCFS.

All jobs processed in a delay cycle arrive when the machine is busy. Then T is replaced by the length of the delay cycle and T_0 is the delay rather than the processing-time of a job which arrives when the machine is idle. The parallel to $E(e^{-zF} \mid \text{busy})$ is directly the Laplace transform associated with flow-time, since a job cannot arrive during an idle period. Using the subscript c to denote a delay cycle, we have

$$\beta_c(z) = \frac{\gamma(z)\big(1 - \eta_0(z)\big)}{E(T_c)\big(\lambda\gamma(z) - \lambda + z\big)} = \frac{(1 - \rho)\gamma(z)\big(1 - \eta_0(z)\big)}{E(T_0)\big(\lambda\gamma(z) - \lambda + z\big)} \tag{18}$$

and

$$E(F_c) = E(P) + \frac{\lambda E(P^2)}{2(1 - \rho)} + \frac{E(T_0^2)}{2E(T_0)}. \tag{18a}$$

The Laplace transform associated with the distribution of waiting-time of jobs processed during a delay cycle is

$$\alpha_c(z) = \frac{1 - \eta_0(z)}{E(T_c)\big(\lambda\gamma(z) - \lambda + z\big)} = \frac{(1 - \rho)\big(1 - \eta_0(z)\big)}{E(T_0)\big(\lambda\gamma(z) - \lambda + z\big)}. \tag{19}$$

8–5 SELECTION DISCIPLINES THAT
ARE INDEPENDENT OF PROCESSING-TIMES

In this section we consider the class of selection disciplines that do not use information relating to the processing-times of waiting jobs. They are grouped together because they all have the same distribution of number of jobs in the system. By the flow equation (Section 2–4), it then follows that they all have the same mean flow-time, although they do have different flow-time distributions. Three particularly important disciplines in this class are the *first-come, first-served rule*, the *last-come, first-served rule*, and the *random discipline*. The latter two disciplines will be discussed in this section.

8–5.1   Distribution of Number of Jobs in the System

Let N represent the number of jobs in the system under steady-state conditions. Specifically, this means that if $N(t)$ is the number of jobs present at time t, the distribution of N is defined by

$$\text{Prob } (N = n) = \lim_{t \to \infty} \text{Prob } \big(N(t) = n\big).$$

We wish to show that the distribution of N is the same for all rules which do not depend on processing-time.

It is difficult to determine the distribution of N directly. Instead, we consider N_k, the number of jobs remaining in the system immediately after the departure of the kth job processed. By a theorem due to Khintchine (see [178] page 186),

$$\text{Prob } (N = n) = \lim_{k \to \infty} \text{Prob } (N_k = n),$$

so that we may investigate the limiting behavior of N_k rather than $N(t)$. Suppose that M_k jobs arrive while the kth job is being processed. Then N_{k+1} is related to N_k by

$$N_{k+1} = \begin{cases} M_{k+1} & \text{for } N_k = 0, \\ M_{k+1} + N_k - 1 & \text{for } N_k > 0. \end{cases}$$

This relationship shows that the sequences of random variables, $N_1, N_2, \ldots$ is a Markov chain for which the transition probabilities, $\text{Prob } (N_{k+1} = i \mid N_k = j)$, are probability statements about the number of jobs arriving during a processing time. The distribution of M_k will be the same for all selection disciplines that are independent of processing-times; it will be a function only of the processing-time distribution and the arrival rate and not of N_k. We conclude that the limiting distribution of N_k is the same for all disciplines in this class.

To determine the distribution of the number of jobs in the system we may use results already derived from the FCFS rule. Let $\xi(x) = E(x^N)$ be the probability-generating function for the distribution of N. Under the FCFS rule all jobs left behind by a departing job must have arrived during its flow-time. Given that a job's flow-time is f, the conditional distribution of the number of remaining jobs is Poisson with mean λf. Using a subscript to denote the FCFS rule, we have

$$\xi(x) = E(x^N) = \int_{f=0}^{\infty} e^{-\lambda f} \sum_{n=0}^{\infty} x^n \frac{(\lambda f)^n}{n!} \, dB(f) \underset{\text{FCFS}}{=} \beta(\lambda - \lambda_x) \underset{\text{FCFS}}{.} \tag{20}$$

When we substitute $(\lambda - \lambda x)$ for z, Eq. (14) results in

$$\xi(x) = \frac{(1 - \rho)(1 - x)\gamma(\lambda - \lambda x)}{\gamma(\lambda - \lambda x) - x}. \tag{21}$$

By differentiation, the first moment of N is

$$E(N) = \rho + \frac{\lambda^2 E(P^2)}{2(1 - \rho)}. \tag{21a}$$

The two terms in this expression represent the mean number of jobs on the machine and the mean number of jobs waiting, respectively. The appearance of the second moment of the processing-time distribution in the equations for mean number of jobs in the system (21a) and mean flow-time (14a) indicates that the amount of congestion is heavily influenced by the variance of the processing-time distribution. If σ^2 is the variance of P, then

$$E(W) = \frac{\lambda}{2(1 - \rho)} ([E(P)]^2 + \sigma^2).$$

If the processing-times distribution is Erlang with mean $E(P)$ and with k phases, the variance of P is $(E(P))^2/k$, and

$$E(W) = \frac{\lambda(E(P))^2}{2(1 - \rho)} (1 + 1/k), \tag{22}$$

so that the mean waiting-time is linear in the reciprocal of the number of phases. Two special cases of importance are $k = 1$, the exponential distribution, and $k = \infty$, constant processing-time. From Eq. (22) we see that the mean waiting-time with exponentially distributed processing-times is twice that for the case of a constant processing-time equal to the mean of the exponential distribution.

For the FCFS rule, there is a converse to Eq. (21):

$$\underset{\text{FCFS}}{\beta(z)} = \xi\left(1 - \frac{z}{\lambda}\right). \tag{23}$$

The dual relationship expressed by Eqs. (20) and (23) allow, by means of differentiation, the moments of N to be expressed in terms of the moments of F, and vice versa for the FCFS discipline. In particular, they imply the flow equation, $E(N) = \lambda E(F)$ for the FCFS case.

When processing-times are exponentially distributed, Eq. (21) allows the results of Section 8-1 on Poisson-exponential queues to be obtained expeditiously. If $\gamma(z) = \mu/(\mu + z)$, then

$$\gamma(\lambda - \lambda x) = \mu/(\mu + \lambda - \lambda x) = 1/(1 + \rho - \rho x).$$

The function $\xi(x)$ then simplifies to

$$\xi(x) = (1 - \rho)/(1 - \rho x) = \sum_{n=0}^{\infty} (1 - \rho)\rho^n x^n,$$

so that

$$\text{Prob } (N = n) = (1 - \rho)\rho^n.$$

8-5.2 The Last-Come, First-Served Discipline

The last-come, first-served (LCFS) discipline selects the latest arrival from queue for processing. It is perhaps initially surprising, but there are many situations in which this is a natural and common discipline. For example, jobs arriving at a machine center may be stacked in such a way that the latest arrival is the most accessible and thus the one selected. One would suspect that this discipline would lead to some very long flow-times, and that the distribution of flow-times under the LCFS rule would have a higher variance than that for the FCFS rule.

For the LCFS rule, the Laplace transform associated with the distribution of waiting-time is easily obtained from the developments of Section 8–3 on busy periods. A job that arrives during a busy period takes precedence over those already waiting. It must wait until the job in process is completed and then it might also have to stand aside for other jobs which arrived later, during the completion of the job in process, and also all other jobs which arrived during their processing, etc. In other words, it must wait until the queue of all later arriving jobs is dissipated. We conclude that the waiting-time for such a job is a type of delay cycle as described in Section 8–3, where the delay part is the random modification of a processing-time.

The Laplace transform associated with the random modification of a processing-time is given by Eq. (4). The Laplace transform associated with a delay cycle is given by Eq. (9). Combining these, we obtain $E(e^{-zW} \mid \text{busy})$, the Laplace transform associated with the waiting-time of jobs that arrive during a busy period:

$$E(e^{-zW} \mid \text{busy}) = \frac{1 - \gamma(z + \lambda - \lambda\eta(z))}{E(P)(z + \lambda - \lambda\eta(z))} = \frac{1 - \eta(z)}{E(P)(z + \lambda - \lambda\eta(z))}.$$

As in the case of FCFS, a job arrives when the machine is busy with probability ρ, and arrives when the machine is idle, incurring a zero waiting-time, with probability $(1 - \rho)$. Thus the unconditional Laplace transform associated with waiting-time under LCFS is

$$\alpha(z) = E(e^{-zW}) = 1 - \rho + \frac{\lambda(1 - \eta(z))}{(z + \lambda - \lambda\eta(z))}. \tag{24}$$

The second moment of waiting-time is

$$E(W^2) = \frac{\lambda E(P^3)}{3(1 - \rho)^2} + \frac{(\lambda E(P^2))^2}{2(1 - \rho)^3}. \tag{24a}$$

The Laplace transform associated with the distribution of flow-time is obtained from $\beta(z) = \alpha(z)\gamma(z)$:

$$\beta(z) = E(e^{-zF}) = (1 - \rho)\gamma(z) + \frac{\lambda\gamma(z)(1 - \eta(z))}{(z + \lambda - \lambda\eta(z))}. \tag{25}$$

The second moment of flow-time is

$$E(F^2) = E(P^2) + \frac{\rho E(P^2)}{1 - \rho} + \frac{\lambda E(P^3)}{3(1 - \rho)^2} + \frac{(\lambda E(P^2))^2}{2(1 - \rho)^3}. \tag{25a}$$

Comparing the expression for $E(W^2)$ above with that for the FCFS rule, Eq. (15b), we see that

$$E_{\text{LCFS}}(W^2) = \frac{1}{1 - \rho} E_{\text{FCFS}}(W^2).$$

While the mean waiting-times are the same, the second moments of waiting-time differ by a factor that depends on ρ but not upon the form of the processing-time distribution. As expected, this relationship implies that the variance of waiting-time or flow-time is less under FCFS than under the LCFS rule. The relationship heavily favors the FCFS rule for high values of utilization; $1/(1 - \rho)$ becomes explosively large for ρ near 1.

8-5.3 The Random Rule

In some queuing situations it is either impossible or undesirable to impose a particular selection discipline. In order to model these cases, we assume that service is rendered according to the random rule. Under this rule, all jobs in queue at the moment a job is to be selected have equal probability of being chosen.

The analysis of the distribution of flow-time under the random discipline is more difficult than for FCFS or LCFS. The usual approach is to write differential-difference equations involving the conditional distributions of waiting-time given that an arriving job finds itself to be one of n jobs waiting. If the system of differential-difference equations can be solved, the distribution of the number of jobs in the system can be used to find the distribution of waiting-time. Examples of this kind of analysis for the special case of exponentially distributed processing-times may be found in Riordan [170] and Morse [143].

For constant processing-times, Burke [23] has given a numerical method for calculating the waiting-time distribution. Kingman [111] has studied the random discipline applied to queues with Poisson arrivals and general processing-time distribution, and has obtained sets of difference equations from which it might be possible to determine the moments of the waiting-time distribution.

Riordan [170] gives formulas by which the first four moments of the waiting-time may be computed when processing-times are exponentially distributed. For that special situation, Riordan's results indicate that

$$E_{\mathrm{RANDOM}}(W^2) = \frac{1}{(1 - \rho/2)} E_{\mathrm{FCFS}}(W^2).$$

One may conjecture that this relationship is also true for any processing-time distribution. Then of the three processing-time-independent disciplines that have been discussed, the FCFS rule would have the least variance of waiting-time, while the LCFS rule would have the greatest.

8-6 NONPREEMPTIVE PRIORITY AND SHORTEST-PROCESSING-TIME DISCIPLINES

The general assumptions underlying all priority disciplines are: that each arriving job belongs to one out of r classes, that the within-class arrival processes are Poisson, and that each class may have its own distribution of processing-times. The classes are assigned class indexes in inverse order of priority; class 1 is the highest-priority class and its members take precedence over jobs of the other classes. This convention of giving lower indexes to the higher-priority classes is somewhat confusing, but universally accepted in the literature. A nonpreemptive priority discipline will select a job from the queue only after the job in process is completely finished. It selects a job from the class with the lowest index, breaking ties by the use of the FCFS rule. In this section we develop the distributions of waiting-times and flow-times for nonpreemptive priority disciplines and study the shortest-processing-time rule as a limiting case of priority class selection.

8–6.1 Nonpreemptive Priority Disciplines

The r classes are indexed by k, $k = 1, 2, \ldots, r$. For class k we have

λ_k = Poisson arrival rate for jobs of class k,

P_k = random variable for processing-time of a job from class k,

$G_k(p)$ = distribution of processing-times for jobs of class k,

$\gamma_k(z)$ = Laplace transform associated with processing-times of jobs of class k,

$\rho_k = \lambda_k E(P_k)$ = utilization factor for jobs of class k.

To analyze the waiting-time under this type of discipline we focus attention on jobs belonging to a particular class, say class k. If it were not for the existence of the other $r - 1$ classes of jobs, class k jobs would constitute a system operating under the FCFS rule with Poisson arrivals at rate λ_k and processing-times drawn from $G_k(p)$. Jobs from other classes disrupt such a system in two ways. First, while a class k job is being processed, higher-priority jobs may arrive and cause the selection of the next class k job to be delayed. Second, an idle period can terminate with the arrival of a job from some other class, causing the ensuing busy period to be a delay cycle rather than a simple busy period for class k. The first type of disruption can be accounted for by an appropriate redefinition of what is meant by the processing-time of a type k job. The second type can be dealt with by using results on delay cycles.

From the point of view of their effect on the progress of class k jobs, it is clear that jobs from classes 1 through $k - 1$ are equivalent and that jobs from classes $k + 1$ through r are equivalent. These jobs will be grouped into two composite classes, denoted by class a (for priority *above* that of class k) and class b (for priority *below* that of class k). From the discussion of aggregation of Poisson streams in Section 8–1, one can see that the arrival processes for these composite classes are Poisson, with rates

$$\lambda_a = \sum_{i=1}^{k-1} \lambda_i \quad \text{and} \quad \lambda_b = \sum_{i=k+1}^{r} \lambda_i. \tag{26a}$$

The symbols and functions for the composite classes are given by

P_a, P_b \qquad random variables for processing-time,

$G_a(p), G_b(p)$ \quad distributions of processing-time,

$$G_a(p) = (1/\lambda_a) \sum_{i=1}^{k-1} \lambda_i G_i(p), \qquad G_b(p) = (1/\lambda_b) \sum_{i=k+1}^{r} \lambda_i G_i(p), \tag{26b}$$

$\gamma_a(z), \gamma_b(z)$ \quad Laplace transforms of processing-times,

ρ_a, ρ_b \qquad utilization factors, and

$$\rho_a = \lambda_a E(P_a), \qquad \rho_b = \lambda_b E(P_b).$$

When a type k job goes onto the machine there are no class a jobs present. While it is being processed, type a jobs may arrive. Before the next type k job can be processed, or until the system is idle as far as type k jobs are concerned, these arriving class a jobs and their successors must be processed. To a type k job waiting in queue the processing-time of a type k job ahead of it consists not of P_k but of P_k plus the time required to get the system free of class a jobs. This time interval will be called *blocking-time*, and will be denoted by P_{ka}, since it plays the role of a processing-time. It is useful to define a class ka job as a task that occupies the machine in a blocking-time. Once some type k job gets access to the machine, all type k jobs in its busy period effectively encounter processing-times of P_{ka} for the previous type k jobs processed.

From this description we see that blocking-time is equivalent to a delay cycle for which the delay portion is a class k processing-time and the jobs processed in the delay busy period are from class a. Denoting the Laplace transform associated with the distribution of blocking-time by $\gamma_{ka}(z)$, from Eq. (9) we have

$$\gamma_{ka}(z) = \gamma_k\big(z + \lambda_a - \lambda_a \eta_a(z)\big), \tag{27}$$

where $\eta_a(z)$ is the solution of

$$\eta_a(z) = \gamma_a\big(z + \lambda_a - \lambda_a \eta_a(z)\big),$$

and represents the Laplace transform associated with the length of a busy period in which only class a jobs are processed. The first moment of blocking-time is

$$E(P_{ka}) = E(P_k)/(1 - \rho_a). \tag{27a}$$

The notion of blocking-time effectively handles the first type of deviation from a FCFS system with only type k jobs arriving. Dealing with the second deviation involves a consideration of how busy periods get started. Some class k jobs will arrive during periods when the machine is idle, and will not have to wait. All others arrive during delay cycles, of which there are three types, depending on the class of job which starts the delay cycle. Any of these three types of delay cycles ends whenever a processing completion leaves the system empty of both class a and k jobs; there may or may not be class b jobs in the system when any one of these delay cycles ends. A type a cycle commences with the arrival of a class a job at an idle machine. Likewise, the arrival of a class k job at an idle machine causes the initiation of a type k cycle. A type b cycle is started every time a class b job goes onto the machine, whether it arrives at an idle machine or whether it has been waiting for the system to clear of class a and k jobs. The three types of cycles are diagrammed in Fig. 8–7. It is important to note that any instant of time during which the machine is busy falls unambiguously into one and only one of these three cycle types. A true busy period consists of at most one cycle which is either type a or type k and possibly several type b cycles.

Equation (19) can be used to obtain the Laplace transform associated with waiting-time conditioned on the type of cycle. When we let the subscript i, $i = a, k, b$, denote the type of delay cycle, Eq. (19) becomes

$$\alpha_{ci}(z) = \frac{(1 - \rho_{ka})(1 - \eta_{0i}(z))}{E(T_{0i})(\lambda\gamma(z) - \lambda + z)}. \tag{28}$$

Here ρ_{ka} equals the arrival rate of type k jobs times the expected blocking-time, or

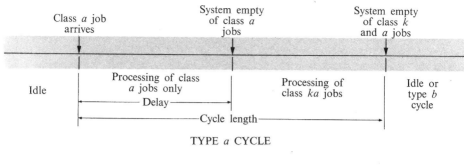

TYPE a CYCLE

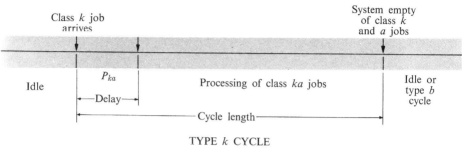

TYPE k CYCLE

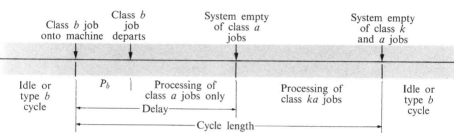

TYPE b CYCLE

Figure 8–7

$\lambda_k E(P_{ka}) = \rho_k/(1 - \rho_a)$. For type a cycles T_{0a} is a busy period involving only class a jobs. For type k cycles T_{0k} is a blocking-time, P_{ka}. For type b cycles T_{0b} is itself a delay cycle in which the delay portion is the processing of a class b job (P_b) and the jobs in the delay busy period are from class a. Equation (9a) is used to obtain $E(T_{0b})$ and Eq. (9) to obtain $\eta_{0b}(z)$. The required substitutions for the use of Eq. (28), and the results, are given in Table 8–1.

These conditional Laplace transforms may be combined by using the probabilities that an arriving class k job finds each type of cycle in progress. Define the system to be in state 0, a, k, or b at an instant in time according as the machine is idle or engaged in a cycle of type a, k, or b. Let π_i be the steady-state probability that the system is in state i, let m_i be the mean length of state i, and let l_i be the mean time between entrances into state i. The required probabilities are equal to the π_i's, which are given by Eq. (6), $\pi_i = m_i/l_i$.

Table 8–1

i, Cycle type	λ	ρ_{ka}	$\gamma(z)$	$\eta_{0i}(z)$
a	λ_k	$\rho_k/(1 - \rho_a)$	$\gamma_{ka}(z)$	$\eta_a(z)$
k	λ_k	$\rho_k/(1 - \rho_a)$	$\gamma_{ka}(z)$	$\gamma_{ka}(z)$
b	λ_k	$\rho_k/(1 - \rho_a)$	$\gamma_{ka}(z)$	$\gamma_b(z + \lambda_a - \lambda_a \eta_a(z))$

i, Cycle type	$E(T_{0i})$	$\alpha_{ci}(z)$
a	$E(P_a)/(1 - \rho_a)$	$\dfrac{1 - \rho_a - \rho_k}{\lambda_k \gamma_{ka}(z) - \lambda_k + z}\ \dfrac{1 - \eta_a(z)}{E(P_a)}$
k	$E(P_{ka})/(1 - \rho_a)$	$\dfrac{1 - \rho_a - \rho_k}{\lambda_k \gamma_{ka}(z) - \lambda_k + z}\ \dfrac{1 - \gamma_{ka}(z)}{E(P_k)}$
b	$E(P_b)/(1 - \rho_a)$	$\dfrac{1 - \rho_a - \rho_k}{\lambda_k \gamma_{ka}(z) - \lambda_k + z}\ \dfrac{1 - \gamma_b(z + \lambda_a - \lambda_a \eta_a(z))}{E(P_b)}$

Let $\rho = \rho_a + \rho_k + \rho_b$. Clearly, ρ is the system utilization and the probability that any arriving jobs have to wait, and $\pi_0 = 1 - \rho$. The mean length of time between successive entrance into the idle state, l_0, is equal to $(1/\lambda)/(1 - \rho)$.

To find l_a and l_k, let $\overline{N}_i(t)$, $i = a, k, b$, be the expected number of busy periods in an interval of length t which are initiated by the arrival of a class i job to the system when it is in the idle state. The rate, r_i, at which such busy periods occur may be taken as

$$r_i = \lim_{t \to \infty} \frac{\overline{N}_i(t)}{t}.$$

These rates are in the same proportion to each other as are the corresponding job-arrival rates because an initiating job will be from class i with probability λ_i/λ. Then

$$r_k = \frac{\lambda_k}{\lambda_a} r_a \quad \text{and} \quad r_b = \frac{\lambda_b}{\lambda_a} r_a.$$

The total rate, r, at which busy periods are initiated is the sum of the individual rates, or

$$r = r_a + r_k + r_b = r_a \left(1 + \frac{\lambda_k}{\lambda_a} + \frac{\lambda_b}{\lambda_a}\right) = r_a \frac{\lambda}{\lambda_a}.$$

This implies that

$$\frac{1}{r_a} = \frac{\lambda}{\lambda_a} \frac{1}{r}.$$

The mean time between the initiation of successive busy periods is l_0, and a limiting result in renewal theory (see [162], page 180) implies that

$$l_0 = \frac{1}{r} \quad \text{and} \quad l_a = \frac{1}{r_a},$$

so that

$$l_a = \frac{\lambda}{\lambda_a} l_0.$$

Since $l_0 = (1/\lambda)/(1 - \rho)$, it follows that $l_a = (1/\lambda_a)/(1 - \rho)$. Similarly, $l_k = (1/\lambda_k)/(1 - \rho)$.

Since every class b job is responsible for initiating a type b cycle, there are on the average λ_b type b cycles per unit of time. From the renewal-theory result just cited, it follows that $l_b = 1/\lambda_b$. The term m_i, for $i = a, k,$ and b, is obtained by the use of Eq. (9a), so that $m_i = E(T_{0i})/(1 - \rho_{ka})$. When we use these values, the state probabilities, π_i, are

$$\pi_0 = 1 - \rho, \qquad \pi_a = \rho_a(1 - \rho)/(1 - \rho_a - \rho_k),$$
$$\pi_k = \rho_k(1 - \rho)/(1 - \rho_a - \rho_k), \qquad \pi_b = \rho_b/(1 - \rho_a - \rho_k).$$

The unconditional Laplace transform, $\alpha_k(z)$, associated with the distribution of waiting-time for a class k job is

$$\alpha_k(z) = E(e^{-zW}k) = \pi_0 + \pi_a \alpha_{ca}(z) + \pi_k \alpha_{ck}(z) + \pi_b \alpha_{cb}(z)$$
$$= \frac{(1 - \rho)(z + \lambda_a - \lambda_a \eta_a(z)) + \lambda_b(1 - \gamma_b(z + \lambda_a - \lambda_a \eta_a(z)))}{\lambda_k \gamma_k(z + \lambda_a - \lambda_a \eta_a(z)) - \lambda_k + z}. \quad (29)$$

The first moment of waiting-time is

$$E(W_k) = \frac{\lambda_a E(P_a^2) + \lambda_k E(P_k^2) + \lambda_b E(P_b^2)}{2(1 - \rho_a - \rho_k)(1 - \rho_a)}. \quad (29a)$$

This result may be expressed in terms of the original r classes by making the following substitutions:

$$\lambda_a E(P_a^2) = \sum_{i=1}^{k-1} \lambda_i E(P_i^2), \qquad \lambda_b E(P_b^2) = \sum_{i=k+1}^{r} \lambda_i E(P_i^2), \qquad \rho_a = \sum_{i=1}^{k-1} \rho_i.$$

With these substitutions, Eq. (29a) becomes

$$E(W_k) = \frac{\sum_{i=1}^{r} \lambda_i E(P_i^2)}{2(1 - \sum_{i=1}^{k-1} \rho_i)(1 - \sum_{i=1}^{k} \rho_i)}. \quad (29a')$$

Suppose that there are r classes of jobs. The overall expected waiting-time or flow-time will be minimized by assigning priorities in inverse order of the class processing-time means. That is, the class indexes should be set so that for $i < j$, $E(P_i) \leq E(P_j)$. More generally, assume that there is a linear cost or other weighting factor, u_j, associated with the waiting-time for members of class j, and that the measure of performance to be minimized is the average cost per job, which is

$$E(W_u) = \sum_{i=1}^{r} \frac{\lambda_j}{\lambda} u_j E(W_j) = \sum_{j=1}^{r} \frac{\lambda_j u_j \sum_{i=1}^{r} \lambda_i E(P_i^2)}{2\lambda(1 - \sum_{j=1}^{j-1} \rho_i)(1 - \sum_{i=1}^{j} \rho_i)}.$$

The optimal priority assignment in this case would be such that

$$\frac{E(P_1)}{u_1} \leq \frac{E(P_2)}{u_2} \leq \cdots \leq \frac{E(P_r)}{u_r}.$$

This result is analogous to Theorem 3–10 (Section 3–7) and is similarly proved by an interchange argument.

If the jobs' processing-times are known when they arrive, the nonpreemptive priority discipline can be used to reduce mean flow-time below that obtainable through the use of a processing-time-independent discipline. In the simplest case two classes would be defined by choosing a number, d, and giving priority to jobs having processing-times less than d. The optimal value of d depends on the processing-time distribution and the arrival rate. The expected waiting-time taken over all jobs is

$$E(W) = \frac{\lambda E(P^2)}{2(1 - \rho)} \left(\frac{1 - \rho G(d)}{1 - \rho_1} \right).$$

The factor in large parentheses represents the ratio of mean waiting-time for the nonpreemptive rule to the mean waiting-time for the FCFS rule. Denoting this ratio by $R(d)$, it is easy to see that $R(d)$ is less than one because $\rho_1 = \lambda G(d)E(P \mid P < d)$, which is less than $\lambda G(d)E(P) = \rho G(d)$.

When we assume that $G(p)$ has a density function, $g(p)$, we may obtain an expression relating the optimal dividing point to the arrival rate for a given processing-time distribution by differentiating $R(d)$ and finding the value of d that makes $R'(d)$ equal to zero. The derivative is

$$\frac{dR(d)}{dd} = \frac{-\rho g(d)\left(1 - \lambda \int_0^d t\, dG(t)\right) + \lambda\, dg(d)\left(1 - \rho G(d)\right)}{\left(1 - \lambda \int_0^d t\, dG(t)\right)^2}.$$

Setting this expression equal to zero implies that d should be chosen so that

$$\frac{d}{E(P)} - 1 + \lambda \left(\int_0^d t\, dG(t) - dG(d) \right) = 0. \tag{30}$$

It can be shown that for a given $G(p)$, d is an increasing function of λ and varies between $E(P)$ and a value so large that class 2 does not exist. To show that d is an increasing function of λ, we differentiate Eq. (30) with respect to λ to obtain

$$\frac{1}{E(P)} \frac{dd}{d\lambda} + \int_0^d t\, dG(t) - dG(d) - \lambda G(d) \frac{dd}{d\lambda} = 0.$$

After integrating by parts and solving for $dd/d\lambda$, we have

$$\frac{dd}{d\lambda} = \frac{\int_0^d G(t)\, dt}{(1/E(P)) - \lambda G(d)},$$

which is greater than zero for $\lambda E(P) = \rho < 1$.

The lower limit of d is obtained by setting λ equal to zero in Eq. (30). The upper limit of d comes from setting λ equal to $1/E(P)$, corresponding to $\rho = 1$ in Eq. (30).

Then Eq. (30) becomes

$$d\big(1 - G(d)\big) + \int_0^d t\, dG(t) - E(P) = d\big(1 - G(d)\big) - \int_d^\infty t\, dG(t) = 0,$$

which is satisfied for d such that $G(d) = 1$.

When processing-times are uniformly distributed on $(0, \theta)$, Eq. (30) becomes

$$\frac{2d}{\theta} - 1 - \frac{\lambda d^2}{2\theta} = 0 \qquad \text{or} \qquad d = \frac{2}{\lambda} - \frac{\sqrt{4 - 2\lambda\theta}}{\lambda}.$$

In terms of $\rho = \lambda\theta/2$, this is

$$d = \frac{2}{\lambda}(1 - \sqrt{1 - \rho}).$$

For exponentially distributed processing-times having mean $1/\mu$, relationship (30) is

$$\mu d - 1 + \frac{\lambda}{\mu}(1 - e^{-\mu d}) - \lambda d e^{-\mu d} - \lambda d(1 - e^{-\mu d}) = 0 \qquad \text{or} \qquad \frac{\mu}{\lambda} - \frac{e^{-\mu d}}{\mu d - 1} = 1.$$

8-6.2 The Shortest-Processing-Time Discipline

The idea of using a nonpreemptive priority discipline to reduce mean waiting-time can be extended by defining $r + 1$ boundary points $(0 = d_0 < d_1 < \cdots < d_r)$, with d_r so large that $G(d_r) = 1$. A job having processing-time P with $d_{j-1} < P \le d_j$ will be a member of class j. The overall mean waiting-time resulting from this way of setting up the classes is

$$E(W) = \frac{\lambda E(P^2)}{2} \sum_{j=1}^r \frac{G(d_j) - G(d_{j-1})}{\big(1 - \lambda\int_0^{d_{j-1}} t\, dG(t)\big)\big(1 - \lambda\int_0^{d_j} t\, dG(t)\big)}. \tag{31}$$

If the number of classes is increased in such a way that $(d_j - d_{j-1})$ becomes arbitrarily small for all j, the selection discipline would always choose the job with the lowest processing-time. This is the queuing version of the shortest-processing-time discipline. The expression for the mean waiting-time under the shortest-processing-time discipline as found by taking the appropriate limit of Eq. (31) is

$$E(W) = \frac{\lambda E(P^2)}{2} \int_{p=0}^\infty \frac{dG(p)}{\big(1 - \lambda\int_{t=0}^{p^-} t\, dG(t)\big)\big(1 - \lambda\int_{t=0}^{p} t\, dG(t)\big)}. \tag{32}$$

If $G(p)$ is continuous, the denominator may be replaced by

$$\left(1 - \lambda\int_0^p t\, dG(t)\right)^2.$$

When preemption is not allowed, the shortest-processing-time discipline is optimal with respect to minimizing mean flow-time. This statement of optimality requires qualification, since it is possible to do better in situations in which the arrival and processing-times of jobs are known in advance of their arrivals. However, the problem of how to make the best utilization of such advance information (presumably by incorporating inserted-idleness) has not been satisfactorily solved (see Section 4-2).

8–6.3 Multiple-Level Nonpreemptive Priority Disciplines

A multiple-level priority system would be the dynamic analog of the multiple-class sequencing model of Section 3–8. The partitioning of jobs into classes can be considered as taking place in two stages, perhaps using two attributes of the jobs. A primary attribute would partition the jobs into classes, and a secondary attribute would partition them into subclasses within each primary class. For example, the primary partition might be into classes of urgent and nonurgent jobs, and the partition within these classes might be according to processing-time. One would then select any urgent job before any nonurgent job, and among urgent jobs, select a short job before a long one.

Of interest is the effect that this superposition of the primary classes has on the basic operation of the secondary partitioning system. For the example above, it is intuitively clear that this multiple-level partitioning will not be as effective as simple partitioning by processing-time in the reduction of mean flow-time or mean system state. One way to examine this question is to compare the performance of a system with an r-class nonpreemptive discipline to a similar system in which each of the r classes is divided into two subclasses in such a way that each subclass has a Poisson arrival process. Let f represent the fraction of the jobs of original class k to be assigned to the preferred subclass of k. The original class k is then replaced by two classes:

$$\text{class } k, \qquad \text{with Poisson arrival rate } f\lambda_k,$$
$$\text{class } k + r, \qquad \text{with Poisson arrival rate } (1 - f)\lambda_k.$$

Let $E(W_k)$ be the expected waiting-time for a class k job under the original rule, and $E(W'_k)$ be the expected waiting-time for a job originally of class k under the two-level rule. From Eq. (29a'), we have

$$E(W_k) = \frac{\sum_{i=1}^{r} \lambda_i E(P_i^2)}{2(1 - \sum_{i=1}^{k-1} \rho_i)(1 - \sum_{i=1}^{k} \rho_i)}$$

and

$$E(W'_k) = \sum_{i=1}^{r} \lambda_i(P_i^2) \left(\frac{f}{2(1 - f\sum_{i=1}^{k-1} \rho_i)(1 - f\sum_{i=1}^{k} \rho_i)} \right.$$
$$+ \frac{1 - f}{2(1 - f\rho - (1 - f)\sum_{i=1}^{k-1} \rho_i)(1 - f\rho - (1 - f)\sum_{i=1}^{k} \rho_i)} \right)$$
$$= \sum_{i=1}^{r} \lambda_i E(P_i^2) \left(\frac{f}{2(1 - f\sum_{i=1}^{k-1} \rho_i)(1 - f\sum_{i=1}^{k} \rho_i)} \right.$$
$$+ \frac{(1 - f)/(1 - f\rho)^2}{2(1 - ((1 - f)/(1 - f\rho))\sum_{i=1}^{k-1} \rho_i)(1 - ((1 - f)/(1 - f\rho))\sum_{i=1}^{k} \rho_i)} \right).$$

This expression of $E(W'_k)$ says that the expected waiting-time of a class k job in a two-level rule is a linear combination of its waiting-time performance in the two systems:

(1) each utilization is $f\rho_i$,   (2) each utilization is $\dfrac{1 - f}{1 - f\rho} \rho_i$.

Table 8-2

Expected Flow-Time for Two-Level SPT Rule*

Values of ρ \ Values of f	0.0	0.2	0.4	0.6	0.8	1.0	FCFS
0.60	1.962	2.054	2.138	2.192	2.174	1.962	2.500
0.70	2.312	2.455	2.598	2.715	2.727	2.312	3.333
0.80	2.882	3.104	3.352	3.603	3.746	2.882	5.000
0.90	4.197	4.562	5.025	5.614	6.306	4.197	10.000
0.95	6.263	6.794	7.503	8.527	10.181	6.263	20.000
0.98	11.283	12.092	13.216	14.969	18.400	11.283	50.000

* Exponentially distributed processing-times, $E(P) = 1$ and f equals fraction of jobs in high priority level.

This relationship is, at the least, intriguing. It can be used to show that performance cannot be improved by a two-level procedure. The $E(W_k)$ is convex in $\sum_{i=1}^{k-1} \rho_i$ and ρ_k, so that for any $0 \le f \le 1$,

$$\frac{f}{2(1 - f\sum_{i=1}^{k-1}\rho_i)(1 - f\sum_{i=1}^{k}\rho_i)} + \frac{1 - f}{2(1 - (1-f)\sum_{i=1}^{k-1}\rho_i)(1 - (1-f)\sum_{i=1}^{k}\rho_i)}$$
$$\ge \frac{1}{2(1 - \sum_{i=1}^{k-1}\rho_i)(1 - \sum_{i=1}^{k}\rho_i)}.$$

Since $(1 - f\rho) \le 1$,

$$\frac{f}{2(1 - f\sum_{i=1}^{k-1}\rho_i)(1 - f\sum_{i=1}^{k}\rho_i)}$$
$$+ \frac{(1 - f)/(1 - f\rho)^2}{2(1 - ((1-f)/(1-f\rho))\sum_{i=1}^{k-1}\rho_i)(1 - ((1-f)/(1-f\rho))\sum_{i=1}^{k}\rho_i)}$$
$$\ge \frac{1}{2(1 - \sum_{i=1}^{k-1}\rho_i)(1 - \sum_{i=1}^{k}\rho_i)}.$$

At the endpoints ($f = 0$ and $f = 1$), $E(W'_k) = E(W_k)$. The performance is, however, usually not symmetric in f about $f = 0.5$, in contrast to the result of Section 3-8. Table 8-2 gives results for a two-level SPT rule.

8-7 PREEMPTIVE PRIORITY DISCIPLINES

Preemptive priority rules operate in a manner similar to that of nonpreemptive priority disciplines, except that a job which arrives while a lower-priority job is on the machine will go onto the machine immediately and the lower-priority job will go back to the head of the queue of its class. Preemption may be necessary when there are emergency classes of jobs, or when the machine is subject to breakdowns.

8–7.1   Preemptive Resume and Preemptive Repeat Disciplines

There are three types of preemptive priority disciplines which will be analyzed. They differ as to the nature of the processing-time requirements of preempted jobs as they return to the machine. If preemption causes no loss so that the processing of a preempted job can be taken up where it left off, the rule is called *preemptive resume*; otherwise it is called *preemptive repeat*, for which we distinguish two varieties. When the variability in processing-times is caused by erratic behavior of the machine, it is reasonable to assume that each time a single job returns to the machine a new processing-time is sampled. But if the within-class processing variability arises from differing requirements of the jobs, then the processing-time for a job is sampled only once, regardless of the number of preemptions experienced. These two cases are called *preemptive repeat with resampling* and *preemptive repeat without resampling*. These three preemptive rules will be analyzed in parallel. The same equation number will be used for parallel expressions, appended by pr, rs, or rw* to differentiate among the disciplines.

The type of analysis is similar to that used for nonpreemptive priorities. Attention is focused on a given class of jobs, say class k, and on how a preemptive system with additional classes differs in operation from a FCFS system with only class k jobs. The first departure is that a job waiting to get on the machine for the first time observes not P_k but some other time interval for a previously selected job from class k to clear the machine. The second departure is that busy periods can start in ways other than by the processing of a class k job.

For a class k job the *waiting-time*, W_k, is the time which elapses between the arrival of the job and the time it first goes onto the machine, and it does not include the time a job must stand aside after it has been preempted. *Residence-time*, P_{rk}, is the time which elapses between the time the job first goes onto the machine and its completion-time. Flow-time is the sum of waiting-time and residence-time. A single interruption in the processing-time of a preempted job will be called a *breakdown* and the length of such an interruption a *breakdown-time*, T_{bk}; a job may encounter several breakdowns before it is completed. *Gross processing-time*, P_{gk}, is the total amount of time that a job actually spends on the machine. For a preemptive resume rule, gross processing-time and processing-time are the same, although they are, in general, different for preemptive repeat rules.

Figure 8–8 illustrates the relationships among these random variables, and shows a job which suffers two preemptions. In this case the gross processing-time is the sum of three separate intervals. If the discipline is preemptive repeat, the first two of these are *wasted processing-times*, P_{wk}, and the final interval during which processing is completed is a *successful processing-time*, P_{sk}.

Residence-time in preemptive rule analysis plays the same role as blocking-time in the analysis of nonpreemptive disciplines. Residence-time will be expressed in terms of its components, breakdown-time and gross processing-time, and then results

---

* The letters pr refer to "preemptive resume," rs refers to "repeat with resampling," and rw refers to "preemptive repeat without resampling."

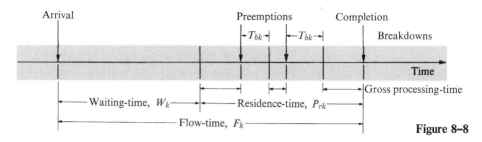

Figure 8–8

on the FCFS rule operating in a delay cycle will be used to determine the flow-time of class k jobs. Since it is convenient to express the final results in terms of gross processing-time, its moments and transform will be developed along with those for residence-time.

When a breakdown starts on a class k job there can be only one higher-priority job in the system—the one which has gained the services of the machine. This interrupting job may come from any class with an index less than k. The breakdown-time consists of the residence-time of this job plus the residence-times of class $k - 1$ jobs until the system is entirely clear of jobs from classes 1 through $k - 1$. Thus breakdown-time and residence-time depend on each other recursively. The residence-time for class 1 jobs determines the breakdown-time for class 2 jobs which determines the residence-time for class 2 jobs, etc. At present we shall avoid this recursive problem and use the symbol T_{bk} for a breakdown-time suffered by class k jobs. The Laplace transform associated with the distribution of T_{bk} will be denoted by $\eta_{bk}(z)$. A breakdown occurs with Poisson intensity $\lambda_1 + \lambda_2 + \cdots + \lambda_{k-1}$. The symbol λ_a will be used for this sum; N will represent the number of preemptions suffered by a job before it is completed.

The residence-time under preemptive resume is the processing-time, P_k, plus the sum of the breakdown-times which interrupt processing, so that we have

$$E(e^{-zP_{rk}} \mid P_k = p, N = n) = e^{-zp}\big(\eta_{bk}(z)\big)^n.$$

The distribution of N, given $P_k = p$, is Poisson with parameter $\lambda_a p$, so that

$$\gamma_{rk}(z) = E(e^{-zP_{rk}}) = \int_{p=0}^{\infty} e^{-zp} \sum_{n=0}^{\infty} \frac{(\lambda_a p \eta_{bk}(z))^n}{n!} e^{-\lambda_a p} \, dG_k(p)$$
$$= \gamma_k(z + \lambda_a - \lambda_a \eta_{bk}(z)). \tag{33, pr}$$

The first two moments of residence-time are

$$E(P_{rk}) = \big(1 + \lambda_a E(T_{bk})\big)E(P_k), \tag{33a, pr}$$

$$E(P_{rk}^2) = \big(1 + \lambda_a E(T_{bk})\big)^2 E(P_k^2) + \lambda_a E(T_{bk}^2)E(P_k). \tag{33b, pr}$$

Under preempt resume, $P_{gk} = P_k$, so that

$$\gamma_{gk}(z) = \gamma_k(z), \tag{34, pr}$$
$$E(P_{gk}) = E(P_k), \tag{34a, pr}$$
$$E(P_{gk}^2) = E(P_k^2). \tag{34b, pr}$$

The residence-time under a preemptive repeat rule with resampling consists of N independent pairs, $P_{wk} + T_{bk}$, followed by an independent P_{sk}. If P_k is used to represent the processing-time of a job drawn on a particular attempt to be processed completely and Y is the interval from the start of processing to the next arrival of a class a job, the joint density of Y and P_k is

$$\text{Prob}\,(y \le Y \le y + \mathrm{d}y,\, p \le P_k \le p + \mathrm{d}p) = \lambda_a e^{-\lambda_a y}\, \mathrm{d}y\, \mathrm{d}G_k(p).$$

The probability that an attempt to complete is successful is the probability that Y is greater than P_k, or

$$\text{Prob}\,(\text{success}) = \int_{p=0}^{\infty} \int_{y=p}^{\infty} \lambda_a e^{-\lambda_a y}\, \mathrm{d}y\, \mathrm{d}G_k(p) = \int_{p=0}^{\infty} e^{-\lambda_a p}\, \mathrm{d}G_k(p) = \gamma_k(\lambda_a).$$

Thus the distribution of N is geometric with parameter $\gamma_k(\lambda_a)$.

The length of a wasted processing-time is Y, under the condition that Y is less than P_k, so that

$$\text{Prob}\,(y \le P_{wk} \le y + \mathrm{d}y) = \frac{\text{Prob}\,(y \le Y \le y + \mathrm{d}y,\, Y < P_k)}{\text{Prob}\,(Y < P_k)}$$

$$= \frac{1}{1 - \gamma_k(\lambda_a)} \int_{p=y}^{\infty} \lambda_a e^{-\lambda_a y}\, \mathrm{d}G_k(p)\, \mathrm{d}y.$$

From this, $\gamma_{wk}(z)$—the Laplace transform associated with the distribution of P_{wk}—is

$$\gamma_{wk}(z) = E(e^{-zP_{wk}}) = \frac{\lambda_a}{1 - \gamma_k(\lambda_a)} \int_{y=0}^{\infty} \int_{p=y}^{\infty} e^{-(z+\lambda_a)y}\, \mathrm{d}G_k(p)\, \mathrm{d}y$$

$$= \frac{\lambda_a}{z + \lambda_a} \frac{1 - \gamma_k(z + \lambda_a)}{1 - \gamma_k(\lambda_a)}.$$

In a similar manner we obtain the Laplace transform associated with the distribution of P_{sk} as

$$\gamma_{sk}(z) = \frac{\gamma_k(z + \lambda_a)}{\gamma_k(\lambda_a)}.$$

Given that a job is preempted n times, its residence-time is the sum of n wasted processing intervals, n breakdown intervals, and one successful processing interval. These $2n + 1$ random variables are all independent, and we may write

$$\gamma_{rk}(z) = \gamma(\lambda_a)\gamma_{sk}(z) \sum_{n=0}^{\infty} \left(\big(1 - \gamma_k(\lambda_a)\big)\gamma_{wk}(z)\eta_{bk}(z) \right)^n$$

or

$$\gamma_{rk}(z) = \frac{(z + \lambda_a)\gamma_k(z + \lambda_a)}{z + \lambda_a - \lambda_a \eta_{bk}(z)\big(1 - \gamma_k(z + \lambda_a)\big)}. \tag{33, rs}$$

The Laplace transform associated with the distribution of gross processing-time, P_{gk}, is obtained by replacing $\eta_{bk}(z)$ by 1 in the above expression. The result is that

$$\gamma_{gk}(z) = \frac{(z + \lambda_a)\gamma_k(z + \lambda_a)}{z + \lambda_a - \lambda_a\big(1 - \gamma_k(z + \lambda_a)\big)}. \tag{34, rs}$$

The first two moments of P_{rk} and P_{gk} are

$$E(P_{rk}) = \big(1 + \lambda_a E(T_{bk})\big)E(P_{gk}), \tag{33a, rs}$$

$$E(P_{rk}^2) = \big(1 + \lambda_a E(T_{bk})\big)E(P_{gk}^2) + 2\lambda_a E(T_{bk})\big(1 + \lambda_a E(T_{bk})\big)\big(E(P_{gk})\big)^2$$
$$+ \lambda_a E(T_{bk}^2)E(P_{gk}), \tag{33b, rs}$$

$$E(P_{gk}) = \frac{1 - \gamma_k(\lambda_a)}{\lambda_a \gamma_k(\lambda_a)}, \tag{34a, rs}$$

$$E(P_{gk}^2) = \frac{2}{\lambda_a^2 \big(\gamma_k(\lambda_a)\big)^2}\big(1 - \gamma_k(\lambda_a) - \lambda_a E(P_k e^{-\lambda_a P_k})\big). \tag{34b, rs}$$

The residence-time under a preemptive repeat rule without resampling consists of N dependent pairs, $P_{wk} + T_{bk}$, followed by P_k. The dependence can be avoided by first working with random variables conditioned on a particular value of P_k and then integrating over the distribution of P_k.

The conditional distribution of the number of preemptions is also geometric:

$$\text{Prob}\,(N = n \mid P_k = p) = (1 - e^{-\lambda_a p})^n e^{-\lambda_a p}.$$

The conditional density of a wasted processing-time is given by

$$\text{Prob}\,(y \leq P_{wk} \leq y + \mathrm{d}y \mid P_k = p) = \frac{\lambda_a e^{-\lambda_a y}}{1 - e^{-\lambda_a p}}\,\mathrm{d}y \qquad \text{for } y < p,$$

from which

$$E(e^{-zP_{wk}} \mid P_k = p) = \frac{\lambda_a}{1 - e^{-\lambda_a p}} \int_{y=0}^{p} e^{-(z+\lambda_a)y}\,\mathrm{d}y$$

$$= \frac{\lambda_a}{z + \lambda_a}\frac{1 - e^{-(z+\lambda_a)p}}{1 - e^{-\lambda_a p}}.$$

Since $P_{sk} = P_k$, then

$$E(e^{-zP_{sk}} \mid P_k = p) = e^{-zp}.$$

When we combine these results, the conditional Laplace transform associated with residence-time is

$$E(e^{-zP_{rk}} \mid P_k = p) = e^{-zp}e^{-\lambda_a p}\sum_{n=0}^{\infty}\left((1 - e^{-\lambda_a p})\eta_{bk}(z)\left(\frac{\lambda_a}{z + \lambda_a}\right)\left(\frac{1 - e^{-(z+\lambda_a)p}}{1 - e^{-\lambda_a p}}\right)\right)^n$$

$$= \frac{(z + \lambda_a)e^{-(z+\lambda_a)p}}{z + \lambda_a - \lambda_a \eta_{bk}(z)(1 - e^{-(z+\lambda_a)p})}.$$

The final step is to integrate the above expression with respect to the distribution of P_k:

$$\gamma_{rk}(z) = \int_{p=0}^{\infty} \frac{(z + \lambda_a)e^{-(z+\lambda_a)p}}{z + \lambda_a - \lambda_a \eta_{bk}(z)\big(1 - e^{-(z+\lambda_a p)}\big)}\,\mathrm{d}G_k(p). \tag{33, rw}$$

Again, the Laplace transform associated with the distribution of P_{gk} is obtained by setting $\eta_{bk}(z)$ to 1 in the above expression. Equation (34, rw), so obtained, is omitted.

The first two moments of P_{rk} and P_{gk} are

$$E(P_{rk}) = \left(1 + \lambda_a E(T_{bk})\right)E(P_{gk}), \tag{33a, rw}$$

$$E(P_{rk}^2) = \left(1 + \lambda_a E(T_{bk})\right)E(P_{gk}^2) + \frac{2E(T_{bk})\left(1 + \lambda_a E(T_{bk})\right)}{\lambda_a} E\left((e^{\lambda_a P_k} - 1)^2\right)$$
$$+ \lambda_a E(T_{bk}^2)E(P_{gk}), \tag{33b, rw}$$

$$E(P_{gk}) = \frac{\gamma_k(-\lambda_a) - 1}{\lambda_a}, \tag{34a, rw}$$

$$E(P_{gk}^2) = (2/\lambda_a)\left(\gamma_k(-2\lambda_a) - \gamma_k(-\lambda_a) - \lambda_a E(P_k e^{\lambda_a P_k})\right). \tag{34b, rw}$$

Once we have determined the Laplace transform for residence-time for each type of preemptive discipline, we can use these to determine the Laplace transform of flow-time by the use of our analysis of FCFS operating in delay cycles. Some class k jobs may arrive when the machine is either idle or is processing jobs of lower priority and these class k jobs will not have to wait. For such jobs, $F_k = P_{rk}$. All others arrive during delay cycles, of which there only two types. A type a cycle is initiated when an arriving class a job finds no class k or a jobs in the system, and ends when there are no class k or a jobs in the system. A type k cycle starts when an arriving class k job finds no class k or a jobs in the system and ends in the same manner as a type a cycle. So far as class k jobs are concerned, the machine is busy only when one of these cycle types is in progress and idle otherwise, even though it may be processing lower-priority jobs. Compared with nonpreemptive priority rules, only the top two cycles of Fig. 8–7 exist under preemptive disciplines, since jobs with priorities lower than class k have no effect on class k jobs.

To obtain the Laplace transform associated with flow-time, conditional on the type of cycle, we use Eq. (18). When we let the subscript i, $i = a, k$, denote the type of delay cycle, Eq. (18) becomes

$$\beta_{ci}(z) = \frac{\left(1 - \lambda_k E(P_{rk})\right)\gamma_{rk}(z)\left(1 - \eta_{0i}(z)\right)}{E(T_{0i})\left(\lambda_k \gamma_{rk}(z) - \lambda_k + z\right)}.$$

For a type a cycle $T_{0a} = T_{bk}$ and for a type k cycle $T_{0k} = P_{rk}$.

The unconditional Laplace transform associated with the distribution of flow-time for class k jobs can be found, as for nonpreemptive priorities, once the π_i's are determined. Define the system to be in state i, $i = 0, a$, or k, at an instant in time according to whether the machine is idle or is engaged in a type a or a type k cycle. As before, these are determined by the use of Eq. (6).

The mean length of the idle state, m_0, is $1/(\lambda_a + \lambda_k)$. The term π_0 is not $1 - \rho$ but $1 - \bar{U}$, since generally with preemption the expected gross processing-time required by a job is higher than $E(P)$. To determine π_0, note that the expected time between idle periods, l_0, is equal to the expected time of the idle period plus the expected time of the cycle which follows, or

$$l_0 = m_0 + \frac{\lambda_a m_a + \lambda_k m_k}{\lambda_a + \lambda_k}.$$

From Eq. (9a) we obtain

$$m_a = E(T_{bk})/\left(1 - \lambda_k E(P_{rk})\right), \qquad m_k = E(P_{rk})/\left(1 - \lambda_k E(P_{rk})\right).$$

We also note from our discussion of nonpreemptive priorities that

$$l_a = \frac{\lambda_a + \lambda_k}{\lambda_a} l_0 \quad \text{and} \quad l_k = \frac{\lambda_a + \lambda_k}{\lambda_k} l_0$$

and the π_i's are

$$\pi_0 = 1 - \bar{U} = \big(1 - \lambda_k E(P_{rk})\big)/\big(1 + \lambda_a E(T_{bk})\big),$$
$$\pi_a = \lambda_a E(T_{bk})/\big(1 + \lambda_a E(T_{bk})\big),$$
$$\pi_k = \lambda_k E(P_{rk})/\big(1 + \lambda_a E(T_{bk})\big).$$

The unconditional Laplace transform associated with the distribution of flow-times for class k jobs is

$$\begin{aligned}
\beta_k(z) &= \pi_0 \gamma_{rk}(z) + \pi_a \beta_{ca}(z) + \pi_k \beta_{ck}(z) \\
&= \frac{\gamma_{rk}(z)\big(1 - \lambda_k E(P_{rk})\big)\big(z + \lambda_a - \lambda_a \eta_{bk}(z)\big)}{\big(1 + \lambda_a E(T_{bk})\big)\big(z + \lambda_k - \lambda_k \gamma_{rk}(z)\big)}.
\end{aligned} \tag{35}$$

The first moments of flow-time and waiting-time are

$$E(F_k) = E(P_{rk}) + \frac{\lambda_k E(P_{rk}^2)}{2\big(1 - \lambda_k E(P_{rk})\big)} + \frac{\lambda_a E(T_{bk}^2)}{2\big(1 + \lambda_a E(T_{bk})\big)}, \tag{35a}$$

$$E(W_k) = E(F_k) - E(P_{rk}) = \frac{\lambda_k E(P_{rk}^2)}{2\big(1 - \lambda_k E(P_{rk})\big)} + \frac{\lambda_a E(T_{bk}^2)}{2\big(1 + \lambda_a E(T_{bk})\big)}. \tag{36a}$$

Equations (33a) and (33b) give $E(P_{rk})$ and $E(P_{rk}^2)$ in terms of the first two moments of P_{gk} and T_{bk}. To express the first moment of flow-time in terms of the moments of gross processing-time, a relation between breakdown-time and gross processing-time is required. This relationship is recursive, as discussed. The breakdown-time for a class $k + 1$ job is either a type a cycle or a type k cycle for a class k job. Letting T_{ca} and T_{ck} denote the lengths of these cycles, we have

$$T_{b,k+1} = \begin{cases} T_{ca} & \text{with probability } \lambda_a/(\lambda_a + \lambda_k), \\ T_{ck} & \text{with probability } \lambda_k/(\lambda_a + \lambda_k). \end{cases}$$

The moments of T_{ca} and T_{ck} can be determined by the use of Eqs. (9a) and (9b). From these, the moments of $T_{b,k+1}$ are

$$E(T_{b,k+1}) = \frac{\lambda_a E(T_{bk}) + \lambda_k E(P_{rk})}{(\lambda_a + \lambda_k)\big(1 - \lambda_k E(P_{rk})\big)}, \tag{37a}$$

$$E(T_{b,k+1}^2) = \frac{\lambda_a E(T_{bk}^2)\big(1 - \lambda_k E(P_{rk})\big) + \lambda_k E(P_{rk}^2)\big(1 + \lambda_a E(T_{bk})\big)}{(\lambda_a + \lambda_k)\big(1 - \lambda_k E(P_{rk})\big)^3}. \tag{37b}$$

Expected waiting-time can be expressed recursively; from Eq. (36a) we have, for class $k + 1$,

$$E(W_{k+1}) = \frac{\lambda_{k+1} E(P_{r,k+1}^2)}{2\big(1 - \lambda_{k+1} E(P_{r,k+1})\big)} + \frac{(\lambda_a + \lambda_k) E(T_{b,k+1}^2)}{2\big(1 + (\lambda_a + \lambda_k) E(T_{b,k+1})\big)}.$$

When we use Eqs. (37a), (37b), and (36a), this reduces to

$$E(W_{k+1}) = \frac{\lambda_{k+1} E(P_{r,k+1}^2)}{2\big(1 - \lambda_{k+1} E(P_{r,k+1})\big)} + \frac{1}{1 - \lambda_k E(P_{rk})} E(W_k). \tag{36a'}$$

When we use Eqs. (33a) and (33b), which give the moments of residence-time in terms of the moments of gross processing-time, and apply Eq. (36a') recursively, we can obtain, after considerable algebraic manipulation, the first moment of flow-time in terms of the moments of gross processing-time. To simplify the expression, let

$$\Lambda_i = \sum_{j=1}^{i} \lambda_j, \quad \omega_i = \lambda_i E(P_{gi}), \quad \kappa_i = \lambda_i E(P_{gi}^2).$$

For each of the three types of preemptive-repeat disciplines the expected flow-times are

$$E(F_k) = \frac{E(P_{gk})}{1 - \sum_{i=1}^{k-1} \omega_i} + \frac{\sum_{i=1}^{k} \kappa_i}{2(1 - \sum_{i=1}^{k-1} \omega_i)(1 - \sum_{i=1}^{k} \omega_i)}, \tag{38, pr}$$

$$E(F_k) = \frac{E(P_{gk})}{1 - \sum_{i=1}^{k-1} \omega_i} + \frac{\sum_{i=1}^{k} \left(\kappa_i (1 - \sum_{j=1}^{i-1} \omega_j) + 2\lambda_i \left(E(P_{gi}) \right)^2 \sum_{j=1}^{i-1} \omega_j \right)}{2(1 - \sum_{i=1}^{k-1} \omega_i)(1 - \sum_{i=1}^{k} \omega_i)}, \tag{38, rs}$$

$$E(F_k) = \frac{E(P_{gk})}{1 - \sum_{i=1}^{k-1} \omega_i}$$
$$+ \frac{\sum_{i=1}^{k} \left(\kappa_i (1 - \sum_{j=1}^{i-1} \omega_j) + 2(\lambda_i/\Lambda_{i-1})(\sum_{j=1}^{i-1} \omega_j) E((e^{\Lambda_{i-1} P_i} - 1)^2) \right)}{2(1 - \sum_{i=1}^{k-1} \omega_i)(1 - \sum_{i=1}^{k} \omega_i)}. \tag{38, rw}$$

By using Eqs. (33a) and (37a), we can show that the first term in Eq. (38) is $E(P_{rk})$, so that the second term represents the first moment of waiting-time, and

$$E(P_{rk}) = \frac{E(P_{gk})}{1 - \sum_{i=1}^{k-1} \omega_i}. \tag{39}$$

The amount of work required of the machine per unit of time is ω_k. Over the r classes the system utilization is then

$$\bar{U} = \sum_{k=1}^{r} \omega_k.$$

For a preemptive-resume rule, $P_{gk} = P_k$, so that $\omega_k = \rho_k$ and $\bar{U} = \rho$. For the other two preemptive rules, $\bar{U} \geq \rho$.

When one compares Eq. (38, pr) with (29a'), one notes that the mean waiting-time in the resume case is the same as it would be for the nonpreemptive priority discipline if the $r - k$ lowest-priority classes did not exist. Further, when there are classes with priority lower than that of class k, the mean waiting-time under preemptive resume is lower than that for nonpreemptive priorities. This should be expected, since under preemption the lower-priority classes can in no way effect the progress of class k jobs, whereas without preemption there is the chance that an arriving class k job may have to wait until some processing is completed on a lower-priority job.

Under the preemptive-repeat disciplines, more congestion results when processing-times for preempted jobs are not resampled. Jobs that have long processing-times are the ones that tend to be preempted. Under a non-resample regime, such a job retains this same long processing-time on its next attempt to be processed, whereas under

Table 8–3

Preemptive Repeat Priorities Without Resampling*

Mean steady-state flow-times averaged over jobs of both classes

Values of λ_1 \\ $\lambda_1 + \lambda_2$	0.30	0.40	0.50	0.60	0.70	0.80	0.90	0.95	0.98	0.99
0.0	1.429	1.667	2.000	2.500	3.333	5.000	10.000	20.000	50.000	100.000
0.005										201.970
0.01								25.652	101.939	
0.02	1.492	1.761	2.141	2.721	3.714	5.809	13.114	34.791		
0.04	1.546	1.850	2.282	2.953	4.136	6.781	18.017	96.276		
0.06	1.600	1.935	2.424	3.157	4.603	7.967	26.758			
0.08	1.637	2.014	2.566	3.454	5.124	9.437	46.413			
0.10	1.667	2.086	2.706	3.723	5.707	11.295	128.628			
0.12	1.687	2.150	2.843	4.005	6.360	13.701				
0.14	1.697	2.204	2.976	4.299	7.094	16.910				
0.16	1.695	2.247	3.104	4.605	7.924	21.360				
0.18	1.682	2.279	3.224	4.925	8.864	27.853				
0.20	1.658	2.296	3.345	5.255	9.934	38.047				
0.22	1.623	2.299	3.434	5.555	11.156	55.973				
0.24	1.578	2.286	3.512	5.944	12.556	94.562				
0.26	1.528	2.255	3.588	6.302	14.168	230.000				
0.28	1.476	2.207	3.637	6.666	16.031					
0.30	1.429	2.140	3.664	7.035	18.196					
0.32		2.057	3.665	7.411	20.726					
0.34		1.957	3.637	7.796	23.710					
0.36		1.852	3.578	8.196	27.276					
0.38		1.747	3.483	8.637	31.630					
0.40		1.667	3.350	9.120	37.143					
0.42			3.176	9.750	44.576					
0.44			2.957	10.695	55.761					
0.46			2.690	12.505	76.324					
0.48			2.372	17.953	134.748					
Saturates at $\lambda_1 =$			0.50	0.50	0.50	0.276	0.113	0.053	0.020	0.010

resampling a preempted job at least has an opportunity to draw a shorter processing-time for its next try.

The difference between these two models with respect to mean flow-time can be tremendous, as shown in Table 8–3. The computations for this table are based on two job classes having identical exponential processing-time distributions, with mean processing-times equal to 1. Each column represents a value of the overall arrival rate, $\lambda_1 + \lambda_2$, and each row represents a value of λ_1. The figures in the body of the table are the mean steady-state flow-time per job averaged over both classes.

The memoryless property of the exponential distribution implies that, in this case, the preemptive-resume rule will yield the same results as the preemptive-repeat rule with resampling. (With exponentially distributed processing-times, P_{gk} and P_k have the same distribution.) Since both processing-time distributions have the same mean, the overall flow-time is independent of the proportion of jobs in class 1, and these two rules have the same overall mean flow-time as would the FCFS discipline with arrival rate $\lambda_1 + \lambda_2$. These values correspond to the first row where λ_1 is 0.0.

For the nonresample rule there are three types of upper bounds on permissible values of λ_1, depending on the magnitude of $\lambda_1 + \lambda_2$. Certainly λ_1 must not be greater than the overall arrival rate. In the present example, this bound is operative for $(\lambda_1 + \lambda_2) \leq 0.5$. A second upper bound is provided by the condition that $E(e^{2\lambda_1 P_2})$ must be finite. When we let $1/\mu_k$ be the mean processing-time for class k jobs, this condition manifests itself here by the requirement that $2\lambda_1 < \mu_2$, or $\lambda_1 < 0.5$. When this bound is exceeded, the variance of the gross processing-times of class 2 jobs becomes infinite, implying that eventually a low-priority job will not be able to get through the system. No matter how small λ_2 is, if it is not zero, there is an upper limit to the class 1 arrival rate that can be sustained.

The third bounding condition is that $\bar{U}$ must be less than 1. For exponentially distributed processing-times, the condition turns out to be $\lambda_1/\mu_1 + \lambda_2/(\mu_2 - \mu_1) < 1$. In the present example, for a fixed value of $\lambda_1 + \lambda_2 = \lambda$, the maximum permissible value of λ_1 is the smaller root of $x^2 - x + 1 - \lambda = 0$. This upper bound becomes limiting for $0.75 < \lambda < 1.0$.

8–7.2 Semi-preemptive Priority and Shortest-Remaining-Processing-Time Disciplines

In Section 8–6 it was suggested that mean flow-time could be reduced below the value obtainable with a processing-time-independent discipline by imposing a nonpreemptive priority discipline with class membership based on processing-time requirements. If the situation were such that preemption in the resume mode is allowable, one might consider carrying out a similar scheme using the preemptive resume discipline. Compared to the nonpreemptive priority rule, this would tend to reduce the waiting-time of the shorter (high-priority) jobs but increase the residence-time of the longer jobs. However, if a preemptive resume rule were used, there would be occasions when a preempted job which is nearing completion had less remaining processing-time than the job that displaced it. To interrupt under these circumstances is not in the best interest of reducing mean flow-time. The *semipreemptive priority discipline* is designed to remedy this defect of preemptive resume by using the amount of processing-time remaining, rather than the original processing-time requirement, to determine class membership. This allows the priority status of a job to improve as its processing progresses.

When one employs the notation of Section 8–6.2, the semipreemptive priority discipline operates in the same manner as the preemptive-resume priority rule except that a job whose remaining processing-time is in the interval (d_{k-1}^{+}, d_k) is considered to be a member of class k. In the limit, as the class intervals are made arbitrarily small, this discipline becomes the *shortest-remaining-processing-time discipline*. Under this rule, a newly arriving job will preempt a job on the machine if the processing-time requirement of the new job is smaller than the remaining processing-time of the job in process. In selecting another job at the moment of a job completion, preference is given to the job with the least remaining processing-time.

When the number of classes is small, the semipreemptive priority discipline may not produce lower overall mean flow-time than the preemptive resume discipline operating with the same classes. But when the number of classes is increased, the semipreemptive rule becomes relatively better, and in the limit the shortest-remaining-processing-time discipline becomes optimal with respect to minimizing overall mean flow-time. This optimality of the shortest-remaining-processing-time discipline does not actually require the arrival process to be Poisson, but this will not be proved here.

In developing the expression for mean flow-time under a semipreemptive priority discipline, we shall consider the mean waiting-time and the mean residence-time separately. The class membership of a job cannot change during its waiting-time, so we may begin by investigating the mean waiting-time of jobs that arrive as members of class k.

For economy of notation, let σ_k be the utilization due to jobs that arrive as members of the first k classes. In the notation of Section 8–6,

$$\sigma_k = \sum_{i=1}^{k} \rho_i = \lambda \int_0^{d_k} t \, dG(t).$$

The numerator of Eq. (29a'), giving the mean waiting-time for class k jobs under the nonpreemptive priority discipline, represents the product of the overall arrival rate and the second moment of the processing-time distribution, and may be rewritten as

$$\sum_{i=1}^{r} \lambda_i E(P_i^2) = \lambda E(P^2) = \lambda \int_0^{\infty} t^2 \, dG(t) = \lambda \int_0^{d_k} t^2 \, dG(t) + \lambda \int_{d_k+}^{\infty} t^2 \, dG(t).$$

When the priority assignment is based on processing-times, the second integral above represents the contribution to $E(P^2)$ made by jobs of lower priority than class k. In the new notation Eq. (29a') becomes

$$E(W_k) = \frac{\lambda \int_0^{d_k} t^2 \, dG(t) + \lambda \int_{d_k+}^{\infty} t^2 \, dG(t)}{2(1 - \sigma_{k-1})(1 - \sigma_k)}. \tag{40}$$

Under both nonpreemptive and semipreemptive priority disciplines, a necessary condition for a class k job to go onto the machine for the first time is that the system be empty of all jobs that were originally from the first $k - 1$ classes and all previous class k arrivals. This shows that the interference of jobs from the first k classes to class k jobs is the same for both disciplines. Under a semipreemptive priority discipline, a job whose original processing requirement is greater than d_k can have no effect on class k jobs until its remaining processing-time is reduced to d_k. At that point it becomes a class k job. From the viewpoint of class k jobs, all lower-priority jobs behave as though they arrive with processing-times of d_k. The arrival process for this new kind of job is no longer Poisson, but that is of no consequence; that the arrival process of class b jobs is Poisson was neither required nor used in the derivations leading to Eq. (29a').

These observations imply that to write an expression for $E(W_k)$ applicable to a semipreemptive priority discipline, we need only replace the second integral in the numerator of Eq. (40) by $(1 - G(d_k))d_k^2$, giving

$$E(W_k) = \frac{\lambda \int_0^{d_k} t^2 \, dG(t) + \lambda (1 - G(d_k)) d_k^2}{2(1 - \sigma_{k-1})(1 - \sigma_k)}.$$

To determine the mean residence-time, suppose that a job requires p units of processing-time while it is in class i status before it can acquire the priority of class $i - 1$ jobs. While the job is in class i status, it is subject to the same interruptions as would a class i job in a preemptive resume system having the same class structure. Then, according to Eq. (39), its mean residence-time in that state is $p/(1 - \sigma_{i-1})$. The total residence-time of a job that arrives as a member of class k is the sum of its residence-times in each priority status from k to 1, and

$$E(P_{rk}) = \frac{E(P_k) - d_{k-1}}{1 - \sigma_{k-1}} + \sum_{i=1}^{k-1} \frac{d_i - d_{i-1}}{1 - \sigma_{i-1}}.$$

The overall mean flow-time is

$$E(F) = \sum_{k=1}^{r} \left(G(d_k) - G(d_{k-1}) \right) \left(E(P_{rk}) + E(W_k) \right),$$

or

$$E(F) = \sum_{k=1}^{r} \left(\frac{\int_{d_{k-1}}^{d_k} t \, dG(t)}{1 - \sigma_{k-1}} + \left(G(d_k) - G(d_{k-1}) \right) \left(-\frac{d_{k-1}}{1 - \sigma_{k-1}} + \sum_{i=1}^{k-1} \frac{d_i - d_{i-1}}{1 - \sigma_{i-1}} \right) \right)$$
$$+ \frac{\lambda}{2} \sum_{k=1}^{r} \left(\frac{\left(G(d_k) - G(d_{k-1}) \right) \left(\int_0^{d_k} t^2 \, dG(t) + d_k^2 (1 - G(d_k)) \right)}{(1 - \sigma_{k-1})(1 - \sigma_k)} \right). \tag{41}$$

The formula for the expected flow-time under the shortest-remaining-processing-time discipline is obtained by taking the limit of this expression as r grows large and $d_k - d_{k-1}$ goes to zero. This results in

$$E(F) = \int_{p=0}^{\infty} \left(\int_{x=0}^{p} \frac{dx}{1 - \lambda \int_0^x t \, dG(t)} \right) dG(p)$$
$$+ \frac{\lambda}{2} \int_{p=0}^{\infty} \left(\frac{\int_{t=0}^{p} t^2 \, dG(t) + p^2 (1 - G(p))}{(1 - \lambda \int_{t=0}^{p} t \, dG(t))(1 - \lambda \int_{t=0}^{p} t \, dG(t))} \right) dG(p). \tag{42}$$

8–8 THE DUE-DATE RULE AND RELATED DISCIPLINES

A research group under Jackson at UCLA has studied the properties of what it terms *dynamic* priority disciplines [94, 97, 99, 71]. Jackson and his group have made an interesting and useful distinction between static and dynamic priorities. Under a *static priority discipline*, the distribution of priority indexes of jobs arriving at a particular queue is stationary, and a given job in queue with a particular priority index faces each new arrival with the same probability that its index will be exceeded by that of the newcomer. Under a *dynamic priority discipline*, the distribution of indexes changes over time, presumably in such a manner that each succeeding arrival has lower and lower probability of taking preference over a job already in queue. However, under a static discipline, a job simply waits in queue until by chance there remains no job with higher priority in the same queue. For a job with a low priority in a heavily loaded system, this can be an exceedingly long wait. Under a dynamic discipline the job's chances of escape improve with age so that extreme values of waiting-time are eliminated.

Operationally, priority is the sum of the job's time of arrival and a quantity called an urgency number. Variations in discipline are obtained by altering the mechanism for assigning urgency numbers. As one limiting case, when the urgency numbers are all identically zero, there is the normal FCFS discipline. At the other extreme, if the range of the urgency numbers is so great as to completely dominate the arrival-times, then one could assign those numbers to follow any nonpreemptive priority discipline. In this sense all priority disciplines are dynamic, but it is useful to reserve this designation for the class of disciplines that lie between the extremes and for which both the arrival-time and the urgency number are significant.

The principal rule of this type is obviously the due-date rule, and Jackson frequently refers to dynamic disciplines as "due-date-like" rules. The urgency number corresponds to "a"—the allowable processing-time (as defined in Section 2–1). The measure of performance has consistently been waiting-time, and the studies have been concerned with the complete distribution of waiting-time, particularly the upper tail. Of particular concern to the UCLA group is the effect of various priority disciplines on the slowest jobs to pass through a system, where much of the other work has been concerned with the average of all jobs that pass through the system. Specifically the work has centered on two conjectures by Jackson [97]. Let $A_u(w)$ be the equilibrium probability that a job which is given urgency number u will wait no longer than w. For $0 \leq f \leq 1$, define

$$w_u(f) = \inf\left(w \mid A_u(w) \geq f\right).$$

This is the waiting-time corresponding to fractile f in the cumulative waiting-time distribution for jobs with urgency number u. For jobs with urgency number u, $w_u(f)$ gives the (minimum) waiting-time which, with probability f, will not be exceeded.

Similarly, for the same system operating under the FCFS discipline, for $0 \leq f \leq 1$, define

$$w(f) = \inf\left(w \mid A_{\text{FCFS}}(w) \geq f\right).$$

Conjecture 1. In any dynamic priority system operating in equilibrium, if u and v are urgency numbers which are actually attained, then

$$\lim_{f \to 1} \left(w_u(f) - w_v(f) \right) = u - v.$$

That is, the difference in waiting-times approaches the difference in urgency numbers in the upper tail of the distribution.

Conjecture 2. In any dynamic priority system operating in equilibrium, there exists a number u^* such that if u is any urgency number which is actually attained, then

$$\lim_{f \to 1} \left(w_u(f) - w(f) \right) = u - u^*.$$

That is, the difference in waiting-time between jobs with urgency numbers u and what would be experienced under FCFS approaches the difference between u and u^* in the upper tail of the distribution. It is further conjectured that u^* is approximately the mean of the distribution of urgency numbers.

An alternative way of considering these conjectures is the following. With "due-date-like" rules one would normally consider lateness, rather than waiting-time, to be the primary measure of performance. Suppose one defines an allowance, a', to be the allowable or preferred flow-time after deducting the inescapable actual processing-time. Then $a_i' = a_i - p_i$. Now the lateness of a job is given by $L_i = W_i - a_i'$. If the urgency number, u, is interpreted to be an allowance of the a' type, then $w_u(f) - u$ is the *lateness* corresponding to fractile f in the cumulative lateness distribution for jobs with urgency number u. The Jackson conjectures then become:

Conjecture 1. In any dynamic priority system operating in equilibrium, if u and v are urgency numbers which are actually attained, then

$$\lim_{f \to 1} \frac{w_u(f) - u}{w_v(f) - v} = 1.$$

Conjecture 2. In any dynamic priority system operating in equilibrium, there exists a number u^* such that if u is any urgency number which is actually attained, then

$$\lim_{f \to 1} \frac{w_u(f) - u}{w(f) - u^*} = 1.$$

Roughly speaking, this suggests that in any dynamic priority system the *maximum lateness* that might be expected for each job is the same, regardless of the urgency number assigned. Stated in another way, given any arbitrary set of due-dates (or mechanism for assigning due-dates) the dynamic priority discipline that sequences according to these due-dates offers each job the same probabilistic bound on lateness.

One might conjecture that for a given mechanism for assigning due-dates, the dynamic priority rule using these due-dates offers the lowest upper bound on lateness that can be achieved. This bound is uniform for each different value of allowance (urgency number). Other priority disciplines might provide a lower bound for certain values of allowance, but this would be offset by greater bounds for other values, so that in particular the maximum lateness bound would be greater than for the dynamic priority discipline. This would be the continuous-case analog of due-date sequencing (Section 3–3) in the static case, which minimizes the maximum lateness. However, the present discipline seems to bear somewhat more resemblance to slack-time sequencing (Section 3–3), which controls the minimum rather than the maximum lateness.

Recall the situation with respect to mean lateness. This is minimized in the finite sequencing case by the SPT rule, and the mean of the distribution of latenesses is minimized in certain continuous single-server cases by the SPT discipline (Sections 3–2 and 8–6.2). We may conjecture that an analogous situation exists with respect to maximum lateness.

Jackson has proved Conjecture 2 (and hence also 1) for a single-server discrete queuing system with geometrically distributed interarrival-times and processing-times [99]. He also has strong empirical evidence in support of the hypothesis for a single-server, Poisson-exponential queuing system [97]. He exhibits these results by plotting the cumulative waiting-time distributions for specific values of urgency number on semilogarithmic coordinates. One such plot (from [97]) is reproduced as Fig. 8–9. On these coordinates the waiting-time distribution under FCFS plots as a straight line. Conjecture 2 says that the graph of a waiting-time distribution for a dynamic priority discipline is asymptotically parallel to the line for FCFS: a separate line for each value of u, but each parallel to the others at the upper tail of the distributions.

The due-date rule operating as a preemptive resume system would always be processing the job with the earliest due-date, interrupting processing of a job when another one arrives if the arriving job's due-date is closer than the due-date of the job in process. If there is only a finite number of arrivals the situation would be that of Section 4–2: a single machine, a finite number of jobs, and intermittent arrivals.

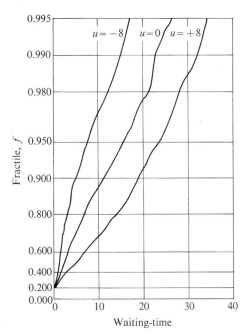

Fig. 8–9. Waiting-time distribution for a dynamic-priority discipline, $\rho = 0.8$ (from Jackson [97]).

In that situation the due-date rule operating with preempt resume does minimize the maximum lateness (this was proved by Jackson [90]). There is as yet no proof that this property (or a probabilistic counterpart) carries over to the continuous, infinite-horizon case. Yet the situation is so similar to that for the shortest-remaining-processing-time discipline that the same result must hold.

A dynamic priority discipline of a multiplicative nature has been analyzed by Kleinrock [112]. There are r classes of jobs, as in a nonpreemptive priority discipline, and each class, k, has associated with it a number b_k, where $b_k \leq b_{k+1}$, $k = 1, 2, \cdots$, $r - 1$. The priority of a job of class k at time t is given by $(t - T)b_k$, where T is the arrival time of the job. The job with the *highest* value for priority is selected for service by the machine. Jobs of the same class are processed in FCFS order, since only the arrival time, T, differs for jobs of the same class. Two jobs of different classes can be ranked differently, depending on the difference between their arrival times and on how long they have waited, as shown below. For example, in the figure below, from time t_1 to time t_2 job I is preferred over job J, but after t_2 job J is preferred.

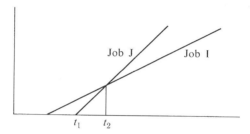

We present Kleinrock's result without proof. It is established in a manner similar to that originally used by Cobham [28] to establish Eq. (8–29a′). If $E(W_k)$ is the expected waiting-time for a job of class k, then,

$$E(W_k) = \frac{\frac{1}{2}\sum_{i=1}^{r} \lambda_i E(P_i^2) + \sum_{i=1}^{r} \rho_i y_{ik} E(W_i)}{1 - \sum_{i=1}^{r} \rho_i z_{ik}}, \tag{43}$$

where

$$y_{ik} = \begin{cases} 1 \text{ if } b_i \geq b_k \\ b_i/b_k \text{ if } b_i < b_k \end{cases} \quad \text{and} \quad z_{ik} = \begin{cases} 0 \text{ if } b_i \leq b_k \\ 1 - b_k/b_i \text{ if } b_i > b_k. \end{cases}$$

If $b_k = b$, $k = 1, \ldots, r$, then Eq. (43) reduces to the expected waiting-time under FCFS, Eq. (15a). On the other hand, if b_k/b_{k+1} is set equal to c_k, $k = 1, \ldots, r - 1$, and b_r is allowed to approach zero, the limiting form of Eq. (43) is equivalent to the result for nonpreemptive priorities, Eq. (29a′), with the class indexes reversed. Thus, in a sense, the discipline proposed by Kleinrock is a generalization of a nonpreemptive priority discipline.

8–9 THE EFFECT OF PROCESSING-TIME-DEPENDENT DISCIPLINES ON FLOW-TIME

Many of the disciplines studied in the preceding sections were motivated by the general principle that giving preference to shorter jobs will tend to reduce mean flow-time and the expected number of jobs in the system. It is worth illustrating just how much can be accomplished by this approach. Since all selection disciplines which do not use information related to processing-times yield the same mean flow-time, the first-come, first-served discipline is a reasonable standard against which the improvement obtained from the other rules may be measured.

The data in Tables 8-4 and 8-6 represent the mean flow-times for a range of arrival rates under six selection disciplines; FCFS, nonpreemptive (NP), semipreemptive (SP)-preempt resume (PR), shortest-processing-time (SPT), and shortest-remaining,

Table 8-4

Comparison of Mean Flow-Time for Various Queuing Disciplines*

Arrival rate	Mean steady-state flow-times					
	FCFS	NP	SP	PR	SPT	SRPT
0.30	1.429	1.377	1.267	1.246	1.359	1.197
0.40	1.667	1.555	1.404	1.372	1.518	1.296
0.50	2.000	1.782	1.578	1.540	1.713	1.425
0.60	2.500	2.088	1.845	1.778	1.962	1.604
0.70	3.333	2.542	2.245	2.147	2.312	1.874
0.80	5.000	3.328	2.970	2.821	2.882	2.352
0.90	10.000	5.284	4.847	4.590	4.197	3.552
0.95	20.000	8.529	8.036	7.646	6.263	5.540
0.98	50.000	16.636	16.046	15.481	11.283	10.494
0.99	100.000	28.283	27.699	26.905	18.449	17.625

* Exponentially distributed processing-times with mean 1.0.

Table 8-5

Optimal Class Dividing Points for Two-Class Rules*

Arrival rate	Nonpreemptive priorities		Semipreemptive priorities		Preemptive resume	
	Boundary point	Prop. in class 1	Boundary point	Prop. in class 1	Boundary point	Prop. in class 1
0.30	1.138	0.679	1.045	0.648	1.323	0.734
0.40	1.200	0.699	1.115	0.672	1.386	0.750
0.50	1.278	0.722	1.200	0.699	1.464	0.769
0.60	1.378	0.748	1.307	0.729	1.561	0.790
0.70	1.514	0.780	1.455	0.767	1.691	0.816
0.80	1.718	0.821	1.675	0.813	1.885	0.848
0.90	2.101	0.878	2.078	0.875	2.245	0.894
0.95	2.524	0.920	2.515	0.919	2.640	0.929
0.98	3.134	0.957	3.133	0.956	3.214	0.960
0.99	3.627	0.973	3.629	0.974	3.686	0.975

* Exponentially distributed processing-times with mean 1.0.

processing-time (SRPT). In both tables, the mean processing-time is equal to 1.0. Table 8–4 is for exponentially distributed processing-times, while Table 8–6 is based on processing-times uniformly distributed over the interval (0, 2). The examples of the three finite class rules (nonpreemptive, preemptive resume, and semipreemptive priorities) are for two classes with the optimal point separating the high- and low-priority classes in each case. These boundary points and the resulting proportions of jobs in the high-priority classes are given in Tables 8–5 and 8–7.

Table 8–6

Comparison of Mean Flow-Time for Various Queuing Disciplines*

Arrival rate	Mean steady-state flow-times					
	FCFS	NP	SP	PR	SPT	SRPT
0.30	1.286	1.262	1.228	1.228	1.245	1.203
0.40	1.444	1.394	1.348	1.348	1.379	1.309
0.50	1.667	1.569	1.512	1.512	1.541	1.451
0.60	2.000	1.816	1.749	1.749	1.766	1.656
0.70	2.556	2.204	2.028	2.028	2.115	1.984
0.80	3.667	2.930	2.849	2.849	2.764	2.610
0.90	7.000	4.949	4.870	4.870	4.575	4.379
0.95	13.667	8.750	8.681	8.681	8.041	7.850
0.98	33.667	19.643	19.592	19.592	18.206	18.007
0.99	66.000	37.300	37.260	37.260	34.993	34.791

* Processing-times uniformly distributed on (0, 2).

Table 8–7

Optimal Class Dividing Points for Two-Class Rules*

Arrival rate	Nonpreemptive priorities		Semipreemptive priorities		Preemptive resume	
	Boundary point	Prop. in class 1	Boundary point	Prop. in class 1	Boundary point	Prop. in class 1
0.30	1.089	0.544	0.864	0.432	0.864	0.432
0.40	1.127	0.564	0.920	0.460	0.920	0.460
0.50	1.172	0.586	0.988	0.494	0.988	0.494
0.60	1.225	0.613	1.070	0.535	1.070	0.535
0.70	1.292	0.646	1.170	0.585	1.170	0.585
0.80	1.382	0.691	1.298	0.649	1.298	0.649
0.90	1.519	0.760	1.480	0.740	1.480	0.740
0.95	1.635	0.817	1.618	0.809	1.618	0.809
0.98	1.752	0.876	1.746	0.873	1.746	0.873
0.99	1.818	0.909	1.816	0.908	1.816	0.908

* Processing-times uniformly distributed on (0, 2).

Compared with the shortest-remaining-processing-time discipline, which is known to produce minimum flow-time, the other processing-time-dependent rules seem to perform quite well in reducing flow-time below the values for FCFS. The effectiveness of the two-class nonpreemptive priority discipline, which is the least sophisticated of the priority rules, shows that there is a great deal of benefit in merely classifying jobs as long or short. This suggests that some kind of processing-time-related discipline would be worthwhile in situations in which low flow-times are desirable, even when accurate estimates of processing-times are not available.

Comparing the nonpreemptive and the preemptive resume priority disciplines, or the shortest-processing-time and the shortest-remaining-processing-time discipline, indicates that preemption, which might be difficult to administer, does not bring about much improvement. It is interesting to note that in both examples there is a cross-over between the two-class preemptive resume priority discipline and the shortest-processing-time rule. At higher utilization, when there are often many jobs competing to get on the machine, the ability to discriminate well between waiting jobs is important. But at low utilization, when there are usually few jobs from which to choose at the moment of a processing completion, fine discrimination is much less important than the ability to preempt a long job when a shorter one arrives.

The two-class preemptive resume discipline may not always be better than the two-class nonpreemptive priority discipline. Given fixed classes, the increase in mean residence-time for class 2 jobs in going from nonpreemptive to preemptive resume priorities is $\rho_1 E(P_2)/(1 - \rho_1)$, while the decrease in mean waiting-time enjoyed by the class 1 jobs would be $\lambda_2 E(P_2^2)/(1 - \rho_1)$. For preemptive resume to be better, it would have to be true that

$$\frac{\lambda_2 \rho_1 E(P_2)}{1 - \rho_1} - \frac{\lambda_1 \lambda_2 E(P_2^2)}{2(1 - \rho_1)} < 0,$$

which implies that the preemptive resume priority discipline is better than the nonpreemptive priority discipline if and only if

$$E(P_1) < \frac{E(P_2^2)}{2E(P_2)} \, .$$

The right-hand side of this inequality is the mean of the random modification of a class 2 processing-time and represents the expected remaining processing-time of a class 2 job as seen by an arriving high-priority job that finds a low-priority job on the machine.

Consider a similar comparison between the two-class semipreemptive and the two-class preemptive resume priority discipline for which jobs with processing-times less than d are assigned to the high-priority class. Compared to the semipreemptive priority discipline, the mean residence-time of class 2 jobs under preemptive resume is greater by $\rho_1 d/(1 - \rho_1)$, while the expected waiting-time of class 1 jobs is less by $\lambda_2 d^2/2(1 - \rho_1)$. For the semipreemptive priority discipline to result in lower overall mean flow-time, it would have to be true that

$$\frac{\lambda_1 \lambda_2 d^2}{2(1 - \rho_1)} - \frac{\lambda_2 \rho_1 d}{1 - \rho_1} < 0 \quad \text{or} \quad E(P_1) > \frac{d}{2} \, .$$

This condition is equivalent to

$$\int_0^d G(t)\,dt < \frac{dG(d)}{2}.$$

The left-hand side of the inequality is the area under the distribution function up to the point d, while the right-hand side is the area of the triangle under the line from the origin to the point $(d, G(d))$. With this interpretation, we may see that since the distribution function for the exponential distribution is convex (second derivative negative), a two-class semipreemptive priority discipline can never be better than a two-class nonpreemptive priority rule when the underlying processing-time distribution is exponential. This conclusion agrees with the data in Table 8–4. When processing-times are uniformly distributed on the interval $(0, 2E(P))$, the condition is satisfied as an identity for any choice of d in $(0, 2E(P))$, and both disciplines produce the same mean flow-time. If processing-times are uniformly distributed over (θ_1, θ_2) with $\theta_1 > 0$, the semipreemptive priority discipline will be better.

Many other typical processing-time distributions, such as those belonging to the Erlang family, have distribution functions that are concave for small values of p and become convex for large values of p. Hence there is some dividing point, p^*, such that for $d < p^*$, the two-class semipreemptive priority discipline is better than the two-class preemptive resume priority discipline, but for $d > p^*$, the situation is reversed. Since the optimal value of d for any of these rules is an increasing function of the arrival rate, in such cases the choice of discipline would depend on λ as well as $G(p)$.

In two-class situations, when the preemptive resume priority rule is better, it would be so because the semipreemptive priority rule does not allow enough preemptions. As soon as a class 2 job has its remaining processing-time worked down to d, it is no longer eligible for preemption. But d is larger than the expected processing-time of a class 1 job that may arrive. Therefore it would be better to allow preemption of class 2 jobs until their remaining processing-times are reduced to some smaller value, d'. But then class 1 jobs with remaining processing-times between d' and d should also be subject to preemption. Thus it would seem that the remedy is to increase the number of classes. The addition of more classes brings us closer to the shortest-remaining-processing-time discipline.

A comparison between two-class nonpreemptive and semipreemptive priority disciplines, both with dividing point d, shows that the semipreemptive rule will be better if

$$E(P_1) < \frac{E(P_2^2) - d^2}{2\bigl(E(P_2) - d\bigr)}.$$

In the special case of exponentially distributed processing-times, it is found that the expected waiting-time under the shortest-processing-time rule is equal to the expected flow-time for the shortest-remaining-processing-time rule multiplied by ρ. That is,

$$E_{\mathrm{SPT}}(W) = \rho E_{\mathrm{SRPT}}(F).$$

Table 8-8

Effect of Variance of Processing-Time Distribution
Under Shortest-Processing-Time Rule*

Phases k	Variance of processing times	Mean flow-time	
		FCFS	SPT
∞	0.000	5.500	5.500
10	0.100	5.590	4.197
5	0.200	6.400	4.089
3	0.333	7.000	4.115
2	0.500	7.750	4.160
1	1.000	10.000	4.197

* Processing-times sampled from k-phase Erlang with
mean 1.0 utilization equal to 0.9.

Even though the two processing-time distributions assumed in the computations
of Tables 8-4 to 8-7 have the same means, the greater variance of the exponential
distribution results in higher levels of congestion under FCFS at all levels of utilization.
In the comparison of the two examples under the shortest-remaining-processing-time
rule, the opposite relationship holds. For all of the other disciplines, there is a cross-
over point in the neighborhood of 80% to 90% utilization such that at low values of
λ the uniform distribution displays less congestion and at the higher values it displays
more.

While the mean flow-time resulting from the use of a processing-time-independent
discipline varies linearly with the variance of the processing-time distribution (with
$E(P)$ held constant), this is not true for the processing-time-independent disciplines.
This is illustrated in Table 8-8. These data are based on an arrival rate of 0.9 and an
Erlang processing-time distribution with mean equal to 1.0 and k phases.

The mean flow-time as a function of the variance for the shortest-processing-time
rule is surprisingly constant. Note especially that the value corresponding to $k = 1$,
the exponential distribution, is less than that for the degenerate distribution ($k = \infty$).
It appears that a discipline which takes advantage of varying processing-time requirements requires a certain amount of variability in the processing-time distribution in
order to be effective. If there are several jobs in the system and there is some variability
in processing-times, the rule has a chance to make a good selection. On the other hand,
too much variability permits an occasional long job to create a high degree of congestion.

With respect to the minimization of variance of flow-time one would intuitively
expect FCFS to be preferable to SPT, but this is not uniformly correct.

Equations (14a) and (14b) are used to get the variance of flow-time under the
FCFS rule. The second moment of flow-time under the SPT rule is derived in a
manner similar to the development of Eq. (32), and the details are omitted. Table 8-9
(based on [184]) gives the variance of flow-time under these two rules for both exponential and 2-phase Erlang processing-time distributions.

Table 8–9

Variance of Flow-Time Under FCFS and SPT*

	Exponential		2-Phase Erlang	
ρ	FCFS	SPT	FCFS	SPT
0.10	1.2345	1.179	0.6180	0.6156
0.20	1.5625	1.482	0.7856	0.7773
0.30	2.0408	1.896	1.031	1.020
0.40	2.666	2.563	1.467	1.420
0.50	4.000	3.601	2.062	2.158
0.60	6.250	5.713	3.265	3.747
0.70	11.111	12.297	5.895	8.042
0.80	25.000	32.316	13.50	25.08
0.90	100.00	222.20	55.06	186.20
0.95	400.00	1596.5	227	1514
0.98	2500.0	22096	1400	19826
0.99	10000	161874	5612	147239

* Mean processing-time = 1.0.

At low utilizations, the SPT rule achieves a slightly lower variance of flow-time than that for the FCFS rule. At higher and more interesting levels of utilization, the FCFS rule is substantially better than SPT. The crossover point between the two rules apparently decreases as the variance of the processing-times decreases.

8–10 HISTORICAL NOTES

The body of literature relating to queuing theory is enormous. Saaty [178] gives a bibliography of over 900 items published up to 1961. In his introductory chapter, Saaty also presents an excellent short history of the development of the field. This section contains collected historical notes concerning the topics covered in Chapter 8.

The earliest work, pioneered by A. K. Erlang in the first three decades of this century, was produced by engineers concerned with the design of telephone switching networks. The technique of differential-difference equations, illustrated in Section 8–1 and in Chapter 10, was the dominant method of analysis.

Section 8–4, treating the distribution of flow-times under the first-come, first-served discipline, is based on a particular decomposition of the length of the busy period into a sequence of dependent random variables. This way of viewing busy periods was suggested by Cobham [28] in his study of the nonpreemptive priority discipline.* It was then used by Avi-Itzhak and Naor [8] to derive the first two moments of the busy period, and then by Avi-Itzhak, Maxwell, and Miller [7] in the derivation given in Section 8–4. Equation (8–14a), giving the mean flow-time resulting from the use of a processing-time-independent discipline in a system with Poisson

---

* A similar notion was employed by Tanner [192]. He in turn attributes the procedure to E. Borel, *C. R. Acad. Sci.*, Paris, **214,** 1942.

arrivals and arbitrary processing-time distribution is the well-known Khintchine-Pollaczek formula.

The ideas of Section 8–5.1 regarding the number of jobs remaining in the system at moments of job completions form the basis of the "method of the imbedded Markov chain," introduced by Kendall [107]. This technique has played a dominant role in the analysis of queues with Poisson arrivals and general processing-times. However, the derivation of Section 8–4 is used in this book to provide equivalent fundamental results.

The last-come, first-served rule of Section 8–5.2 has been studied by Wishart [202], while the Random discipline of Section 8–5.3 has had a somewhat frustrating history. Several references are given in the text and others are indicated in Saaty ([178], page 23).

The original paper on priority disciplines was by Cobham [28] in which Eq. (29a') was derived for the mean waiting-time under the nonpreemptive priority discipline. The extension to the shortest-processing-time discipline was made by Phipps [163]. Equation (29), giving the Laplace transform for the distribution of waiting-time, was first derived by Kesten and Runnenberg [109]. The derivation given here is more elementary, and is a culmination of the preceding material in Chapter 8.

There have been several important papers dealing with preemptive priority disciplines. White and Christie [201] examined the case of two classes, Poisson arrivals and exponential processing-times under the resume regime. Heathcote [74] studied the time-dependent distribution of numbers of jobs in the system under similar conditions. R. G. Miller [138] gave important results for the preemptive resume rule with general processing-times. Although a distinction between the resume and repeat modes of preemption was recognized in the White and Christie paper, it appears that the difference between the two types of preemptive repeat disciplines was not recognized, at least in the formulation of analytic models, until the papers of Gaver [59] and Avi-Itzhak [5]. Some of the terminology used in Section 8–7.1 is borrowed from the latter.

The shortest-remaining-processing-time discipline (Section 8–7.2) was first studied by Schrage [183], and the paper by Miller and Schrage [137] is an expanded version of the material presented here.

The work on the due-date rule is attributable to Jackson. He provided the fundamental results [94, 97, 99] and stimulated the work of others [71, 112].

CHAPTER 9

SINGLE-SERVER QUEUING
SYSTEMS WITH SETUP CLASSES

One significant limitation to the queuing models of Chapter 8 is the exclusion of setup considerations. The processing-times could, of course, be defined to represent both setup and actual processing, but this is effective only if the setup is entirely dependent on the particular job and unaffected by the presence of other jobs and their position in sequence. It also excludes the rather realistic model in which the setup portion of the machine-time must be repeated if preempted, but the processing portion of machine-time can be resumed from the point of preemption. It is the purpose of this chapter to explicitly separate setup and processing-times and study the very different and interesting behavior of queuing systems under these circumstances. The analysis becomes much more difficult and the results are very modest, but they afford some useful and practical insight and certainly suggest areas in which further work could be done.

The central concept of the chapter is that of a *setup class*. Each job arriving at a system is a member of one such class. The machine is set up to perform work on a particular class of jobs; that is, setup-time is required between jobs of different classes, but not between jobs of the same class. Chapter 8 could be considered as representing either of the limiting cases: each job has a unique setup class, or all jobs belong to the same setup class. Except for Section 9–1, Chapter 9 assumes that the number of classes is less than the number of jobs, but greater than one. (In fact the results are largely limited to the case of two classes.) In general one could allow the setup-time to depend on the class which preceded on the machine as well as the class intended to go on the machine, in a manner analogous to the traveling-salesman problem of Section 4–1, but the algebra is tedious and the work just has not been done. Clearly, for the case with only two classes, sequence-dependent class setup-times are not relevant.

The first section of the chapter is a more general treatment of the preemption question than was provided in Chapter 8: setup-time is preempt repeat, processing-time is preempt resume. The results of Chapter 8 are simply special cases of this analysis.

The other three sections are concerned with selection disciplines. The first is the familiar first-come, first-served, in which the occurrence of setup-time is entirely determined by the arrival process. The second is a discipline in which one processes jobs of a single class as long as possible before shifting to another class.

The notation of Chapter 8 is retained, with the addition of the symbols S for setup-times, $K(s)$ for the distribution function of setup-time, and $\kappa(z)$ for the Laplace transform of $K(s)$.

9–1 PREEMPTIVE RESUME PRIORITIES WITH PREEMPTIVE REPEAT SETUP-TIMES FOR EACH JOB

Assume that there are r classes of jobs, with class k having preemptive priority over all classes with indexes less than k. Each time a job is started on the machine, whether initially or after having been preempted, setup must take place. (The setup is also subject to preemption.) The residence-time of a job of class k consists of a complete setup, S_{ck}, followed by some processing, possible breakdown-times, resetup, further processing, etc., until the total required processing of the job is completed. A typical residence-time of a job of class k, together with the definition of the random variables of concern, is given in Fig. 9–1. A complete setup consists of wasted setup-times, S_{wk}, followed by breakdown-times, T_{bk}, and is terminated by a successful setup-time, S_{sk}. The gross setup-time, S_{gk}, is the sum of the complete setup-times required.

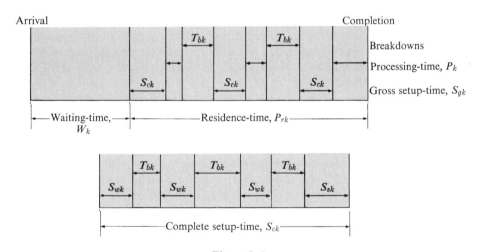

Figure 9–1

After the first complete setup-time has elapsed, the system behaves in the same manner as a preemptive resume system with a breakdown interval of $T_{bk} + S_{ck}$. Using Eq. (8–33, pr) for the preemptive resume rule, we obtain

$$\gamma_{rk}(z) = \kappa_{ck}(z)\gamma_k\big(z + \lambda_a - \lambda_a\eta_{bk}(z)\kappa_{ck}(z)\big). \tag{1}$$

Effectively, a complete setup-time in this situation is analogous to residence-time under the preemptive repeat system of Section 8-7.1. With appropriate substitution we can use Eq. (8–33, rs) for the resample case and Eq. (8–33, rw) for the nonresample case:

$$\kappa_{ck}(z) = \frac{(z + \lambda_a)\kappa_k(z + \lambda_a)}{z + \lambda_a - \lambda_a \eta_{bk}(z)\big(1 - \kappa_k(z + \lambda_a)\big)}, \tag{2, rs}$$

$$E(S_{gk}) = \frac{1 - \kappa_k(\lambda_a)}{\lambda_a \kappa_k(\lambda_a)}, \tag{3a, rs}$$

$$E(S_{gk}^2) = \frac{2}{\lambda_a^2\big(\kappa_k(\lambda_a)\big)^2}\big(1 - \kappa_k(\lambda_a) - \lambda_a E(S_k e^{-\lambda_a S_k})\big), \tag{3b, rs}$$

$$E(S_{ck}) = \big(1 + \lambda_a E(T_{bk})\big)E(S_{gk}), \tag{2a, rs}$$

$$E(S_{ck}^2) = \big(1 + \lambda_a E(T_{bk})\big)E(S_{gk}^2)$$
$$+ 2\lambda_a E(T_{bk})\big(1 + \lambda_a E(T_{bk})\big)\big(E(S_{gk})\big)^2 + \lambda_a E(T_{bk}^2)E(S_{gk}), \tag{2b, rs}$$

$$\kappa_{ck}(z) = \int_{s=0}^{\infty} \frac{(z + \lambda_a)e^{-(z+\lambda_a)s}}{z + \lambda_a - \lambda_a \eta_{bk}(z)\big(1 - e^{-(z+\lambda_a)s}\big)}\, dK_k(s), \tag{2, rw}$$

$$E(S_{gk}) = \frac{\kappa_k(-\lambda_a) - 1}{\lambda_a}, \tag{3a, rw}$$

$$E(S_{gk}^2) = (2/\lambda_a)\big(\kappa_k(-2\lambda_a) - \kappa_k(-\lambda_a) - \lambda_a E(S_k e^{\lambda_a S_k})\big), \tag{3b, rw}$$

$$E(S_{ck}) = \big(1 + \lambda_a E(T_{bk})\big)E(S_{gk}), \tag{2a, rw}$$

$$E(S_{ck}^2) = \big(1 + \lambda_a E(T_{bk})\big)E(S_{gk}^2)$$
$$+ (2/\lambda_a)\big(E(T_{bk})\big)\big(1 + \lambda_a E(T_{bk})\big)E(e^{\lambda_a S_k} - 1)^2 + \lambda_a E(T_{bk}^2)E(S_{gk}). \tag{2b, rw}$$

Equations (8–37a), (8–37b), and (8–36a′), which recursively define the moments of waiting-time and breakdown-time, still apply. These equations, together with the other equations for moments required to solve for the first moments of waiting- and flow-time, are listed below:

$$E(F_k) = E(W_k) + E(P_{rk}), \tag{4}$$

$$E(T_{b,k+1}) = \frac{\lambda_a E(T_{bk}) + \lambda_k E(P_{rk})}{(\lambda_a + \lambda_k)\big(1 - \lambda_k E(P_{rk})\big)}, \tag{5a}$$

$$E(T_{b,k+1}^2) = \frac{\lambda_a E(T_{bk}^2)\big(1 - \lambda_k E(P_{rk})\big) + \lambda_k E(P_{rk}^2)\big(1 + \lambda_a E(T_{bk})\big)}{(\lambda_a + \lambda_k)\big(1 - \lambda_k E(P_{rk})\big)^3}, \tag{5b}$$

$$E(P_{rk}) = E(S_{ck}) + E(P_k)\big(1 + \lambda_a\big(E(S_{ck}) + E(T_{bk})\big)\big), \tag{1a}$$

$$E(P_{rk}^2) = E(S_{ck}^2) + 2E(S_{ck})E(P_k)\big(1 + \lambda_a\big(E(S_{ck}) + E(T_{bk})\big)\big)$$
$$+ E(P_k^2)\big(1 + \lambda_a\big(E(S_{ck}) + E(T_{bk})\big)\big)^2$$
$$+ \lambda_a E(P_k)\big(E(S_{ck}^2) + 2E(S_{ck})E(T_{bk}) + E(T_{bk}^2)\big), \tag{1b}$$

$$E(W_{k+1}) = \frac{\lambda_{k+1}E(P_{r,k+1}^2)}{2\big(1 - \lambda_{k+1}E(P_{r,k+1})\big)} + \frac{E(W_k)}{1 - \lambda_k E(P_{rk})}. \tag{6}$$

For class 1, $\lambda_a = 0$ and $T_{bk} = 0$, so that Eq. (6) gives $E(W_1)$ directly, and $E(F_1) = E(W_1) + E(S_1) + E(P_1)$, since $S_{c1} = S_1$. Then, recursively, given $E(T_{bk})$, $E(T_{bk}^2)$, $E(P_{rk})$, and $E(P_{rk}^2)$, Eqs. (5a) and (5b) give $E(T_{b,k+1})$ and $E(T_{b,k+1}^2)$. These in turn allow $E(S_{ck+1})$ and $E(S_{ck+1}^2)$ to be determined. From these, one can find $E(P_{r,k+1})$ and $E(P_{r,k+1}^2)$, and then one can compute $E(W_{k+1})$ and $E(F_{k+1})$.

9–2 ALTERNATING-PRIORITY DISCIPLINE WITHOUT SETUP-TIME

Consider a two-class priority discipline in which sequence within each class is on a first-come, first-served basis. Suppose that the class priorities alternate so that the higher priority is assigned to that class which includes the job currently being processed. The operation of this discipline is simply to continue to process jobs of a particular class—in arrival order—until no further jobs of that class are waiting and available for assignment. Assuming that there are one or more jobs of the other class waiting at that point, the earliest of these is selected for processing and the high-priority status is shifted to the class now being processed. The effect is to alternate between the two classes, processing a run of jobs of one class, then a run of the other. The length of run is a random variable depending on the overall utilization of the machine and the balance between the two classes.

In general, one could consider such a discipline with k classes, but the discipline must then include a rule to determine which of the $k - 1$ classes is to be processed next when the queue for the incumbent class is exhausted. With only two classes there is obviously no such choice, and the analysis here is limited to this simpler situation.

In order to investigate the steady-state flow-time for this discipline, consider two types of busy periods depending on the class affiliation of the initiating job. A busy period is further subdivided into *phases* which are unbroken time-intervals during which only one class of job is processed. The length of a phase is denoted by T_{ikj}, where the subscripts have the following interpretations:

i-class receiving service during the phase ($i = 1, 2$);
k-serial numbers of the type i phases in the busy period ($k = 1, 2, \ldots$);
j-class of job initiating the busy period ($j = 1, 2$).

Figure 9–2 illustrates the sequence of phases in each type of busy period.

Type 1 busy period

Phases	T_{111}	T_{211}	T_{121}	T_{221}	T_{131}	$\ldots$	$T_{2(k-1)1}$	T_{1k1}	T_{2k1}	$T_{1(k+1)1}$
Cycles	T_{11}	T_{21}		T_{31}			T_{k1}		$T_{(k+1)1}$	

Type 2 busy period

Phases	T_{212}	T_{112}	T_{222}	T_{122}	$\ldots$	$T_{2(k-1)2}$	$T_{1(k-1)2}$	T_{2k2}	T_{1k2}	
Cycles	T_{12}		T_{22}			$T_{(k-1)2}$		T_{k2}		

Figure 9–2

Because of the obvious symmetry in the problem, it is necessary to analyze only the flow-time for class 1 jobs, and the results for class 2 jobs are obtained by interchanging subscripts.

The central idea of the analysis is to consider a busy period as being composed of a sequence of delay cycles. The busy-period portion of such a delay cycle is a phase in which class 1 jobs are processed, and the delay portion, except for the first cycle of a type 1 busy period, is the preceding phase during which class 2 jobs are processed. Let T_{kj} be the length of the kth cycle in a type j busy period. Then

$$T_{11} = T_{111},$$

$$T_{k1} = T_{2(k-1)1} + T_{1k1}, \qquad k = 2, 3, \ldots,$$

$$T_{k2} = T_{2k2} + T_{1k2}, \qquad k = 1, 2, \ldots.$$

The length of a busy period initiated by a class j job will be denoted by T_j, where

$$T_j = \sum_{k=1}^{\infty} T_{kj}.$$

We shall define both singly and doubly subscripted system states by considering the system to be in state 0, 1, or 2 according to whether it is idle or engaged in a type 1 or a type 2 busy period, and by saying that the system is in state kj if a cycle indexed by kj is in progress. If the system is in state kj, it is also in state j. For each state s, whether singly or doubly subscripted, let π_s, m_s, and l_s have the usual meanings of steady-state probability, mean persistence-time, and mean recurrence-time.

Let $\eta_{ikj}(z) = E(e^{-zT_{ikj}})$ and let $E(e^{-zF_1}| s)$ be the conditional Laplace transform associated with the distribution of flow-times of class 1 jobs, given that the job arrives while the system is in state s. Then

$$E(e^{-zF_1} \mid 0) = \gamma_1(z),$$

and by substitution into Eq. (8–18), we have

$$E(e^{-zF_1} \mid 11) = \frac{\gamma_1(z)\big(1 - \gamma_1(z)\big)}{m_{11}\big(\lambda_1\gamma_1(z) - \lambda_1 + z\big)}, \tag{7}$$

$$E(e^{-zF_1} \mid k1) = \frac{\gamma_1(z)\big(1 - \eta_{2(k-1)1}(z)\big)}{m_{k1}\big(\lambda_1\gamma_1(z) - \lambda_1 + z\big)}, \tag{8}$$

$$E(e^{-zF_1} \mid k2) = \frac{\gamma_1(z)\big(1 - \eta_{2k2}(z)\big)}{m_{k2}\big(\lambda_1\gamma_1(z) - \lambda_1 + z\big)}. \tag{9}$$

The state probabilities, π_{kj}, required to weight these conditional Laplace transforms are obtained in a manner similar to those for the priority-class schemes of Chapter 8. We have

$$m_0 = \frac{1}{\lambda_1 + \lambda_2}, \qquad m_1 = \frac{E(P_1)}{1 - \rho}, \qquad m_2 = \frac{E(P_2)}{1 - \rho},$$

$$l_0 = \frac{1}{(\lambda_1 + \lambda_2)(1 - \rho)}, \qquad l_1 = \frac{\lambda_1 + \lambda_2}{\lambda_1} l_0, \qquad l_2 = \frac{\lambda_1 + \lambda_2}{\lambda_2} l_0,$$

$$\pi_0 = 1 - \rho, \qquad \pi_1 = \rho_1, \qquad \pi_2 = \rho_2.$$

Given that the system is in state j, the conditional probability that it is in state kj is m_{kj}/m_j, so that

$$\pi_{kj} = \frac{m_{kj}}{m_j} \pi_j = \pi_0 \lambda_j m_{kj} = (1 - \rho)\lambda_j m_{kj}. \tag{10}$$

The unconditional Laplace transform for the distribution of class 1 flow-times is then

$$\beta_1(z) = E(e^{-zF_1}) = \pi_0 E(e^{-zF_1} \mid 0) + \sum_{j=1}^{2} \sum_{k=1}^{\infty} \pi_{kj} E(e^{-ZF_1} \mid kj)$$

$$= \frac{(1 - \rho)\gamma_1(z)}{\lambda_1\gamma_1(z) - \lambda_1 + z} \left(z + \lambda_1 \sum_{k=1}^{\infty} (1 - \eta_{2k1}(z)) + \lambda_2 \sum_{k=1}^{\infty} (1 - \eta_{2k2}(z)) \right). \tag{11}$$

The first moment of the mean flow-time is obtained by differentiation as

$$E(F_1) = (1 - \rho)\left(E(P_1) + \frac{2(1 - \rho_1)E(P_1) + \lambda_1 E(P_1^2)}{2(1 - \rho_1)^2} \right.$$

$$\times \left(\rho_1 + \lambda_1 \sum_{k=1}^{\infty} E(T_{2k1}) + \lambda_2 \sum_{k=1}^{\infty} E(T_{2k2}) \right)$$

$$\left. + \frac{1}{2(1 - \rho_1)} \left(\lambda_1 E(P_1^2) + \lambda_1 \sum_{k=1}^{\infty} E(T_{2k1}^2) + \lambda_2 \sum_{k=1}^{\infty} E(T_{2k2}^2) \right) \right). \tag{11a}$$

The sum of the means of type 2 phases in a type 1 cycle required in Eq. (11a) is the total expected amount of time spent serving class 2 jobs in a type 1 cycle, so that

$$\sum_{k=1}^{\infty} E(T_{2k1}) = \rho_2 m_1 = \frac{\rho_2 E(P_1)}{1 - \rho}. \tag{12}$$

Similarly,

$$\sum_{k=1}^{\infty} E(T_{2k2}) = E(P_2) + \rho_2 m_2 = \frac{(1 - \rho_1)E(P_2)}{1 - \rho}. \tag{13}$$

To evaluate the sums of second moments, we use Eqs. (8–7a), (8–7b), (8–8a), and (8–8b). Since T_{111} is really a simple busy period,

$$E(T_{111}^2) = \frac{E(P_1^2)}{(1 - \rho_1)^3},$$

while T_{1k1} for $k > 1$ is the busy-period portion of a delay cycle for which the delay is $T_{2(k-1)1}$, so that

$$E(T_{1k1}^2) = \frac{\lambda_1 E(P_1^2)}{(1 - \rho_1)^3} E(T_{2(k-1)1}) + \frac{\rho_1^2}{(1 - \rho_1)^2} E(T_{2(k-1)1}^2) \qquad \text{for } k = 2, 3, \ldots.$$

Summing the last equation over the indicated range of k, adding the previous equation, and then substituting Eq. (12) results in

$$\sum_{k=1}^{\infty} E(T_{1k1}^2) = \frac{(1 - \rho_2)E(P_1^2)}{(1 - \rho_1)^2(1 - \rho)} + \frac{\rho_1^2}{(1 - \rho_1)^2} \sum_{k=1}^{\infty} E(T_{2k1}^2).$$

A second equation is produced from the relationship

$$E(T_{2k1}^2) = \frac{\lambda_2 E(P_2^2)}{(1 - \rho_2)^3} E(T_{1k1}) + \frac{\rho_2^2}{(1 - \rho_2)^2} E(T_{1k1}^2), \qquad \text{for } k = 1, 2, \ldots,$$

which, when it is summed over $k = 1, 2, \ldots$, and when

$$\frac{(1 - \rho_2)E(P_1)}{1 - \rho}$$

is then substituted for $\sum_{k=1}^{\infty} E(T_{1k1})$, yields

$$\sum_{k=1}^{\infty} E(T_{2k1}^2) = \frac{\lambda_2 E(P_1)E(P_2^2)}{(1 - \rho_2)^2(1 - \rho)} + \frac{\rho_2^2}{(1 - \rho_2)^2} \sum_{k=1}^{\infty} E(T_{1k1}^2).$$

These two simultaneous linear equations in the sums of the second moments are solved to produce

$$\sum_{k=1}^{\infty} E(T_{1k1}^2) = \frac{(1 - \rho_2)^3 E(P_1^2) + (\lambda_2/\lambda_1)\rho_1^3 E(P_2^2)}{(1 - \rho)^2((1 - \rho_1)(1 - \rho_2) + \rho_1\rho_2)}, \tag{14}$$

$$\sum_{k=1}^{\infty} E(T_{2k1}^2) = \frac{\rho_2^2(1 - \rho_2)E(P_1^2) + (\lambda_2/\lambda_1)(1 - \rho_1)^2\rho_1 E(P_2^2)}{(1 - \rho)^2((1 - \rho_1)(1 - \rho_2) + \rho_1\rho_2)}. \tag{15}$$

By interchanging subscripts in Eq. (14), we have

$$\sum_{k=1}^{\infty} E(T_{2k2}^2) = \frac{(1 - \rho_1)^3 E(P_2^2) + (\lambda_1/\lambda_2)\rho_2^3 E(P_1^2)}{(1 - \rho)^2((1 - \rho_1)(1 - \rho_2) + \rho_1\rho_2)}. \tag{16}$$

To obtain $E(F_1)$ now is a matter of substituting Eqs. (12), (13), (15), and (16) into (11a). Thus

$$E(F_1) = E(P_1) + \frac{\lambda_1 E(P_1^2)}{2(1 - \rho_1)} + \frac{\lambda_1\rho_2^2 E(P_1^2) + \lambda_2(1 - \rho_1)^2 E(P_2^2)}{2(1 - \rho_1)(1 - \rho)((1 - \rho_1)(1 - \rho_2) + \rho_1\rho_2)}, \tag{17}$$

and, by symmetry,

$$E(F_2) = E(P_2) + \frac{\lambda_2 E(P_2^2)}{2(1 - \rho_2)} + \frac{\lambda_1(1 - \rho_2)^2 E(P_1^2) + \lambda_2\rho_1^2 E(P_2^2)}{2(1 - \rho_2)(1 - \rho)((1 - \rho_1)(1 - \rho_2) + \rho_1\rho_2)}. \tag{18}$$

One can conclude from these results that such a system has some rather curious properties. If both classes have the same processing-time distribution, then the sequencing mechanism is entirely independent of processing-time and the overall mean flow-time is the same as that obtained under simple FCFS (Section 8–4). If both classes also have the same distribution of interarrival-times, then the model is completely symmetric with respect to class, and each class has the same mean flow-time. However, as one introduces asymmetry into the model, the results are not what might be expected. While maintaining the same overall arrival rate, if arrivals to class α are more frequent than arrivals to class β, this discipline tends to favor α in the sense that the mean flow-time for jobs in α is less than the overall mean flow-time and, of course, less than the mean for jobs in β. Moreover, if jobs of α and β are equally frequent and class α is favored by its jobs having processing-times with a smaller mean than those for jobs of β, the result is just the opposite: the mean flow-time is less for β than for α. The jobs of class α clear the machine rapidly once they attain access, but must wait for long periods while jobs of class β are served. A nonstochastic analogy might help one to understand this phenomenon. Suppose that two identical water tanks are being continuously filled at the same rate. Each has an outlet with a shutoff valve. The two tanks empty into a common drain and the rule is that one or the other, but not both, of the tanks is to be draining at any instant. The procedure is to leave the outlet open on one tank until that tank is empty, then switch to the other until it is empty, etc. It is interesting to discover that if the outlet capacities of the two tanks are different, the average water level will be *higher* in the tank with the *greater* outlet capacity.

The overall mean flow-time in an asymmetric case can be either better or worse than that for simple FCFS sequencing. When processing-times are exponentially distributed with means $1/\mu_\alpha$ and $1/\mu_\beta$, respectively, with $\mu_\alpha > \mu_\beta$, the alternating-priorities discipline gives smaller overall mean flow-time than FCFS if and only if

$$\frac{1 - \rho_\alpha}{1 - \rho_\beta} > \frac{\mu_\beta}{\mu_\alpha} \quad \text{or} \quad (\mu_\alpha - \mu_\beta) > \lambda_\alpha - \lambda_\beta.$$

The alternating-priority discipline is in one respect optimal, in that for two classes of jobs this discipline results in fewer phases than any other discipline that does not permit inserted idle-time. This means that the frequency of changeover from one class to another is minimized and this becomes most important when setup-times are introduced in Section 9–4.

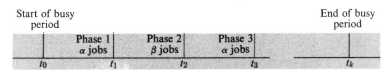

Start of busy period

End of busy period

Phase 1	Phase 2	Phase 3	
α jobs	β jobs	α jobs	
t_0	t_1	t_2	t_3	t_k

Fig. 9–3. Phases of a type α busy period for alternating-priority discipline.

This property is easily shown. Suppose that the alternating-priority rule is applied in a two-class system. Consider a busy period that begins with the arrival of an α job at t_0, as shown in Fig. 9–3. Phase 1, consisting of the processing of α jobs, ends at t_1; phase 2, the processing of β jobs, ends at t_2, etc., until the busy period ends at t_k with the termination of the kth phase. Now suppose that another discipline, **R**, different from alternating priority, was used during this busy period. Under **R**, phase 1 ends at t_1', phase 2 ends at t_2', etc., until the final jth phase ends at t_j'. Observe that $t_k = t_j'$, since the length of a busy period is independent of discipline so long as the discipline does not permit the insertion of idle-time. In general, there would be a different number of phases under the two rules, so that $k \neq j$. What must be shown is that, in fact, $k \leq j$.

Note that either rule starts with the α job that initiates the busy period and that under either rule all the odd-numbered phases consist of α jobs and the even-numbered phases consist of β jobs. Now compare the nature of phase 1 under the two rules. Under the alternating-priority rule, phase 1 ends when there are simply no more α jobs available for processing. The machine must either change over to β or go idle. Since the latter is prohibited, the processing of a β job is begun. The discipline $\mathbf{R}$ can differ from alternating priorities in the order in which jobs are processed within phase 1 (but this would not affect the total length of the phase) or by terminating phase 1 when there is still α work to do. What $\mathbf{R}$ cannot do is extend phase 1 beyond t_1, for there simply is not any α work available at t_1. The conclusion is that for any $\mathbf{R}$, $t_1' \leqq t_1$. By a similar argument, $\mathbf{R}$ cannot extend phase 2 beyond t_2 so that $t_2' \leqq t_2$. Continuing in this manner, $t_i' \leqq t_i$ for $1 \leqq i \leqq k$, so that $\mathbf{R}$ could divide the busy period into more than, but not fewer than, k phases.

If the insertion of deliberate idle-time in a busy period is permitted, then the situation is rather different. One must then consider rules that prolong an α phase to await the arrival for further α work even though there is β work waiting to be processed. This is analogous to the contrast between delay and nondelay schedules in Section 6–6 and appears again in the experimental investigation in Section 11–3. In theory, this sort of inserted idle-time is useful and one can construct examples to show that it is necessary to achieve a good schedule. In practice it is exceedingly difficult to contrive a procedure that is sufficiently frugal with inserted idle-time, and results are often disastrous.

The distribution of the number of changeovers can be studied by numerical methods. Let $E(N_{11})$ be the expected number of type 1 phases in a type 1 busy period. Having exactly k type 1 phases in a busy period corresponds to the event

$$\{T_{1k1} > 0, T_{1(k+1)1} = 0\}.$$

The probability that a type 1 phase is of zero length is the probability that there are no class 1 arrivals during the preceding type 2 phase. Letting $H_{1kj}(t)$ be the distribution function of T_{ikj},

$$\text{Prob}\,(T_{1k1} = 0) = \int_0^\infty e^{-\lambda_1 t}\,dH_{2(k-1)1}(t) = \eta_{2(k-1)1}(\lambda_1).$$

This probability includes the event $\{T_{2(k-1)1} = 0\}$. The probability that there are exactly k type 1 phases in a type 1 busy period is

$$\text{Prob}\,(N_{11} = k) = \text{Prob}\,(T_{1(k+1)1} = 0) - \text{Prob}\,(T_{1k1} = 0)$$
$$= \eta_{2k1}(\lambda_1) - \eta_{2(k-1)1}(\lambda_1).$$

From this, one obtains

$$\sum_{k=1}^q k\,\text{Prob}\,(N_{11} = k) = \eta_{2q1}(\lambda_1) + \sum_{k=1}^{q-1} \big(\eta_{2q1}(\lambda_1) - \eta_{2k1}(\lambda_1)\big).$$

If the system is unsaturated, $\lim_{q\to\infty}\eta_{2q1}(\lambda_1) = 1$. Taking the limit in the equation above produces

$$E(N_{11}) = 1 + \sum_{k=1}^\infty \big(1 - \eta_{2k1}(\lambda_2)\big). \tag{19}$$

A similar derivation shows that the mean number of type 2 phases in a type 1 busy period is

$$E(N_{21}) = \sum_{k=1}^{\infty} \left(1 - \eta_{1k1}(\lambda_2)\right). \tag{20}$$

The remaining expectations, $E(N_{12})$ and $E(N_{22})$, may be written from consideration of symmetry. These sums can be appoximated numerically for particular cases by computing $\eta_{111}(\lambda_2)$, $\eta_{211}(\lambda_1)$, $\eta_{121}(\lambda_2)$, ... iteratively from Eq. (8–8).

When there are more than two classes of jobs, there are many ways to choose the next class to receive service at the end of a phase. One method would be to use a fixed cyclic sequence starting with the class of the job that initiates the busy period. A class would simply be passed over in a particular cycle if no jobs of that class were in queue when its turn arrived. Another method would be to assign a priority ranking to the classes so that at the end of a phase the class of lowest index having jobs in the queue would be selected. The type of analysis used above for two classes can be adapted for either of these two models, but even for three classes the algebra involved in obtaining the counterpart of Eqs. (14) and (15) becomes monumental.

9-3 FIRST-COME, FIRST-SERVED WITH SETUP CLASSES

The FCFS rule operating with setup-time requirements and setup classes was first studied by Gaver [59]. In his model, setup is not incurred when the initial job of a busy period is of the same class as the terminal job of the previous busy period. Gaver gives procedures for determining the expected waiting-time for each class of a two-class system. However, a much simpler analysis yields the overall waiting-time and utilization for the special case of a completely symmetric r-class system.

Let there be r setup classes and let jobs be processed in FCFS order, regardless of their setup class. We assume that $\lambda_i = \lambda/r$ for $i = 1, \ldots, r$, and that $G_i(p) = G(p)$ and $K_i(s) = K(s)$ for $i = 1, \ldots, r$. Then the machine-time requirements of a typical job, P_m, are given by

$$P_m = \begin{cases} P & \text{with probability } 1/r, \\ S + P & \text{with probability } (r-1)/r. \end{cases}$$

The random variable P_m plays the role of a processing-time in the analysis of the FCFS rule in Section 8–4. The system utilization is $\lambda E(P_m)$, or

$$\overline{U} = \rho + \left(\frac{r-1}{r}\right) \lambda E(S).$$

The Laplace transform and first moment of waiting-time are obtained from Eqs. (8–15) and (8–15a):

$$\alpha(z) = E(e^{-zW}) = \frac{(1 - \overline{U})z}{\lambda \gamma_m(z) - \lambda + z}, \tag{21}$$

$$E(W) = \frac{\lambda}{2(1 - \overline{U})} \left(E(P^2) + \left(\frac{r-1}{r}\right)\left(2E(P)E(S) + E(S^2)\right)\right). \tag{21a}$$

The expected flow-time is $E(W) + \overline{U}/\lambda$.

The solution to the nonsymmetric problem is complicated by the necessity of determining the r joint probabilities: Prob (machine is idle and the last job was class k). These joint probabilities sum to $1 - \bar{U}$, which is easily found, but the individual probabilities are difficult to obtain. For the nonsymmetric case in which there are two classes the utilization is

$$\bar{U} = \rho + \frac{\lambda_1 \lambda_2}{\lambda} \left(E(S_1) + E(S_2) \right). \tag{22}$$

9-4 ALTERNATING-PRIORITY DISCIPLINES WITH SETUP-TIMES

The introduction of setup-times lengthens the phases but does not change the character or properties of the alternating-priority discipline. It remains true that this rule minimizes the number of phases (assuming inserted idle-time is not permitted). It also minimizes the total setup-time and has a rather remarkable automatic compensation for setup-times that makes it essentially impervious to setup saturation.

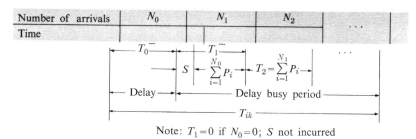

Note: $T_1 = 0$ if $N_0 = 0$; S not incurred

Figure 9-4

The analysis of flow-time distribution is similar to that in Section 9-2. It will be assumed that every phase, including the first phase of a busy period, starts with a setup-time, the distribution of which depends on the class of jobs to be processed within the phase. The first phase of a type 1 busy period is a delay cycle for which the delay portion is the sum of a type 1 setup-time and a class 1 processing-time. These two random variables are independent, and the analog of Eq. (7) is

$$E(e^{-zF_1} \mid 11) = \frac{\gamma_1(z)\big(1 - \kappa_1(z)\gamma_1(z)\big)}{m_{11}\big(\lambda_1 \gamma_1(z) - \lambda_1 + z\big)} .$$

The counterparts of Eqs. (8) and (9), which describe the flow-times of class 1 jobs that arrive during other phases, are obtained by retracing the derivations of Eqs. (8–10) and (8–18) with appropriate modifications to account for setup-times. With a possible setup at the outset of interval number 1, Fig. 8–5 is changed to the situation shown in Fig. 9–4. The setup will change the distribution of T_1 and the flow-time of jobs which arrive in interval 0 and are processed in interval 1.

The Laplace transform associated with the distribution of flow-time for jobs that arrive during interval 0 is obtained in a manner similar to that for Eq. (8–12):

$$E(e^{-zF} \mid j = 0) = \frac{\kappa(z)\gamma(z)\big(\eta_0(\lambda - \lambda\gamma(z)) - \eta_0(z)\big)}{E(T_0)(\lambda\gamma(z) - \lambda + z)} .$$

The analog of Eq. (8–12) is written in this form since it is not true that $\eta_1(z) = \eta_0(\lambda - \lambda\gamma(z))$ for this situation. To obtain $\eta_1(z)$ in terms of $\eta_0(z)$, we retrace the steps leading to Eq. (8–10) for $j = 1$:

$$E(e^{-zT_1} \mid T_0 = t, N_0 = n) = \begin{cases} 1 & \text{for } n = 0, \\ \kappa(z)\big(\gamma(z)\big)^n & \text{for } n > 0. \end{cases}$$

Integrating to eliminate the conditioning distributions, we obtain

$$\eta_1(z) = \big(1 - \kappa(z)\big)\eta_0(z) + \kappa(z)\eta_0\big(\lambda - \lambda\gamma(z)\big).$$

For $j > 1$, Eqs. (8–10) and (8–12) still hold, as well as the rest of Section 8–4. Therefore

$$E(e^{-zF}) = \frac{E(T_0)}{E(T_c)} E(e^{-zF} \mid j = 0) + \frac{\gamma(z)\big(1 - \eta_1(z)\big)}{E(T_c)\big(\lambda\gamma(z) - \lambda + z\big)}$$

$$= \frac{\gamma(z)}{E(T_c)\big(\lambda\gamma(z) - \lambda + z\big)} \big(\kappa(z)\big(\eta_0(\lambda - \lambda\gamma(z)) - \eta_0(z)\big) + 1 - \eta_1(z)\big).$$

Substituting $\eta_1(z)$, given above, into this equation gives the analog of Eq. (8–18) as

$$\beta_c(z) = \frac{\gamma(z)}{E(T_c)\big(\lambda\gamma(z) - \lambda + z\big)} \big(1 - \eta_0(\lambda) + \kappa(z)\big(\eta_0(\lambda) - \eta_0(z)\big)\big). \tag{23}$$

This equation gives the equations comparable to Eqs. (8) and (9) as

$$E(e^{-zF_1} \mid k1) = \frac{\gamma_1(z)\big(1 - \eta_{2(k-1)1}(\lambda_1) + \kappa_1(z)\big(\eta_{2(k-1)1}(\lambda_1) - \eta_{2(k-1)1}(z)\big)\big)}{m_{k1}\big(\lambda_1\gamma_1(z) - \lambda_1 + z\big)},$$

$$E(e^{-zF_1} \mid k2) = \frac{\gamma_1(z)\big(1 - \eta_{2k2}(\lambda_1) + \kappa_1(z)\big(\eta_{2k2}(\lambda_1) - \eta_{2k2}(z)\big)\big)}{m_{k2}\big(\lambda_1\gamma_1(z) - \lambda_1 + z\big)}.$$

The flow-time of a class 1 job that arrives during an idle period is the sum of a type 1 setup interval and its own processing-time, so that

$$E(e^{-zF_1} \mid 0) = \kappa_1(z)\gamma_1(z).$$

The unconditional Laplace transform associated with the distribution of flow-times for class 1 jobs is again found by using the state probabilities to weight the conditional transforms. It can be shown that Eq. (10), $\pi_{kj} = \pi_0\lambda_j m_{kj}$, holds in the setup case. This gives

$$\beta_1(z) = \frac{\pi_0\gamma_1(z)}{\lambda_1\gamma_1(z) - \lambda_1 + z} \bigg(\lambda_1 - \lambda_1\kappa_1(z) + z\kappa_1(z) + \lambda_1 \sum_{k=1}^{\infty} \big(1 - \eta_{2k1}(\lambda_1)\big)$$

$$+ \lambda_1\kappa_1(z) \sum_{k=1}^{\infty} \big(\eta_{2k1}(\lambda_1) - \eta_{2k1}(z)\big)$$

$$+ \lambda_2 \sum_{k=1}^{\infty} \big(1 - \eta_{2k2}(\lambda_1)\big)$$

$$+ \lambda_2\kappa_1(z) \sum_{k=1}^{\infty} \big(\eta_{2k2}(\lambda_2) - \eta_{2k2}(z)\big)\bigg).$$

When we use the factors $E(N_{ij})$ to represent the sums which defined them in Section 9–2, the mean flow-time for class 1 jobs is

$$
\begin{aligned}
E(F_1) = \pi_0 \Bigg(& E(P_1) + E(S_1) + \frac{2(1 - \rho_1)E(P_1) + \lambda_1 E(P_1^2)}{2(1 - \rho_1)^2} \\
& \times \Bigg(\rho_1 + \lambda_1 \sum_{k=1}^{\infty} E(T_{2k1}) + \lambda_2 \sum_{k=1}^{\infty} E(T_{2k2}) \\
& + \lambda_1 E(N_{11})E(S_1) + \lambda_2 E(N_{12})E(S_1) \Bigg) \\
& + \frac{1}{2(1 - \rho_1)} \Bigg(\lambda_1 E(P_1^2) + 2\rho_1 E(S_1) \\
& + \lambda_1 \sum_{k=1}^{\infty} E(T_{2k1}^2) + \lambda_2 \sum_{k=1}^{\infty} E(T_{2k2}^2) \\
& + 2\lambda_1 E(S_1) \sum_{k=1}^{\infty} E(T_{2k1}) + 2\lambda_2 E(S_1) \sum_{k=1}^{\infty} E(T_{2k2}) \\
& + \lambda_1 E(N_{11})E(S_1^2) + \lambda_2 E(N_{12})E(S_1^2) \Bigg) \Bigg).
\end{aligned}
\tag{24}
$$

To complete the solution for $E(F_1)$, one must determine the values of π_0, the expected number of class 1 setups in each type of busy period, and the sums of the moments of the type 2 phase lengths. The first step is to calculate numerically the values of the $E(N_{ij})$'s from Eqs. (19) and (20) derived in Section 9–2. Counterparts of Eqs. (8–7) and (8–8), (8–8a), and (8–8b) with adjustments for setup intervals are required.

The first phase of a busy period initiated by the arrival of a class j job is a delay cycle for which the delay portion is the sum of a type j setup interval and a class j processing-time. It follows that the Laplace transform describing the distribution of the length of the first phase of a type j busy period is

$$
E(e^{-zT_{j1j}}) = \eta_{j1j}(z) = \eta_j(z)\kappa_j\big(z + \lambda_j - \lambda_j\eta_j(z)\big),
$$

where $\eta_j(z)$ is the solution of

$$
\eta_j(z) = \gamma_j\big(z + \lambda_j - \lambda_j\eta_j(z)\big).
$$

For subsequent phases, consider again the problem of the delay cycle with setup-time which was formulated to obtain Eq. (23). Let $\eta_b(z)$ be the Laplace transform describing the distribution of the length of the busy-period portion of a cycle, including the setup-time if a setup is required. The probability that there are n jobs in queue at the end of a setup-time, given the length of a delay portion of the cycle and the length of the setup, is, for $n > 0$,

$$
\begin{aligned}
\text{Prob } (N = n \mid T_0 = t, S = s) &= \sum_{x=1}^{n} \frac{(\lambda t)^x}{x!} \frac{(\lambda s)^{n-x}}{(n-x)!} e^{-\lambda(t+s)} \\
&= \left(\frac{(\lambda(t + s))^n}{n!} - \frac{(\lambda s)^n}{n!} \right) e^{-\lambda(t+s)}.
\end{aligned}
$$

On the other hand, the conditional probability that T_b is zero is $e^{-\lambda t}$. Therefore

$$E(e^{-zT_b} \mid T_0 = t, S = s) = e^{-\lambda(t+s)} \sum_{n=1}^{\infty} \frac{(\lambda\eta(z))^n}{n!} \left((t+s)^n - s^n\right)e^{-zs} + e^{-\lambda t}$$

$$= e^{-\lambda(t+s)}\left(e^{\lambda(t+s)\eta(z)} - e^{\lambda s\eta(z)}\right)e^{-zs} + e^{-\lambda t}.$$

Integration with respect to the conditioning distributions gives

$$\eta_b(z) = \eta_0(z) + \kappa\left(z + \lambda - \lambda\eta(z)\right)\left(\eta_0(\lambda - \lambda\eta(z)) - \eta_0(\lambda)\right). \tag{25}$$

The subscripts 0 and b represent the two phases in a cycle, so that, for example,

$$\eta_{1k1}(z) = \eta_{2(k-1)1}(z)$$
$$+ \kappa_1\left(z + \lambda_1 - \lambda_1\eta_1(z)\right)\left(\eta_{2(k-1)1}(\lambda_1 - \lambda_1\eta_1(z)) - \eta_{2(k-1)1}(\lambda_1)\right).$$

Differentiation of Eq. (25) yields the moments of a phase length as a function of the moments of the length of the previous phase. The sums appearing in Eq. (24) may be computed by the method used to derive Eqs. (12), (13), (15), and (16). The sums of first moments of the type 2 phases are

$$\sum_{k=1}^{\infty} E(T_{2k1}) = \frac{\rho_2}{1 - \rho_1 - \rho_2}\left(E(P_1) + E(N_{11})E(S_1)\right) + \frac{1 - \rho_1}{1 - \rho_1 - \rho_2}E(N_{21})E(S_2),$$
$$\tag{26}$$

$$\sum_{k=1}^{\infty} E(T_{2k2}) = \frac{1 - \rho_1}{1 - \rho_1 - \rho_2}\left(E(P_2) + E(N_{22})E(S_2)\right) + \frac{\rho_2}{1 - \rho_1 - \rho_2}E(N_{12})E(S_1).$$
$$\tag{27}$$

The sums of second moments of type 1 and type 2 phases in a type 1 cycle are obtainable as the solutions to the pair of equations

$$\sum_{k=1}^{\infty} E(T_{1k1}^2) - \frac{\rho_1^2}{(1 - \rho_1)^2} \sum_{k=1}^{\infty} E(T_{2k1}^2)$$

$$= \frac{(1 - \rho_2)E(P_1^2)}{(1 - \rho_1)^2(1 - \rho)} + \frac{2(1 - \rho_2)E(P_1)E(S_1)}{(1 - \rho_1)(1 - \rho)} + \frac{2\rho_1\rho_2(E(S_1))^2}{(1 - \rho_1)^2(1 - \rho)}E(N_{11})$$

$$+ \frac{E(S_1^2)}{(1 - \rho_1)^2}E(N_{11}) + \frac{\lambda_1 E(P_1^2)}{(1 - \rho_1)^2(1 - \rho)}\left(E(N_{11})E(S_1) + E(N_{21})E(S_2)\right)$$

$$+ \frac{2\rho_1 E(S_1)E(S_2)}{(1 - \rho_1)(1 - \rho)}E(N_{21}),$$

and

$$\sum_{k=1}^{\infty} E(T_{2k1}^2) - \frac{\rho_2^2}{(1 - \rho_2)^2} \sum_{k=1}^{\infty} E(T_{1k1}^2)$$

$$= \frac{\lambda_2 E(P_2^2)E(P_1)}{(1 - \rho_2)^2(1 - \rho)} + \frac{2\rho_2 E(P_1)E(S_2)}{(1 - \rho_2)(1 - \rho)} + \frac{2\rho_1\rho_2(E(S_2))^2}{(1 - \rho_2)^2(1 - \rho)}E(N_{21})$$

$$+ \frac{E(S_2^2)}{(1 - \rho_2)^2}E(N_{21}) + \frac{\lambda_2 E(P_2^2)}{(1 - \rho_2)^2(1 - \rho)}\left(E(N_{11})E(S_1) + E(N_{21})E(S_2)\right)$$

$$+ \frac{2\rho_2 E(S_1)E(S_2)}{(1 - \rho_2)(1 - \rho)}E(N_{11}).$$

The expected length of a busy period can be found from Eqs. (26) and (27), so that the probability of the idle state can be determined from

$$\pi_0 = \frac{1/(\lambda_1 + \lambda_2)}{1/(\lambda_1 + \lambda_2) + E(\text{length of busy period})}$$

$$= \frac{1 - \rho_1 - \rho_2}{1 + \big(\lambda_1 E(N_{11}) + \lambda_2 E(N_{12})\big)E(S_1) + \big(\lambda_1 E(N_{21}) + \lambda_2 E(N_{22})\big)E(S_2)}.$$

The most interesting property of the alternating-priority discipline with setup-times is that the system remains unsaturated and the expected waiting-time is finite as long as the setup-times are finite and $\rho < 1$. It is sufficient to show that, under these conditions, the expected length of any phase is finite, since no job remains in the system for more than two phases.

The expected length of the first phase in a type 1 busy period is

$$E(T_{111}) = \frac{E(P_1) + E(S_1)}{1 - \rho_1},$$

and from Eq. (25) we have

$$\left. \frac{-\mathrm{d}\eta_b(z)}{\mathrm{d}z} \right|_{z=0} = \frac{\rho E(T_0)}{1 - \rho} + \frac{E(S)}{1 - \rho}\big(1 - \eta_0(\lambda)\big).$$

Starting with the first of these equations and applying the second one recursively, we find that

$$E(T_{1k1}) = \frac{E(P_1) + E(S_1)}{1 - \rho_1}r^{k-1} + \frac{E(S_1)}{1 - \rho_1}\sum_{i=0}^{k-2}r^i\big(1 - \eta_{2(k-i-1)1}(\lambda_1)\big)$$

$$+ \frac{\rho_1 E(S_2)}{(1 - \rho_1)(1 - \rho_2)}\sum_{i=0}^{k-2}r^i\big(1 - \eta_{1(k-i-1)1}(\lambda_2)\big),$$

where

$$r = \frac{\rho_1\rho_2}{(1 - \rho_1)(1 - \rho_2)}.$$

Since the $\eta(\lambda)$ terms are Laplace transforms, their values are all between zero and one, and the right-hand side would not be made smaller if they were all to be replaced by zero. For $r < 1$, the modified expression would still be finite, and in particular,

$$\lim_{k \to \infty} E(T_{1k1}) \leq \frac{(1 - \rho_2)}{1 - \rho}E(S_1) + \frac{\rho_1}{1 - \rho}E(S_2).$$

Similar statements can be made about $E(T_{2k1})$, etc. Finally, r is less than one if and only if ρ is less than one.

Table 9–1 gives a numerical example of the effect of setup-times on overall mean flow-time and system utilization. It is interesting to observe that the mean flow-time as a function of the mean setup-time, in the range of the data, is very nearly linear. The values of system utilizations show how quickly the utilization approaches 1.0, even though for $\rho < 1$ the system is not saturated. For FCFS, Eq. (21) shows that for most of these conditions the system would be saturated.

Gaver [59] has determined the Laplace transform and the first two moments of the residence-time for low-priority jobs under the following two-class preemptive resume rule with setup-times. An arriving class 1 job may interrupt a class 2 processing-time

Table 9–1

Mean Flow-Time and Utilization for Alternating-Priority Discipline with Setup-Times*

Mean of setup distribution	Mean steady-state flow-times averaged over jobs of both classes				
	$\lambda = 0.5$	0.7	0.8	0.9	0.95
0.0	2.000	3.333	5.000	10.000	20.000
1.0	3.739	5.826	8.369	15.915	30.943
2.0	5.706	8.522	11.912	21.960	41.981
3.0	7.732	11.228	15.446	27.981	53.0
4.0	9.768	13.925	18.967	33.991	64.0
5.0	11.803	16.612	22.479	40.0	
6.0	13.833	19.293	25.987	46.0	
7.0	15.857	21.970	29.491		
8.0	17.878	24.644	33.0		
9.0	19.894	27.316	36.5		

Mean of setup distribution	$\overline{U}$, utilization				
	$\lambda = 0.5$	0.7	0.8	0.9	0.95
0.0	0.5000	0.7000	0.8000	0.9000	0.9500
1.0	0.7390	0.8954	0.8954	0.9476	0.9943
2.0	0.8531	0.9565	0.9824	0.9960	0.9990
3.0	0.9106	0.9790	0.9928	0.9988	0.9998
4.0	0.9421	0.9888	0.9967	0.9996	0.9999
5.0	0.9606	0.9935	0.9983	0.9998	
6.0	0.9721	0.9960	0.9991	0.9999	
7.0	0.9796	0.9974	0.9995		
8.0	0.9847	0.9983	0.9997		
9.0	0.9883	0.9988	0.9998		

* Symmetric problem with two classes of jobs: (1) Exponential processing-time, $\mu_\alpha = \mu_\beta = 1.0$. (2) Exponentially distributed setup-time, common for both classes. (3) Poisson arrivals, $\lambda_\alpha = \lambda_\beta = \lambda/2$.

or setup interval. The processing already accomplished for an interrupted job is not lost, but interrupted setups must be started again from the beginning. After an interruption, a class 2 setup must be successfully completed before the machine may again process class 2 jobs, but repeated setups for the same low-priority job are not resampled. Gaver's expression for the system utilization is

$$\overline{U} = 1 + \rho_2 - \frac{1 - \rho_1}{(1 + \lambda_1 E(S_1))\kappa_2(-\lambda_1)}.$$

Table 9–2

The Effect of Unequal Arrival Rate on the Alternating-Priority
Discipline with Setup-Times*

	Mean flow-time		
λ_α	Overall	α jobs	β jobs
0.00	11.354	—	11.354
0.10	16.327	52.071	11.859
0.20	19.123	36.964	14.025
0.30	20.944	29.233	16.799
0.40	21.847	24.055	20.080
0.45	21.960	21.960	21.960

* Two classes of jobs: (1) Exponential processing-time, $\mu_\alpha = \mu_\beta = 1.0$. (2) Exponentially distributed setup-time, common for both classes, mean $= 2.0$. (3) Poisson arrivals, $\lambda_\alpha + \lambda_\beta = \lambda = 0.9$.

For the special case of exponentially distributed setup-times,

$$\kappa_2(-\lambda_1) = E(e^{\lambda_1 S_2}) = \left(1 - \lambda_1 E(S_2)\right)^{-1}.$$

The superior system capacity under the alternating-priority discipline compared with Gaver's preemptive resume system and FCFS rule can be illustrated by the condition of Table 9–1. Using the last two equations for the preemptive resume rule and Eq. (22) for FCFS with $r = 2$, we may calculate $E(S^*)$, the value of the mean setup-time for which saturation occurs; $E(S^*) = 2(1 - \lambda)/\lambda$ for both disciplines. The values of $E(S^*)$ corresponding to conditions of Table 9–1 are

λ	0.5	0.7	0.8	0.9	0.95,
$E(S^*)$	2.000	0.875	0.500	0.222	0.105.

Consider the effect of varying λ_1 and λ_2 while keeping all other conditions, including λ, the same as in the computation above. For FCFS, the value of $E(S^*)$ increases as the imbalance in arrival rates increases (since the factor, $\lambda_1\lambda_2$, in Eq. (22) with $r = 2$ is maximized by having the two arrival rates equal under the constraint of a fixed sum). However, under Gaver's preemptive resume discipline, the ability of the system to tolerate setups increases with the proportion of low-priority jobs. Under the alternating-priority discipline, imbalance serves to reduce congestion as measured by the mean overall flow-time. This effect is shown in Table 9–2.

CHAPTER 10

MULTIPLE-SERVER QUEUING MODELS

Job-shops and flow-shops can be visualized as queuing networks in which there are arrangements of multiple-servers with customers visiting more than one server before discharge from the system. Unfortunately there are not extensive theoretical results for such systems. The results are principally concerned with identifying the special conditions under which the individual machines are essentially independent and their individual queues can be analyzed separately by methods such as those of Chapters 8 and 9. In general, these conditions are very demanding and restrictive. They require that:

1) The input process be Poisson.

2) The routing of a job—the determination of which machines are to be visited and in what order—be entirely independent of the state of the system.

3) The processing-times be exponentially distributed. (Some slight generalization is permitted.)

4) The order in which jobs are sequenced on a particular machine be independent of (a) the processing-times, (b) the subsequent routing of the jobs, and (c) the knowledge of specific future job arrivals to the machine.

From our point of view, the fourth of these is the most limiting, since it means that the available results are simply not useful in the comparison of interesting sequencing procedures. In fact, a harsh critic could conclude that there are no *network* queuing results. Conditions have been identified under which the individual machines do not behave as an interrelated network and hence can be analyzed as single-machine systems. However, any time the machines behave in a truly interrelated manner there are virtually no applicable theoretical results.

There are three principal types of arrangement of servers that have been considered in the queuing literature. Multiple-channel or parallel-channel queues have a number of identical servers working in parallel to provide a single type of service. This corresponds to a machine group of several interchangeable machines. Tandem queues, or queues in series, correspond to the flow-shop of Chapter 5. Finally, the general-network case in which each customer visits a particular subset of servers, in an order that may be peculiar to that customer, corresponds to the job-shop of Chapter 6. The most important results for these cases are summarized in Sections 10-1, 10-3, and 10-4, respectively.

The analysis of queuing models which allow the input streams for machine groups to be composed partly or entirely of the output streams from other machine groups depends on three key results concerning properties of the Poisson-exponential queuing process. The first two of these are the preservation of the Poisson properties of Poisson streams subject to aggregation and probabilistic branching. These were discussed in Section 8–1. The third property, to be examined in Section 10–2, is that the output of a machine group is Poisson in nature if (1) the input is Poisson, (2) the processing-time is exponential, and (3) the sequencing discipline is independent of the processing-times of the jobs.

10–1 STATE-DEPENDENT COMPLETION RATES; MULTIPLE-CHANNEL QUEUES

Suppose that the input to a machine group is a Poisson stream with rate λ, but that the job-completion rate is dependent on the number of jobs at the group. The queuing process is a generalized birth-death process with Poisson input. For this process, if $N(t)$ represents the number of jobs at the machine group at time t, then

$$\text{Prob}\left(N(t + \Delta t) = n + 1 \mid N(t) = n\right) = \lambda \, \Delta t, \qquad n = 0, 1, \ldots,$$
$$\text{Prob}\left(N(t + \Delta t) = n - 1 \mid N(t) = n\right) = \mu_n \, \Delta t, \qquad n = 1, 2, \ldots.$$

Let p_n be the steady-state probability of n jobs in the system. The steady-state versions of the differential-difference equations that represent the process are equivalent to the equations

$$\lambda p_n = \mu_{n+1} p_{n+1}, \qquad n = 0, 1, \ldots. \tag{1}$$

Applying Eq. (1) recursively shows that

$$p_n = p_0 \frac{\lambda^n}{\prod_{i=1}^{n} \mu_i}. \tag{2}$$

In this expression, p_0 is determined from the requirement that $\sum_{i=0} p_i = 1$, and, for nonsaturation, p_0 must be greater than zero.

Multiple-channel queuing systems are an important special case of this model. Here we assume Poisson input, m identical machines which draw jobs from a common queue, and exponentially distributed processing-time at each machine with mean $1/\mu$. The sequencing procedure must be independent of the jobs' processing-times (say, FCFS). To analyze this case, we specialize the results from the generalized birth-death process by taking $\mu_n = n\mu$ for $n \le m$ and $\mu_n = m\mu$ for $n > m$. Equation (2) becomes

$$p_n = \frac{1}{n!} \left(\frac{\lambda}{\mu}\right)^n p_0, \qquad n = 0, 1, \ldots, m,$$

$$p_n = \frac{m^m}{m!} \left(\frac{\lambda}{m\mu}\right)^n p_0, \qquad n = m + 1, m + 2, \ldots,$$

$$p_0 = \left(\sum_{n=0}^{m-1} \frac{(\lambda/\mu)^n}{n!} + \frac{(\lambda/\mu)^m}{m!(1 - \lambda/m\mu)}\right)^{-1}.$$

The machine group utilization is $\rho = \lambda/m\mu$, which is the amount of processing-time arriving per unit time, (λ/μ), divided by the processing capability per unit time. This can also be established by showing that ρ is the average number of machines

Table 10–1

Expected Flow-Time for Multiple-Channel Queues*

m \ ρ	0.30	0.40	0.50	0.60	0.70	0.80	0.90	0.95	0.98
1	0.429	0.667	1.000	1.500	2.333	4.000	9.000	19.000	49.000
2	0.659	0.952	1.333	1.875	2.745	4.444	9.474	19.487	49.495
3	0.930	1.294	1.737	2.332	3.249	4.989	10.054	20.083	50.100
4	1.216	1.660	2.174	2.831	3.800	5.586	10.690	20.737	50.764
5	1.509	2.040	2.630	3.354	4.382	6.216	11.362	21.428	51.466
6	1.805	2.427	3.099	3.895	4.984	6.871	12.061	22.146	52.194
7	2.103	2.818	3.576	4.448	5.602	7.544	12.780	22.885	52.944
8	2.402	3.212	4.059	5.009	6.231	8.231	13.514	23.639	53.710
9	2.701	3.608	4.546	5.578	6.781	8.929	14.261	24.407	54.489
10	3.001	4.006	5.036	6.152	7.517	9.637	15.019	25.186	55.280
15	4.500	6.001	7.511	9.072	10.829	13.277	18.924	29.202	59.356
20	6.000	8.000	10.004	12.036	14.218	17.024	22.857	33.353	63.570

* Fixed arrival rate, $\lambda = 1$.

being used per unit time, divided by m. This is

$$\frac{1}{m} \left(\sum_{n=0}^{m} n p_n + m \sum_{n=m+1}^{\infty} p_n \right) = \frac{\lambda}{m\mu}.$$

The expected flow-time of a job, $E(F)$, is given by

$$E(F) = \frac{1}{\lambda} \sum_{n=0}^{\infty} n p_n = \frac{\mu(\lambda/\mu)^m}{(m-1)!(m\mu - \lambda)^2} p_0 + \frac{1}{\mu}.$$

Values of $E(F)$ are tabulated in Tables 10–1 and 10–2 for various values of ρ and m. Table 10–1 is based on a fixed arrival rate of $\lambda = 1$, so that for each entry μ is set equal to $(m\rho)^{-1}$. In Table 10–2, the processing rate of each machine is taken to be 1.0, and λ is made equal to $m\rho$.

There are no convenient analytic results for one machine group with more than one machine under any weaker set of assumptions regarding dispatching, although results under weaker distribution assumptions do exist [38].

An interesting variation of the multiple-server queue occurs when waiting lines develop before each of the machines, due to the choice of machines by arriving jobs. Queuing at supermarket checkout stands and bank tellers' counters are examples in which the machine to process a particular job is selected by the job at the time of arrival at the shop.

In actual situations, jobs may switch back and forth between individual queues, a phenomenon called *jockeying*. If an arriving job selects its machine randomly (and independently of the status of the several machines) and jockeying is not allowed, each queue-machine behaves as if it were a single-machine system. If the machine selection is based on more rational grounds, such as the number in queue before each machine

Table 10–2

Expected Flow-Time for Multiple-Channel Queues*

m \ ρ	0.30	0.40	0.50	0.60	0.70	0.80	0.90	0.95	0.98
1	1.429	1.667	2.000	2.500	3.333	5.000	10.000	20.000	50.000
2	1.099	1.190	1.333	1.563	1.961	2.778	5.263	10.256	25.253
3	1.033	1.078	1.158	1.296	1.547	2.079	3.724	7.047	17.041
4	1.013	1.038	1.087	1.179	1.357	1.746	2.969	5.457	12.950
5	1.006	1.020	1.052	1.118	1.252	1.554	2.525	4.511	10.503
6	1.003	1.011	1.033	1.082	1.187	1.431	2.234	3.885	8.877
7	1.001	1.006	1.022	1.059	1.143	1.347	2.029	3.441	7.718
8	1.001	1.004	1.015	1.044	1.113	1.286	1.877	3.110	6.851
9	1.000	1.002	1.010	1.033	1.091	1.240	1.761	2.855	6.178
10	1.000	1.001	1.007	1.025	1.074	1.205	1.669	2.651	5.641
15	1.000	1.000	1.001	1.008	1.031	1.106	1.402	2.049	4.038
20	1.000	1.000	1.000	1.003	1.016	1.064	1.275	1.755	3.243

* Fixed processing rate, $\mu = 1$.

or the expected total processing-time of the jobs in the queues, then the analysis of the system becomes very difficult.

Koenigsberg [113] has studied the case of two dissimilar exponential servers in which a newly arriving job goes to the machine with the shorter queue. A second version of the model allows jockeying when the difference in queue lengths is greater than one. Other investigations of such systems known to the authors are the experimental work of Evans [49] and the analytic development of Kingman [110].

10–2 THE OUTPUT OF A POISSON-EXPONENTIAL QUEUING SYSTEM

In this section we shall establish that the output stream of jobs emerging from a Poisson-exponential queuing system is a Poisson process. This is done first for the single-machine queue and then for the generalized birth-death process, proving the statement for the case of multimachine queues.

Let the time interval elapsing between two successive job completions be denoted by the random variable U having distribution function $D(u)$ and Laplace transform $\delta(z)$, while the random variable V with distribution function $Z(v)$ and Laplace transform $\zeta(z)$ is associated with the interarrival intervals. The notation relating to processing-times $(P, G(p),$ and $\gamma(z))$ is retained from Chapters 8 and 9.

Theorem 10–1 Assume that $E(P) = 1/\mu$ and $E(V) = 1/\lambda$ and that the P's and V's are independent in every sense. For a nonsaturated, single-machine queuing system operating in the steady state, the departure intervals are independently distributed with the distribution function

$$D(u) = 1 - e^{-\lambda u}, \qquad u \geq 0,$$

1) if and only if $Z(v) = 1 - e^{-\lambda v}$, $v \geq 0$,
2) if and only if $G(p) = 1 - e^{-\mu p}$, $p \geq 0$,
3) if and only if the dispatching procedure is independent of the set of processing-times of the jobs waiting for selection.

Proof. Since only a single-machine system is being considered, the probability is $\rho = \lambda/\mu$ that a departing job does not leave the system in the idle state, in which case $U = P$. If a departing job leaves no other jobs behind, the next interdeparture interval is the sum of an interarrival interval and a processing-time, so that with probability $1 - \rho$, $U = V + P$. Then

$$D(u) = \rho G(u) + (1 - \rho)\int_0^u G(u - v)\, dZ(v).$$

In terms of Laplace transforms this is

$$\delta(z) = \rho\gamma(z) + (1 - \rho)\gamma(z)\zeta(z). \tag{3}$$

The hypothesis that P and V are exponentially distributed with parameters μ and λ, respectively, implies that $\gamma(z) = \mu/(\mu + z)$ and $\zeta(z) = \lambda/(\lambda + z)$, so that

$$\delta(z) = \frac{\rho\mu}{\mu + z} + \frac{(1 - \rho)\mu\lambda}{(\mu + z)(\lambda + z)} = \frac{\lambda}{\lambda + z},$$

which shows that U is exponentially distributed with mean $1/\lambda$.

When we again substitute into Eq. (3), the assumptions that $\delta(z) = \lambda/(\lambda + z)$ and that $\gamma(z) = \mu/(\mu + z)$ imply that $\zeta(z) = \lambda/(\lambda + z)$; when $\zeta(z) = \delta(z) = \lambda/(\lambda + z)$, then $\gamma(z)$ must necessarily be $\mu/(\mu + z)$.

On the other hand, $\rho\gamma(z) + (1 - \rho)\gamma(z)\zeta(z) \neq \lambda/(\lambda + z)$, if $\gamma(z)$ and $\zeta(z)$ are not both Laplace transforms of exponential distributions. This proof is tedious and is omitted (it involves a detailed comparison of the coefficients of the power series expansion of $\lambda/(\lambda + z)$).

That the lengths of successive departure intervals are independent follows from the Markovian nature of the Poisson-exponential queuing process and from the independence of successive processing-times as selected by the dispatching procedure when there is more than one job in queue. Consider the $(n - 1)$th and nth departure intervals, separated by the departure of the nth job. The length of the nth departure interval can depend on the previous history of the system only by being dependent on the number of jobs present at the moment of the nth job completion. But it will be shown below that the length of the $(n - 1)$th departure interval and the number of jobs present when the nth departure occurs are independent.

Finally, if the dispatching procedure is not independent of processing-times, the time elapsing between two successive job completions occurring within a busy period will not be exponentially distributed. It will be governed by some other distribution which is dependent on the number of jobs present at the earlier completion.

The following theorem shows that the output from the generalized birth-death process, hence the efflux from a multiple-channel machine center, is Poisson and supplies the proof of independence missing from Theorem 10-1. The proof follows a procedure due to Burke [24], omitting some of the finer details. In proving the theorem, we shall employ the following notation:

U = length of a randomly selected departure interval.

$J(t)$ = number of jobs at the machine group t time units after the most recent job completion.

$J(U)$ = number of jobs remaining at the machine group immediately after a job completion.

$F_n(t)$ = Prob $(J(t) = n$ and $U > t)$.

$F(t) = \sum_{n=0}^{\infty} F_n(t)$ = Prob $(U > t)$, the complement of the marginal distribution function of U.

The symbols p_n, λ, and μ_n are as defined in Section 10–1.

Theorem 10–2 For the generalized birth-death process operating in the steady state and unsaturated, the departure intervals are independent and exponentially distributed with mean $1/\lambda$.

Proof. If the process is operating in the steady state, it can be argued (Section 8–5.1) that

$$F_n(0) = p_n.$$

Using the technique of differential-difference equations, we have

$$F_0(t + \Delta t) = (1 - \lambda \Delta t)F_0(t),$$
$$F_n(t + \Delta t) = (1 - \lambda \Delta t)(1 - \mu_n \Delta t)F_n(t) + \lambda \Delta t F_{n-1}(t), \qquad n = 1, 2, \ldots.$$

These yield the differential equations

$$\frac{dF_0(t)}{dt} = -\lambda F_0(t), \qquad \frac{dF_n(t)}{dt} = -(\lambda + \mu_n)F_n(t) + \lambda F_{n-1}(t), \qquad n = 1, 2, \ldots.$$

This system of equations, along with the boundary condition at $t = 0$, is satisfied by the solution

$$F_n(t) = p_n e^{-\lambda t}, \qquad n = 0, 1, \ldots.$$

Since the p_n's sum to one, it follows that $F(t) = e^{-\lambda t}$, or

$$\text{Prob } (U \leq t) = 1 - e^{-\lambda t}.$$

The next step is to establish the independence of $J(U)$ and U. The joint distribution of $J(U)$ and U is obtained from

$$\text{Prob } (J(U) = n, t < U < t + \Delta t) = \mu_{n+1} \Delta t F_{n+1}(t) = \mu_{n+1} p_{n+1} e^{-\lambda t} \Delta t.$$

Substituting Eq. (1) for p_{n+1} gives

$$\text{Prob } (J(U) = n, t < U < t + \Delta t) = p_n \lambda e^{-\lambda t} \Delta t$$
$$= \text{Prob } (J(U) = n) \, \text{Prob } (t < U < t + \Delta t).$$

The independence of successive departure intervals then follows from the remarks made in the proof of Theorem 10–1.

10-3 QUEUES IN SERIES

As defined in Chapter 5, a flow-shop is one in which the machine groups may be numbered so that every job routing specifies an increasing sequence of machine-group identifiers. The input to a machine group consists of inputs from the external source or from machine groups with lower indexes. It is also possible for a job to exit from the shop after being processed on any machine group. Given that λ_k is the input rate to the kth machine group from outside the shop, ϵ_k is the rate at which the job exits from the shop for the kth machine group, and $\Lambda_k = \sum_{j=1}^{k} \lambda_j - \sum_{j=1}^{k-1} \epsilon_j$ is the rate at which the job enters the kth machine group, we have the situation shown in Fig. 10-1.

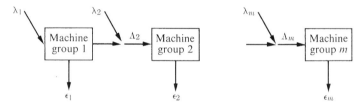

Figure 10-1

Assuming that in this type of flow-shop a job's routing is probabilistically determined when an operation is completed at a machine group (with some probability, say h_k, that it exits from the shop after it is processed at machine group k), and assuming that the m input streams are Poisson in nature and that the processing-times are exponentially distributed (not necessarily with the same distribution over machine groups), then this series of machine groups can be analyzed as though each were an independent machine group with input rate Λ_k.

Formally, let $\bar{l} = (l_1, l_2, \ldots, l_m)$ be a vector representing the number of jobs at each of the m machine groups, $p(\bar{l})$ be the probability of observing the state $\bar{l}$, and $p_k(l_k)$ be the probability that the number of jobs at machine group k is l_k. The result is that

$$p(\bar{l}) = \prod_{k=1}^{m} p_k(l_k), \tag{4}$$

where $p_k(l_k)$ is obtained from the generalized birth-death process with constant parameter of Λ_k.

This result follows immediately from the results of Sections 8-1 and 10-2. The input to the kth machine group consists of the aggregation of Poisson arrivals from the outside combined with a random selection of jobs emerging from machine groups with lower indexes. If the outputs from the preceding machine groups are Poisson, then the total input to the kth machine group is also Poisson. By the conclusions of Section 10-2, we know that the assumptions of exponentially distributed processing-times and processing-time-independent dispatching at the kth machine group will also be Poisson. This reasoning applied to each machine group in sequence proves Eq. (4). A rigorous development of this result has been given by Reich [167].

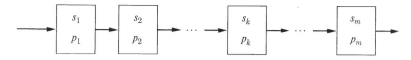

Figure 10–2

Flow-shops with constant processing-time at a machine group and with every job visiting every machine group have been studied by Avi-Itzhak [6] and by Friedman [55]. The situation is shown in Fig. 10–2, where s_k is the number of servers in machine group k and p_k is the known and constant processing-time at group k.

The first machine group must have a queue size which will accommodate all arriving jobs, but the allowable queue sizes in front of the other machine groups may be arbitrary. If a job is finished at machine group k and the queue before group $k + 1$ is full, then the machine on which it was processed cannot be used until the job can be put into the queue of group $k + 1$.

If only the number in the system is of interest, there is no need to consider other than a FCFS discipline, since processing-times are constant. However, it is possible that the arrangement of machine groups could affect the waiting-time of a job, especially the time that could be spent at a machine group when the queue of the following machine group is full. Interest has centered on conditions under which the ordering of the machine groups will not affect the flow-time of a job. The conditions under which it is known that a job's flow-time is independent of the machine group arrangement are:

1) Arbitrary input of jobs to the first machine group.

2) (a) Same number of machines per machine group and arbitrary allowable intermediate queue sizes (Avi-Itzhak [6]). (b) An arbitrary number of machines per machine group and unlimited intermediate queue sizes (Friedman [55]).

Conditions which are intermediate to extremes of 2(a) and 2(b) exist in great numbers. The proofs used by Avi-Itzhak and Friedman are quite dissimilar, so that there appears to be no immediate hope of extending the conditions under which flow-time order independence holds for intermediate conditions, yet attempts by the authors at simple counterexamples have been fruitless.

10–4 GENERAL QUEUE NETWORKS

Certainly the most important analytic result in the analysis of job shops is the decomposition theorem due to Jackson [100]. This gives sufficient conditions under which a general network of queues may be treated as an aggregation of independent queues. In essence, the Jackson result is that (1) *if* the input to the shop is Poisson, (2) *if* the routing of jobs is determined by a probability transition matrix, (3) *if* the processing rates are exponential (possibly depending on the number of jobs at the machine group), and (4) *if* the dispatching rule is independent of a job's routing and processing-times, *then* each machine group behaves, in a probabilistic sense, as an independent machine group of the type covered in Section 10–1.

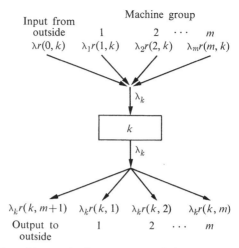

Figure 10–3

It should be noted that this result gives only a sufficient condition for decomposition; the corresponding necessary condition is not known. Moreover, most of the interesting dispatching procedures do not satisfy the conditions for decomposition.

Our proof of the decomposition theorem is similar to that given in Jackson [100]. For a given machine group, say k, the general model is as shown in Fig. 10–3. The parameters indicated on the diagram and used in the proof are the following.

λ = input rate to shop,

λ_k = input rate to kth machine group, $k = 1, 2, \ldots, m$,

$r(j, k)$ = element of routing transition matrix.

These elements are as follows:

$r(0, k)$ = Prob (group k is required for first operation),

$r(k, m + 1)$ = Prob (if machine k appears on routing at operation J, it is last operation of the routing),

$r(j, k)$ = Prob (if operation J requires group j, operation $J + 1$ requires group k), and

$r(k, k) = 0$.

The λ_k's must satisfy $\lambda_k = \lambda r(0, k) + \sum_{j=1}^{m} \lambda_j r(j, k)$.

Let

$\bar{l} = (l_1, l_2, \ldots, l_m)$ be the state vector indicating that there are l_k jobs at machine group k ($k = 1, 2, \ldots, m$),

$\mu(k, l_k)$ be the completion rate at group k when there are l_k jobs at group k,

$p(\bar{l})$ be the steady-state probability of the state vector $\bar{l}$,

$p_k(l_k)$ be the steady-state marginal probability of the state l_k at machine group k.

Theorem 10–3 (Jackson, 1963 [100]) If the conditions above are satisfied, then

$$p(\bar{l}) = \prod_{k=1}^{m} p_k(l_k), \tag{5}$$

where

$$p_k(l_k) = p_k(0) \prod_{i=1}^{l_k} \frac{\lambda_k}{\mu(k, i)}. \tag{6}$$

Proof. This result is established by use of the technique of differential-difference equations. In a small increment of time, the system state $\bar{l}$ may:

1) remain the same,

2) have an arrival from the outside to group k,

3) have a departure to the outside from group k, or

4) have a completion at group j which then goes to group k.

To facilitate writing down the steady-state equations, let

$$\bar{l}(k^-) = (l_1, l_2, \ldots, l_{k-1}, l_k - 1, l_{k+1}, \ldots, l_m),$$
$$\bar{l}(k^+) = (l_1, l_2, \ldots, l_{k-1}, l_k + 1, l_{k+1}, \ldots, l_m),$$
$$\bar{l}(j^+, k^-) = (l_1, l_2, \ldots, l_j + 1, \ldots, l_k - 1, \ldots, l_m).$$

The steady-state equations are

$$\left(\lambda + \sum_{k=1}^{m} \mu(k, l_k) \right) p(\bar{l}) = \lambda \sum_{k=1}^{m} r(0, k) p(\bar{l}(k^-))$$

$$+ \sum_{k=1}^{m} \mu(k, l_k + 1) r(k, m + 1) p(\bar{l}(k^+))$$

$$+ \sum_{j=1}^{m} \sum_{k=1}^{m} \mu(j, l_j + 1) r(j, k) p(\bar{l}(j^+, k^-)). \tag{7}$$

The left-hand side of Eq. (7) is the steady-state probability of being in state $\bar{l}$ multiplied by the total rate at which the system tends to move out of state $\bar{l}$. The right-hand side contains one term for every state from which state $\bar{l}$ can be reached in a single transition. Each of these terms consists of the appropriate state probability multiplied by the rate of transition from the given state to state $\bar{l}$.

The actual proof consists of showing that Eq. (5), written in terms of Eq. (6), satisfies Eq. (7). This can be verified by direct substitution. When the substitution is made, each term on both sides contains the factor $\prod_{k=1}^{m} p_k(0)$, and this cancels. It is interesting to note the manner in which the terms of the right-hand side match with the terms of the left-hand side. To do this, each term on the right is adjusted to a coefficient times $p(\bar{l})$. For example, from Eq. (6), we have

$$p_k(l_k - 1) = p_k(0) \prod_{i=1}^{l_k - 1} \frac{\lambda_k}{\mu(k, i)} = \frac{\mu(k, l_k)}{\lambda_k} p(l_k),$$

so that substitution into Eq. (5) gives

$$p(\bar{l}(k^-)) = \frac{\mu(k, l_k)}{\lambda_k} p(\bar{l}).$$

Then

$$\lambda \sum_{k=1}^{m} r(0, k) p(\bar{l}(k^-)) = p(\bar{l}) \lambda \sum_{k=1}^{m} r(0, k) \frac{\mu(k, l_k)}{\lambda_k},$$

and similarly,

$$\sum_{k=1}^{m} \mu(k, l_k + 1) r(k, m + 1) p(\bar{l}(k^+)) = p(\bar{l}) \sum_{k=1}^{m} \lambda_k r(k, m + 1),$$

$$\sum_{j=1}^{m} \sum_{k=1}^{m} \mu(j, l_j + 1) r(j, k) p(\bar{l}(j^+, k^-)) = p(\bar{l}) \sum_{j=1}^{m} \sum_{k=1}^{m} \frac{\lambda_j}{\lambda_k} \mu(k, l_k) r(j, k).$$

Canceling $p(\bar{l})$ we have

$$\lambda + \sum_{k=1}^{m} \mu(k, l_k) = \lambda \sum_{k=1}^{m} \frac{r(0, k) \mu(k, l_k)}{\lambda_k}$$

$$+ \sum_{k=1}^{m} \lambda_k r(k, m + 1) + \sum_{j=1}^{m} \sum_{k=1}^{m} \frac{\lambda_j \mu(k, l_k) r(j, k)}{\lambda_k}.$$

Cancellation of terms arises from the following:

1) No input to the shop but an output from the shop,

$$\sum_{k=1}^{m} \lambda_k r(k, m + 1) = \sum_{k=1}^{m} \lambda_k \left(1 - \sum_{j=1}^{m} r(k, j)\right) = \sum_{k=1}^{m} \lambda_k - \sum_{k=1}^{m} \sum_{j=1}^{m} \lambda_k r(k, j)$$

$$= \sum_{k=1}^{m} \lambda_k - \sum_{j=1}^{m} (\lambda_j - \lambda r(0, j)) = \lambda \sum_{k=1}^{m} r(0, k) = \lambda.$$

2) No completion in the shop but no exit from the shop,

$$\lambda \sum_{k=1}^{m} \frac{r(0, k) \mu(k, l_k)}{\lambda_k} + \sum_{j=1}^{m} \sum_{k=1}^{m} \frac{\lambda_j \mu(k, l_k) r(j, k)}{\lambda_k}$$

$$= \sum_{k=1}^{m} \frac{\mu(k, l_k)}{\lambda_k} \left(\lambda r(0, k) + \sum_{j=1}^{m} \lambda_j r(j, k)\right) = \sum_{k=1}^{m} \mu(k, l_k).$$

CHAPTER 11

EXPERIMENTAL INVESTIGATION OF THE CONTINUOUS JOB-SHOP PROCESS

An alternative to the algebraic and probabilistic methods of the previous chapters, as a means of studying questions of sequencing, would simply be to try various procedures in a real shop and compare actual performances. In principle, one could in this way avoid the simplifying assumptions necessary for the abstract methods and obtain results that were unquestionably realistic. However, the price paid for this realism would be a loss of generality. Furthermore, one could not be sure that results obtained in this way would be generally useful even in the particular shop in which the experiments were performed. Moreover, the difficulties involved in adequately controlling such an experiment, and the costs that would be incurred, have effectively precluded the conduct of any significant amount of research by this method.

However, the advent of the digital computer made it possible to conduct an experimental investigation by simulating the action of jobs, machines, and scheduling procedures with a computer program. Indicators, counters, lists, and files within computer storage can effectively represent the state of a job-shop process, and a program can be written to cause this state to change over time under the action of a particular scheduling procedure. This possibility apparently occurred to a number of different investigators at about the same time (1952–1954), although, since some of the earliest results were never published, it is difficult to establish the chronological order of the investigations. However, a group at General Electric (Evandale); Jackson, Nelson, and Rowe at UCLA [93, 101]; and Baker and Dzielinski at IBM [10] were among the earliest. The UCLA and IBM work greatly influenced later investigations.

The early "job-shop simulators" were programmed in absolute machine language, and then later in symbolic assembly language. Since this was a nontrivial programming task, the models were necessarily simple. By the early 1960's several programming languages intended specifically for this type of simulation had appeared [114] and the investigation of more complex models and scheduling procedures became practical. These programs are also very demanding in both storage and execution-time requirements; very large-scale scientific machines are required to produce significant results. (For example, in one study an IBM 7090 processed only 300 to 700 jobs per minute, depending on the complexity of the scheduling procedure, in a SIMSCRIPT model of a nine-machine shop [32].)

The scheduling problem has actually been approached from two sides by simulated experimentation. On one side are the investigators who have attempted to extend the theoretical results described in the preceding chapters. On the other side are those people faced with actual problems in real shops who have attempted to pretest procedures before installing them in the shop. There has been at least modest success in both cases. On the theoretical side some interesting extensions in dimension and measure of performance have been explored and some comparatively complex procedures have been tested. Of course, one cannot prove theorems experimentally; one can only reject or fail to reject hypotheses. One cannot establish the optimality of a particular procedure, but one can use experimentation to develop progressively more powerful procedures.

Because very few of the simulation studies of real shops have been published, it is difficult to summarize accomplishment in this area. One published report [119] and several others with which we have had contact have produced no great surprises. Results are, in general, consistent with the more abstract experimental work. There is no evidence to suggest that the use of actual shop data and dimensions significantly alters the comparative performance of key procedures. The greatest difficulty in work of this sort is the collection and disciplining of data to describe a sequence of jobs and the characteristics of the shop. In the future, as computer-based production control systems become more common, this handling of data will become less onerous. It is likely that the simulation of actual shops for scheduling purposes will become a routine technique. In fact, such simulation will probably become an integral part of an on-line computer-based production control system. The computing system that already contains a current-status description of the shop will be asked to project ahead, by what is essentially the simulation of the shop operation, so that one can evaluate the effect of a scheduling decision. During periods when the shop is not in actual operation (e.g., third shift or weekend), this system can be used for less immediate and more systematic explorations.

The results presented in this chapter are derived entirely from abstract simulation studies; no real shops or actual data were involved. All these results have been extracted from our own recent investigations. The factors that led to this selection were familiarity, accessibility, and comparable experimental conditions; there is no intent to depreciate the significance of other studies.*

11-1 EXPERIMENTAL CONDITIONS AND PROCEDURES

All the results of the following sections were attempts to measure equilibrium or steady-state performance of a dynamic, randomly routed job-shop process. Except for the models described in Section 11-4, each was a *simple* one-resource job-shop process. Recall from Section 1-2 that this means that:

1) Each machine is continuously available for assignment, without partition into shifts or days, and without intermittent unavailability (breakdown).

2) Jobs are simple sequences of operations; no assembly.

---

* In particular, the investigations of Bakker [12], Carroll [25], LeGrande [119], Nanot [147], and Nelson [154] are recommended for consideration.

3, 4) Only one machine capable of performing a given operation.

5) No preemption.

6) A job can be in process on at most one operation at any point in time; no overlap scheduling.

7) Each machine can handle at most one operation at a time.

Additional restrictions are:

8) No setup time for operations.

9) Instantaneous transfer to next machine (or queue) after an operation has been completed.

Processing-times were random variables obtained from an exponential distribution with mean equal to one. The random variables were observed prior to the "arrival" of the job so that the processing-times were known in advance and this information was available to the scheduling procedure. Interarrival intervals were also exponentially distributed. These random variables were also observed in advance, but information as to the time of arrival of the next job was not made available to the scheduling procedure.

Within a given series of tests, individual runs differed only in the discipline used to select jobs from queue whenever a machine became available for assignment. The simulator always selected, from among those in queue, the job with minimum priority value. Runs differed in that different priority assignment subroutines were substituted. Identical sets of jobs, arriving at exactly the same instants in time, were presented to different selection disciplines and the aggregate performance recorded. In each case, however, recording did not begin until the process had been running long enough to at least approach steady-state conditions, and recording was terminated before the actual end-of-run. The sample sizes reported do not include the additional jobs required for the "run-in" and "run-out."

Priority rules (selection procedures) are defined and results are tabulated in Appendix C. These data are summarized and discussed in the following sections.

11-2 REDUCTION OF MEAN QUEUE
LENGTH AND WORK-IN-PROCESS INVENTORY

For a system that consists of a single machine and queue, there are, in essence, two ways of measuring work-in-process inventory. (a) One can count the jobs waiting (including the job being processed); this is the conventional measure used in queuing literature. (b) One can total the processing-times of the jobs waiting in order to obtain the "work content" of the queue. In a job-shop, when each job entails a sequence of operations to be performed, one can still consider the aggregate number of jobs in the system. However, there are several ways one can generalize the measurement of the work content.

1) p_r, *work remaining*. The sum of the processing-times of all operations not yet completed or in process for all jobs in the shop.

2) p, *total work content*. The sum of the processing-times of all operations of all jobs in the shop.

3) p_c, *work completed*. The sum of the processing-times of all completed operations of all jobs in the shop. *Work completed* is equal to *total work content* minus *work remaining*.

4) p_q, *imminent operation work content*. The sum of the processing-times of the particular operations for which jobs are waiting in queue.

There are differing situations in which each of these measures is relevant. For example, for the flight-line maintenance activity at an Air Force base, the objective is clearly to minimize the number of jobs (aircraft) in the shop. In civilian industry, one would ordinarily be more concerned with value of inventory than with simple count of jobs, so that one of the work-content measurements would be appropriate. In different circumstances one might be more interested in the work completed because of the investment that it represented, or one might be more interested in the work remaining as a measure of backlog or real congestion in the shop.

Similarly, there are various ways in which single-queue selection disciplines can be generalized in a network of queues. For example, the shortest-processing-time concept could be implemented in several ways:

SPT   *Shortest-processing-time*. Job priority equals processing-time of the imminent operation.

LWKR   *Least work remaining*. Job priority equals the sum of the processing-times for all operations not yet performed.

TWORK   *Total work*. Job priority equals the sum of the processing-times of all operations of the job.

Since in these experiments all processing-times were obtained from a common distribution, the total processing-time and the number of operations of a job are correlated attributes. The following rule would have the same intent as LWKR, and would certainly be easier to implement:

FOPNR   *Fewest operations remaining*. Job priority equals the number of operations remaining to be performed on the job.

The rule antithetical to each of these procedures (Section 3–5) was also tested, in one case with surprising results:

MWKR   *Most work remaining*. Job priority equals minus the sum of the processing-times for all operations not yet performed.

All the above procedures are based in some way on processing-time, which has been shown to be sufficient information for optimal sequencing in either the static or dynamic single-machine case. However, it seems reasonable to expect that in a network there are other types of information—in particular, certain types of status information—that could be used to reduce queue lengths and inventory. For example, it would be rather pointless to select from queue a job with the shortest operation if, when that operation is completed, that job will move on to a machine that already has a large queue of work. Instead one could select for immediate assignment the job which would subsequently move on to the machine with the least backlog of work. One such procedure is:

WINQ   *Work in next queue.* Job priority equals the sum of the imminent operation processing-times of the other jobs in the queue that this job will next enter. A job waiting for its last operation has priority of zero.

The measure of "downstream congestion" used in WINQ is, of course, only approximate, since this congestion is determined at the moment the immediate selection is to be made; by the time the job actually arrives at the next queue the situation may have changed. In fact, it is quite likely to have changed, because a short queue will, in effect, attract work from every other machine in the shop, each such decision being made in ignorance of the other decisions. A potential improvement would be to include in the work-content of the downstream queue any job that is now being processed on another machine but which will arrive in queue before the subject job arrives. Thus a queue would be credited with work that is committed to it, and this would tend to reduce overreaction of the system. This rule is:

XWINQ   *Expected work in next queue.* Job priority equals the sum of the imminent operation processing-times of the other jobs in the queue that this job will next enter. Queue is considered to include jobs now on other machines that will arrive before the subject job. A job waiting for its last operation has priority of zero.

Two standard disciplines, for purposes of comparison, are:

RANDOM   *Random.* Job priority is a random number, determined at time of entry to a particular queue.

FCFS   *First-come, first-served.* Jobs are removed from a particular queue in the same order as they enter.

The performance of each of these rules is shown in Part I of Table 11–1.* RANDOM and FCFS are, of course, equivalent rules, since both are entirely independent of processing-time, and the model is "Jackson-decomposable" under these rules (Section 10–4). The simple SPT rule performs very impressively. This is not surprising for the jobs in queue, but it is interesting that SPT appears better than LWKR with respect to average work remaining, and better than TWORK with respect to average total work. The contrast between LWKR and MWKR is interesting: LWKR has less than half as many jobs in queue, but these represent more than three times the uncompleted work content as compared with MWKR. Imminent-operation work content is interesting only in that there clearly are differences between the rules. Recall that, for a single-server queue, this measure is entirely independent of queue discipline (Section 8–3).

---

* This study was conducted at the RAND Corporation [32].

Table 11-1
Comparison of Inventory Rules in a Simple, Symmetric, Random-Routed Job-Shop of 9 Machines*

Selection disciplines	No. of jobs in queue, $\bar{N}_q$	Average over sample period			Imminent-operation work content, $\bar{p}_a$
		Work remaining, $\bar{p}_r$	Work completed, $\bar{p}_c$	Total work, $\bar{p} = \bar{p}_r + \bar{p}_c$	
I					
SPT	23.25	297	248	545	62.54
LWKR	47.52	989			51.81
FOPNR	52.23	1019			51.83
MWKR	109.97	276			69.07
TWORK	82.91	323	264	587	65.16
WINQ	40.43	421	380	801	39.16
XWINQ	34.03	379	330	709	41.26
RANDOM	59.42	554	526	1080	58.39
FCFS	58.87	559	519	1078	58.29
II					
0.97 (SPT) + 0.03 (WINQ)	22.83	294	245	539	59.98
0.96 (SPT) + 0.04 (XWINQ)	22.67	293	243	536	58.40
III					
SPT/TWORK	42.39	244	191	435	68.60
SPT/WKR	62.94	220			73.79

* Sample size = 8700 jobs; actual utilization = 88.4%; mean operation time = 1.0 (from Appendix C-3).

Certainly the most interesting result is the performance of WINQ and XWINQ. In a single-server system any selection discipline that is independent of processing-time is equivalent to FCFS (Section 8–5). Since WINQ and XWINQ are both independent of processing-time, and clearly are not equivalent to FCFS, they indicate the identification of an additional type of information, available in a network, that is capable of reducing queue length. Status information, such as that used by WINQ and XWINQ, is comparatively difficult to provide, either for a simulation model or for an actual production control system. During the RAND study, a simulation run using XWINQ took almost three times as long as a comparable run under SPT. Actual implementation of such a procedure would require a real-time status-reporting system, whereas all the other rules are completely *local* in their information require-ments. The other rules require only certain attributes of the particular job—which in fact are constant and can be precomputed and printed on a route sheet—and the only computing that has to be done at the time of action is the selection of the job with minimum value.

One goal of the RAND study was to devise a procedure that used information about both processing-time and status in an attempt to reduce queue length below what could be accomplished using information about either separately. Linear combinations of the basic rules were tested, with different runs made for different values of the weighting coefficient. The runs that resulted in minimum average queue length are shown in Part II of Table 11–1. The weighting coefficient also serves as a scaling factor, since the first term in the priority is a single processing-time and the second term is a sum of processing-times, so that a coefficient of 0.96 represents something closer to a five-to-one emphasis on SPT than the apparent twenty to one. In effect, this composite procedure usually makes the same selection as would SPT, particularly when the differences between the processing-times of the jobs in the queue are large. But when the processing-times are nearly equal the rule considers where the jobs will go when they leave the current machine and favors those that will either move on to an uncongested machine or leave the shop. The improvement in performance under the composite procedure—approximately $2\frac{1}{2}\%$ as compared with SPT—is certainly not significant from any practical point of view, particularly in light of the additional information required and the consequent difficulty of implementing such a rule. Whether or not the difference in performance between SPT and the combined rule is statistically significant is a difficult and interesting question, which is discussed at some length in the original report [32].

Another method of combining the basic rules is by means of ratios. This proved interesting and effective with respect to the work-content measures of performance. For a given job, the ratio of imminent-operation processing-time to the sum of all operation processing-times exhibited the lowest values of average total work and aver-age work completed of all rules tested (Part III, Table 11–1). The ratio of imminent-operation processing-time to the sum of remaining processing-times on the job minimized the average work remaining. The relative importance of numerator and denominator in these rules could be altered by raising the denominator to a power. Different runs were made with different values of this exponent, with the results shown

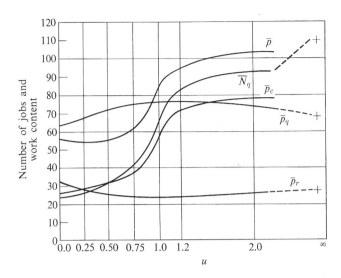

Fig. 11–1. Performance of SPT/(WKR)^u priority in a simple, symmetric, random-routed job-shop of 9 machines. Sample size = 8700 jobs; actual utilization = 88.4%; mean operation time = 1.0 (from Appendix C-3).

in Fig. 11–1. At one extreme—with an exponent of zero—this rule is equivalent to SPT. At the other extreme it approaches the performance of MWKR. It is interesting to note that the various measures react differently to changes in weighting.

11–2.1 Truncation of the Shortest-Processing-Time Rule

An immediate practical objection to any suggestion of actual use of procedures based on SPT lies in the long holding times experienced by jobs with large processing-times. The general belief that SPT has a high variance of flow-time is, in fact, erroneous. For example, the mean and variance of the flow-time of the 8700 jobs in this test was as follows:

	Mean	Variance
SPT	34.02	2,318
RANDOM	74.70	10,822
FCFS	74.43	5,739

The only rule that exhibited lower variance (1565) than SPT was a variation of FCFS in which priority was determined by the order in which the jobs arrived at the shop, rather than by their arrival at individual queues. This result has been independently confirmed by Nanot [147].

Nevertheless, since individual jobs do experience waiting-times that could be prohibitively long, modifications of SPT that would prevent these long waiting-times were considered. One method of truncating SPT is simply to place a limit on the waiting-time of each job in a given queue. When the waiting-time equals or exceeds this limit, the job is selected next, regardless of processing-time. The results for runs with different limiting values were the following (from Appendix C-3).

Limit on wait in each queue	$\overline{N}_q$	
0	58.87	(equivalent to FCFS)
4	55.67	
8	53.50	
16	44.20	
32	32.85	
∞	23.25	(equivalent to SPT)

To evaluate these results, recall that the average processing-time is 1.0 time units. The average waiting-time for a job in an individual queue is 2.78 time units under SPT and 7.27 time units under FCFS. Thus it appears that SPT is quite sensitive to this type of truncation. For example, 32 time units is approximately 12 times the average waiting-time, yet setting a limit at this level results in the loss of 41% of SPT's advantage over FCFS.

A second method of modifying SPT is to use it intermittently to relieve overloaded queues. Consider FCFS to be the normal procedure, which is used until the number of jobs in queue reaches a certain limit. At that point SPT selection is used *for that one queue* until the number of jobs falls back below the given limit. When this method was used, the results were the following (from Appendix C-3).

Number of jobs in queue that causes switch to SPT	$\overline{N}_q$	
1	23.25	(equivalent to SPT)
5	29.49	
9	38.67	
∞	58.87	(equivalent to FCFS)

This seems like a promising possibility. Considering that the average individual queue length under FCFS is 6.54, invoking SPT when the queue reaches 9 seems reasonable, and yields 57% of the advantage that might be obtained from pure SPT.

11-2.2 SPT Sequencing with Multiple Classes

A series of tests were made in which a certain fraction of the jobs were considered *preferred*. The priority rule selected the preferred job with shortest processing-time, if any preferred jobs were present in queue; otherwise the straightforward SPT was used. The performance for different fractions of preferred jobs is illustrated in Fig. 11-2.

Similar multi-class procedures were considered for the $n/1$ case in Section 3-8 and for a single-server queuing system in Section 8-6.3. The present results are similar to those for the single-server case, both in the lack of symmetry with respect to the class partition and in the degree of degradation of SPT performance.

Conditions	Reference	Least-favorable partition	Maximum loss of SPT advantage
$n/1$	Figure 3–9	50% preferred	50%
Single-server	Table 8–2		
queue	90% utilization	80% preferred	36%
Nine-machine			
job-shop	Figure 11–2	90% preferred	26%

In the last two cases the least-favorable partition is not exact; it is simply the partition with the poorest performance of those observed. These results suggest that one can single out a small fraction of jobs in the shop for preferential treatment without serious penalty, but to designate a similar fraction as filler jobs—to be processed only if nothing else is waiting—results in a significant increase in the mean number of jobs in the shop.

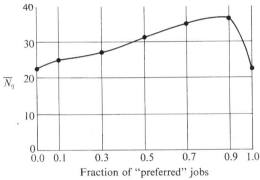

Fraction of "preferred" jobs

Fig. 11–2. Two-class SPT sequencing in a simple, symmetric, random-routed job-shop of 9 machines. Sample size = 8700 jobs; actual utilization = 88.4% (from Appendix C-3).

11–2.3 SPT Sequencing with Incomplete Information

SPT procedures obviously require some *a priori* information about processing-times, and in actual practice exact knowledge of processing-times is rarely available until after the operation has been scheduled and performed. The effect on sequencing performance of basing decisions on imperfect information was explored for the $n/1$ case in Section 3–6 and corresponding tests were conducted for the nine-machine network.

These tests used exactly the same set of jobs and processing-times as the runs previously described but the selection of jobs from queue was based on estimates of the processing-times rather than actual times. The estimates were obtained by multiplying a scaled random number by the actual processing-time. This meant that in one run the estimates of a given operation were equally likely to be any value between 90% and 110% of the actual value; in a second run each value between 0 and 200% of the actual was equally likely.

The effect on shortest-processing-time performance can be tabulated in the following manner.

Type of estimate	$\overline{N}_q$	
Perfect	23.25	(equivalent to SPT)
90–110% of actual	23.23	
0–200% of actual	27.13	
Independent of actual	59.42	(equivalent to RANDOM)

These results are very encouraging. When the errors of estimate were less than 10% there was no apparent degradation of performance. When errors were as much as 100%, SPT still retained 90% of its advantage relative to RANDOM sequencing. (The claim is sometimes made by industrial engineers that a precision of estimate of plus or minus 10% is within the capability of good, experienced time-study men. Whether or not this is true, plus or minus 100% should be possible in many situations.)

A further test assumed that one could classify operations as short or long depending only on whether the processing-time was less than or greater than the mean of all operation processing-times, and that even then there was a chance of misclassification. The priority rule was to select short operations in preference to long, with FCFS ordering within each of these classes. The results were as follows:

Procedure	$\overline{N}_q$	
Perfect short-long classification	35.29	
Short-long with 25% of operations misclassified	44.99	
All operations considered short	58.87	(equivalent to FCFS)

Even this crude two-way classification, with 25% of the operations misclassified, still retained 40% of the advantage of perfect-information SPT.*

To date there is no evidence to suggest that SPT sequencing is highly sensitive to quality of processing-time estimates and hence no reason to avoid its use on that basis.

11–3 SEQUENCING AGAINST DUE-DATES

Of the various measures of performance that have been considered in research on sequencing, certainly the measure that arouses the most interest in those who face practical problems of sequencing is the satisfaction of preassigned job due-dates. Equipment utilization, work-in-process inventory, and job flow-time are all interesting and more or less important, but the ability to fulfill delivery promises on time undoubtedly dominates these other considerations. However, the lateness or tardiness of a job depends on both its completion-time and its due-date, so that an investigation of these measures of performance must consider the manner in which due-dates are determined as well as the efficacy of procedures intended to enforce those dates.

---

* The analysis of a similar two-class procedure (without misclassification) for single-server queuing systems (Tables 8–4 and 8–5) shows that for 90% utilization it would be optimal to partition the jobs so that 87.8% were "short" rather than to use the mean as the dividing point. In a single-server system the optimal short-long procedure sacrifices only 19% of the SPT advantage over FCFS sequencing.

11–3.1 Assignment of Due-Dates

Scheduling procedures of sufficient power to enforce a completely arbitrary set of due-dates do not yet exist. If they did, then it would not be necessary to worry about the manner in which the due-dates are determined and one could proceed as in the earlier chapters. However, it will soon become apparent that the procedures currently known are not this powerful and it is necessary to consider the reasonableness of the due-dates as an important condition of the problem.

In some situations due-dates are exogenous to the scheduling organization; they are set by some independent external agency and announced upon arrival of the job. They are a fixed and given attribute of a job, not unlike processing-times or routing. On the other hand, even when a due-date represents a contract with an external agency, those situations in which the due-date is arbitrarily set without some participation by the organization responsible for satisfying the date are relatively infrequent. A more typical situation is: the producing agency proposes a date, which is adjusted according to the consumer's need and competitive forces, and then the producer and consumer agree on a final due-date. In many situations the consumer is actually another unit of the same organization as the producing agency. In these cases due-dates provide a means of coordinating the activities of many different units of the same organization, and are no less important simply because they are, in a sense, internal. For example, due-dates may be assigned to jobs in several fabrication shops by a central production-planning department whose task is to synchronize the arrival of these jobs on an erection floor where they will be assembled into a single machine. This could be the most demanding sort of due-date, with extreme penalties associated with tardiness.

The point of the external-versus-internal discussion is its effect on the inherent reasonableness or attainability of due-dates. Presumably an internal agency has some knowledge of the capability of the shop and of the scheduling procedure that will be employed, and has some interest in establishing due-dates that can be met. The question is closely related to the prediction of individual job flow-times. If one could exactly predict the flow-time of various jobs under a certain priority procedure one could, if due-dates were under internal control, assign an allowance equal to the flow-time so that the completion-time of each job would be exactly equal to its due-date. However, since flow-time depends not only on characteristics of the individual job and the particular priority rule in use, but also on the nature and status of the other jobs coexistent in the shop during the time that the given job is present, perfect prediction is, for all practical purposes, not attainable. The problem becomes one of finding a combination of due-date assignment procedure and priority sequencing procedure such that:

a) the natural flow-time of a particular job can be estimated with some precision,

b) one can set a due-date based on both this natural passage time and exogenous considerations of relative urgency,

c) the priority procedure can react to this due-date to accelerate or retard a job with respect to its natural rate of progress.

Four different methods of assigning due-date were considered in the RAND study.* In each case exactly the same set of jobs was used, with the same arrival times. The individual job allowances, and hence the due-dates, varied from one run to the next. However, in each different run the *average allowance* remained the same so that, on the average, the different types of due-dates were equally difficult to attain. All that differed was the job-to-job variation. The allowances were determined in the following ways:

TWK　*Total-work due-dates.* The allowance for flow-time (the difference between due-dates and arrival time) was 9 times the sum of the processing-times. That is, a job could spend a total of 8 times its total processing-time waiting in queue before it became tardy.

NOP　*Number-of-operation due-dates.* The allowance for flow-time was proportional to the number of operations.

CON　*Constant-allowance due-dates.* Each job received exactly the same allowance (78.8 time units).

RDM　*Random-allowance due-dates.* Each job was assigned an allowance at random. (Allowances were uniformly distributed between 0 and 157.6 time units.)

The TWK and NOP methods were intended to represent two different ways of assigning reasonable and attainable due-dates, providing a greater allowance for a job with more processing time or a larger number of operations. The CON due-dates are representative of common practice, in which a standard lead-time is quoted. The RDM due-dates were intended to represent completely arbitrary deadlines assigned by an external agency.

The scheduling procedures tested against these different due-dates were suggested by procedures that are actually employed in industry. The basic rules were:

DDATE　*Due-date.* Job priority equals its due-date.

OPNDD　*Operation due-date.* The initial allowance is divided equally among the several operations of the job and due-dates are assigned to each operation. The job priority in a given queue equals the due-date of the imminent operation.

SLACK　*Slack-time.* Job priority equals the time remaining before the job due-date minus all remaining processing time for the job.

S/OPN　*Slack per operation.* Job priority equals the ratio of job slack-time to the number of operations remaining.

The SPT and FCFS procedures were used for controls.

The performance of these procedures is described in Appendix C-3 and summarized in Table 11-2.† It was apparent that employing the rules using due-date

---

* For additional work on this topic see Bakker [12].
† These data are selected from the complete results. For more detail see [32] or [31].

Table 11–2

Comparison of Due-Date Rules in a Simple,
Symmetric, Random-Routed Job-Shop of 9 Machines*

Priority rule		Type of due-date† TWK	NOP	CON	RDM
FCFS	(a)	74.4	74.4	74.4	74.4
	(b)	−4.5	−3.9	−4.4	−4.9
	(c)	41.1	33.5	75.7	87.9
	(d)	0.448	0.399	0.337	0.412
DDATE	(a)	63.7	70.6	72.5	72.9
	(b)	−15.5	−7.7	−6.3	−6.4
	(c)	20.8	23.0	39.6	45.8
	(d)	0.177	0.267	0.439	0.487
OPNDD	(a)	69.0			
	(b)	−9.9			
	(c)	120.0			
	(d)	0.104			
SLACK	(a)	65.8			
	(b)	−13.1			
	(c)	20.8			
	(d)	0.220			
S/OPN	(a)	66.1	72.9	73.7	74.0
	(b)	−12.8	−5.4	−5.1	−5.3
	(c)	15.0	15.0	33.4	41.5
	(d)	0.037	0.216	0.481	0.525
SPT	(a)	34.0	34.0	34.0	34.0
	(b)	−44.9	−44.3	−44.8	−45.3
	(c)	53.7	54.1	48.2	66.5
	(d)	0.050	0.062	0.110	0.198

* Sample size = 8700 jobs; actual utilization = 88.4%; mean operation time = 1.0 (from Appendix C-3).
† Table entries: (a) F, average job flow-time, (b) L, average job lateness, (c) standard deviation of lateness distribution, (d) f_t, fraction of jobs tardy.

information provided a substantial improvement over FCFS in the reduction of the proportion of jobs that did not meet the due-date in this test. This was due in part to a decrease in the mean lateness, but largely to a reduction in the variance of the distribution. The S/OPN rule, which selects the job with the least slack per operation remaining, is the most complicated of these due-date rules; however, it appeared to give somewhat better performance. When this rule was used, the jobs exhibited the smallest value of lateness variance, which is probably the key measure, and also the proportion of jobs with positive lateness under this rule was smaller than under any of the other basic due-date rules.

It is interesting to note that under the SPT rule there is almost as small a proportion of jobs tardy, and yet SPT accomplishes this by an entirely different strategy and *without ever looking at a due-date.* SPT depends entirely on a reduction of the mean lateness to offset an increase in lateness variance and still reduce the area of the positive tail of the distribution.

The effect of the difference in approach between SPT and S/OPN was clearly illustrated by tests using a second and third set of jobs that had slightly higher utilization. The percentage of jobs tardy when TWK due-dates were used was as follows:

	Job set 1: utilization = 88.4%	Job set 2: utilization = 90.4%	Job set 3: utilization = 91.9%
FCFS	44.8%	57.7%	67.5%
S/OPN	3.7%	30.9%	54.8%
SPT	5.0%	6.1%	7.3%

Although the allowances under TWK due-dates were slightly greater, in proportion to the increased processing-times in the second and third job sets, the congestion in the shop increased more rapidly than the allowances. The result was that the due-dates for the second set of jobs were more difficult to achieve than for the first set, and those for the third set were even harder. In each case S/OPN exhibited the most compact (low-variance) distribution, but increasing fractions of this distribution had positive values. The high-variance, highly skewed distribution under SPT was not as greatly affected by this displacement of the mean, and the proportion of jobs tardy remained small. One might conclude on the one hand that SPT performance is much less sensitive to variations in shop load than the other procedures, and on the other hand that TWK due-dates adjust for differences in job length, but do not adequately compensate for differences in shop load.

Detailed examination of the test results indicates another important difference between the SPT and S/OPN approaches to due-date satisfaction. Even under conditions for which the two procedures were similar in aggregate number of jobs tardy (job set 1, TWK due-dates), there were entirely different patterns of which jobs were tardy and when. Under the S/OPN rule, virtually all the tardy jobs occurred in several short periods when, by chance, the utilization and congestion were higher than average for the test. There were long intervals under S/OPN during which all jobs were completed before their due-date. Under SPT the tardy jobs occurred more uniformly throughout the test run. It would seem that a composite procedure that employed an SPT strategy during periods of relatively high congestion and shifted to S/OPN when the peaks were passed would perform very well; such a procedure is described in Section 11–3.2.

A linear combination of SPT and S/OPN was tested with different relative weights. Equal weighting (0.5 SPT + 0.5 S/OPN) produced the best overall performance, which resulted in a mean lateness of -21.3, a standard deviation of 22.8, and only 1.52% of the jobs tardy.

It is interesting to compare the different methods of assigning due-dates for the different rules, remembering that in each case the same set of jobs was involved, and that on the average each kind of due-date was equally easy to achieve. Under FCFS it did not really matter how the due-dates were assigned; the performance was mediocre in any event. Under DDATE, SPT, or S/OPN, the sequence TWK, NOP, CON, and RDM represented progressively poorer ways of assigning due-dates. The SPT rule, which did not consider the due-date in its selections, was clearly less sensitive to the method of assigning due-date than the other procedures, but still behaved best with the reasonable TWK and NOP due-dates. The procedures that used due-date information—DDATE and S/OPN—were strongest when the due-dates were reasonable and bore some relationship to the job characteristics. They did not exhibit impressive ability to enforce an arbitrary set of due-dates. In fact, when either CON or RDM due-dates were used, FCFS provided better performance than either DDATE or S/OPN. It should be noted, however, that the lateness variance was still substantially smaller with DDATE and S/OPN, regardless of which type of due-date was employed—and this is an important characteristic. An increase in the mean allowance would presumably benefit these rules with compact distributions much more than FCFS or SPT, and the proportion of jobs with positive lateness could be made satisfactorily small.

11-3.2 A State-Dependent Due-Date Procedure

The RAND study described in the previous section began consideration of composite rules in which several different factors were considered in determining the priority of jobs, but in which the relative weighting of these factors was constant throughout a given run. A later study [158] considered a dynamic composite rule in which the relative weight of several factors was varied, depending on conditions at the time a decision was to be made. In effect, this permitted different priority rules to be in operation for different machines at a given point in time, and different rules to be employed for a given machine at different points in time. In fact, the entire investigation concerned a single such rule, with different runs to explore the effect of changing values of several control parameters.

The previous work suggested that three factors were important in meeting due-dates:

1) Some function of job *due-date*, to pace the progress of individual jobs and reduce the variance of the lateness distribution.

2) Consideration of *processing-time*, to reduce congestion and to get jobs through the shop as quickly as possible.

3) Some *foresight*, to avoid selecting a job from queue which, when the imminent operation is completed, will move on into a queue which is already congested.

The problem was to determine how to mix these considerations, depending on the conditions that exist in the particular queue and throughout the shop at the moment that a selection must be made. It was arbitrarily decided to base the priority on a due-date term that would always be present, and then add varying weights of the other two considerations, depending on conditions.

After preliminary tests, the particular due-date function that was chosen was what might be called "operation-slack." When the job arrived at the shop, operation due-dates were assigned to each individual operation in the same way that the due-date was assigned to the job. For example, if in a particular run each job were allowed four times the sum of its processing-times for queuing in the assignment of its due-date, then each individual operation would be assigned an operation due-date that allowed four times the processing-time of that particular operation for queuing at that one machine.* The tardiness of the job was determined only by the job due-date (which was equivalent to the operation due-date for the last operation) but these interim operation due-dates provided a means for assessing the job's progress through the shop. Given that

$o_{i,j}$ represents the operation due-date of the jth operation of job i and that

$p_{i,j}$ represents the processing-time of the jth operation of job i,

then the operation-slack at time t would be equal to $o_{i,j} - p_{i,j} - t$. However, since the value of t is the same for each operation in the queue from which the selection is to be made, and since the selection will depend only on relative values of priority, t can be omitted and operation-slack defined as: $s_{i,j} = o_{i,j} - p_{i,j}$.

The second term in the priority was of the form: $B \cdot p_{i,j}$, and the weighting coefficient B was defined as

$$B = b \cdot [\textstyle\sum p_{i,j}]^r,$$

where b and r are constants for a particular run and the summation is over all the operations in the queue from which selection is to be made. When r is set equal to 0, the summation is, of course, inoperative and there is a simple weighting, depending on the value of b, between the processing-time and the operation-slack, similar to rules that were explored in the RAND study. When $r > 0$, then the processing-time of the operation is more heavily weighted in a congested queue than in a relatively empty one.

The third term was of the form $h \cdot W_{nq}$, where h is a weighting coefficient, constant throughout a run, and W_{nq} is the ratio of the sum of the processing times in the next queue this job will visit to the sum of the processing times in all the other queues in the shop. For a job that will move on to an empty queue, $W_{nq} = 0$, or for the last operation on a job W_{nq} is defined to be zero.

In aggregate the priority of the jth operation of job i at time t was given by:

$$Z_{i,j}(t) = s_{i,j} + B \cdot p_{i,j} + h \cdot W_{nq} = o_{i,j} - p_{i,j} + b \cdot [\textstyle\sum p_{i,j}]^r \cdot p_{i,j} + h \cdot W_{nq},$$

where:

$b, r, h,$ are constants assigned for a particular run that select one particular rule out of a parametric family of rules;

$p_{i,j}$ the processing time, is a characteristic of the individual operation, assumed known in advance;

$o_{i,j}$ the operation due-date, is a characteristic of the individual operation, determined by the job due-date and $p_{i,j}$;

$\sum p_{i,j}, W_{nq}$ are variables that describe the status of the shop at time t.

---

* A study by Orkin [159] indicated that this proportional distribution of the job allowance into individual operations was preferable to the equal division used in the RAND study.

Each time a machine completed an operation and two or more jobs were waiting for assignment to this machine, then a priority value was computed for each of the competing jobs and the job with *minimum* value was selected. This method favored selection of jobs which were running out of a slack time and/or had short processing-time and/or would move on to a relatively empty queue.

The tests were performed under conditions very similar to those of the RAND study. There were eight machines instead of nine; the sample size was 6000 jobs; processing-times were small integers with a geometric distribution instead of exponentially distributed random variables; and the actual average utilization turned out to be 86% instead of 88%. But these are all minor points. It was quickly apparent that the performance of this composite rule was better than the RAND rules to a degree that meant that a more taxing set of due-dates would be required if interesting comparisons were to be made. TWK due-dates were used, with an allowance for queuing of four times the sum of the processing-times as compared with eight times for the RAND study. The average congestion in the shop was slightly reduced, due to the lower average utilization, but the due-dates were nevertheless substantially more difficult to achieve. In order to compare conditions and the effect of the tighter due-dates, several of the RAND rules were tested again. The comparison in percentage of jobs tardy is shown below.

| | Queue-time allowance/processing-time | |
Rule	8/1 (RAND)	4/1
DDATE	17.7%	49.8%
OPNDD (equally spaced		
operation due-dates)	10.4%	61.7%
S/OPN	3.7%	54.3%
SPT	5.0%	10.8%

Again it was apparent that SPT, with only the tail of a highly skewed lateness distribution having positive value, is least affected by the tightening of due-dates.

This study concentrated on *tardiness* rather than *lateness* as a measure of performance. The mean tardiness $\bar{T}$, the fraction of jobs tardy f_t, and the conditional mean of those jobs with nonzero tardiness ($\bar{T}_c = \bar{T}/f_t$) were reported, along with the mean and standard deviation of lateness and flow-time. A portion of the performance data is reproduced in Appendix C–4.

In general, the results showed the composite due-date procedure to be remarkably powerful and relatively insensitive to the setting of the control parameters. For reasonably wide ranges of values of the control parameters (b, r, h), this rule yielded:

a) Mean lateness and fraction tardy approaching that of SPT.

b) Variance of lateness less than any of the simple rules.

c) Variance of flow-time comparable to that of SPT, and better than the simple due-date-based rules.

d) Conditional mean tardiness slightly better than the simple due-date-based rules and approximately one-third that of SPT.

e) Mean tardiness approximately one-half that of SPT.

For example, compare the performance of the composite rule with the SPT and operation-slack rules from which it is constructed:

	$\overline{T}$	f_t	$\overline{T}_c$	$\overline{L}$	Std. dev., L	$\overline{F}$	Std. dev., F
SPT	12.1	0.108	112.1	−49.0	106	107.8	147
Operation-slack, $b = 0, r = 0, h = 0$	21.9	0.422	51.9	−1.2	60.5	155.6	160
Composite, $b = 0.30, r = 1, h = 0$	6.11	0.118	51.9	−37.3	56.2	119.5	148

When the rules were judged on the basis of tardiness, experimenters found that use of the composite rule enabled them to maintain almost the same small fraction of tardy jobs as use of SPT, but with much less tardiness for those jobs that did not complete by their due-date. The composite rule was as satisfactory in this respect as pure operation-slack. The product of these two measures gives a value of mean tardiness 50% better than either contributing rule. When the rules were judged on the basis of lateness, experimenters found that use of the composite rule enabled them to reduce the mean of the distribution almost as much as the use of SPT, while at the same time giving a variance less than that obtained by using the operation-slack rule alone.

One could offer SPT as an alternative to simple due-date rules with considerable confidence that it would reduce the fraction of jobs that were tardy, the mean lateness, and even the mean tardiness. However, under very general conditions one could not deny the price that would be exacted in terms of lateness variance and conditional mean tardiness. The type of composite rule described here appears to offer these same improvements without obvious ill effects. In these tests the composite rule was

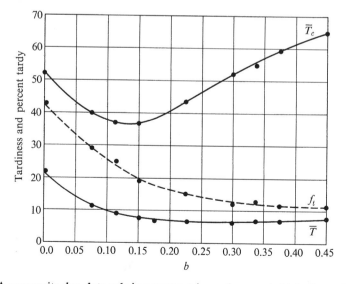

Fig. 11–3. A composite due-date rule in a symmetric random-routed job-shop of 8 machines. Sample size = 6000 jobs; actual utilization = 86%; $r = 1$; $h = 0$ (from Appendix C–4).

uniformly more powerful than any of the simple due-date rules for any measure of performance, and for all practical purposes more powerful than SPT as well.

Figure 11-3 shows the performance of the composite rule as a function of the relative balance of the first two factors—the operation-slack and the operation processing-time. Beginning at the left, one has a pure operation-slack rule. An increase in the abscissa represents increased weight on processing time in determining priority. The fraction of jobs that are tardy decreases monotonically from one extreme to the other, while the conditional mean tardiness finds a minimum that is significantly less than the value at either extreme. The mean tardiness, being the product of these two, has a minimum value that is less than that for either extreme rule.

There are six rules whose performance summarizes the results of these tests:

	$\overline{T}$	f_t	$\overline{T}_c$	$\overline{L}$
(1) SPT	12.1	0.108	112.1	−49.0
(2) DDATE	35.2	0.498	70.7	18.2
(3) Operation-slack,				
$b = 0, r = 0, h = 0$	21.9	0.422	51.9	−1.2
(4) Composite,				
$b = 15, r = 0, h = 0$	6.52	0.152	42.8	−38.7
(5) Composite,				
$b = 0.3, r = 1, h = 0$	6.11	0.118	51.9	−37.3
(6) Composite				
$b = 0.3, r = 1, h = 160$	5.43	0.120	45.3	−39.4

The SPT rule is an illogical but necessary standard, since, although it ignores due-date information completely, it has smaller mean lateness and fraction tardy than any other rule tested. DDATE represents the simplest possible way of using due-date information. Operation-slack is a more sophisticated way of using due-date information and in this study represented the best of the simple due-date rules. Although operation-slack is based on the same idea as the OPNDD rule of the RAND study, two minor changes greatly improved its performance:

1) Allocating queuing allowance among operations in proportion to processing-time rather than equally.

2) Considering operation-slack rather than operation due-date directly.

Rule (4) is an unusually successful composite of (1) and (3), somehow preserving the best features of both. Rule (5) uses shop-status information to control the relative weight given due-date and processing-time considerations, and rule (6) uses additional status information to avoid moving jobs into already congested queues.

The first four of these rules can have the priority value for each individual operation precomputed before the job arrives at the shop. None of these priorities depends on shop status, and none changes with elapsed time. Rules (5) and (6) would require a real-time communication-computing system for implementation, and the improvement in performance would be too modest to justify the cost of such a system.

11-4 INVESTIGATIONS OF MORE COMPLEX JOB-SHOP MODELS

In each of the investigations described previously, a model of a simple job-shop process was employed in order to concentrate attention on a comparison of scheduling procedures. With this as a background, several studies have been performed to explore the effect of relaxing various of the restrictions of the simple job-shop model. Two of these have been concerned with types of flexibility that are present in real shops. In actual practice there is often some latitude in the order in which the operations of a job have to be performed, and some choice as to which machine in the shop can be used to perform a given operation. Both these types of flexibility tend to reduce congestion and queue lengths; neither was permitted in the previous studies. A third study considered a shop in which some of the operations were assembly operations requiring the combining of several jobs. In this case, rather than restricting all jobs to simple linear strings of operations, the experimenters used a tree-structured job-routing. The existence of assembly operations appeared to have marked effect on the relative performance of the simple sequencing rules and required the development of new types of sequencing procedures.

Each of these investigations was based directly on the previous RAND study and was conducted in such a manner that direct comparison of results is possible.

11-4.1 Flexibility in Machine Selection

In many actual job-shops, the scheduler has some freedom to bypass congested machines by assigning operations to an alternate machine which is less heavily loaded. This may be advantageous, even when the alternate machine is somewhat less efficient in performing the operation. Lacking this and other kinds of flexibility, the simple job-shop models have tended to overstate the queuing that would occur for given load conditions, and consequently have probably overstated the differences between scheduling rules.

Wayson [198] conducted a study of the effect of flexibility in machine selection; his study was carried out under the same conditions as the RAND study, but with slightly higher machine utilization. The various choices that were available for machine specifications were described by a square matrix whose number of rows was equal to the number of different types of machine. An element x_{ij} gives the probability that an operation nominally assigned to a machine of type i could also be done on a machine of type j. (In practice, of course, a particular operation either can or cannot be performed on a given machine, but this probability model provided a mechanism in the simulation to allow machine j to serve as an alternate for some, but not all, of the operations on machine i.) The element x_{ii} is obviously 1 for all i. Given that $x_{ij} = 0$ for all $i \neq j$, then one has the strict machine assignment that has been used in previous studies. The value $x_{ij} = 1$ for $i \neq j$ means that any operation nominally assigned to a machine of type i could be processed on a machine of type j. The value $x_{ij} = 0.5$ would mean that in the long run one-half of the i-operations could be processed on j. In the simulation a random number was drawn for each i-operation to determine whether or not it could be done on j. For example, each of the following might be a matrix for a five-machine shop.

$$A: \begin{bmatrix} 1 & 0 & 0 & 0 & 0 \\ 0 & 1 & 0 & 0 & 0 \\ 0 & 0 & 1 & 0 & 0 \\ 0 & 0 & 0 & 1 & 0 \\ 0 & 0 & 0 & 0 & 1 \end{bmatrix} \qquad B: \begin{bmatrix} 1 & 1 & 0 & 0 & 0 \\ 0 & 1 & 1 & 0 & 0 \\ 0 & 0 & 1 & 1 & 0 \\ 0 & 0 & 0 & 1 & 1 \\ 1 & 0 & 0 & 0 & 1 \end{bmatrix}$$

$$C: \begin{bmatrix} 1 & 0.5 & 0.2 & 0 & 0 \\ 0 & 1 & 0.5 & 0.2 & 0 \\ 0 & 0 & 1 & 0.5 & 0.2 \\ 0.2 & 0 & 0 & 1 & 0.5 \\ 0.5 & 0.2 & 0 & 0 & 1 \end{bmatrix} \qquad D: \begin{bmatrix} 1 & 0 & 0 & 0 & 0 \\ 0 & 1 & 0.5 & 0 & 0 \\ 0 & 0.5 & 1 & 0.3 & 0 \\ 0.1 & 0.1 & 0.1 & 1 & 0.1 \\ 1 & 1 & 1 & 1 & 1 \end{bmatrix}$$

The matrix A is the matrix for strict assignment; B provides a single fixed alternative for each machine; C provides a major and minor alternative for each machine; and D has a different alternative pattern for each machine; in particular, operations nominally assigned to machine 5 can always be done on any other machine in the shop. Wayson considered only matrices with two symmetries:

$$\sum_{j=1}^{m} x_{ij} = K \quad \text{for all } i, \qquad \sum_{i=1}^{m} x_{ij} = K \quad \text{for all } j,$$

where K is a constant between 1 and m. Examples A, B, and C possess this symmetry; D does not. The value $K - 1$ is a measure of the number of alternate machines available for each nominal machine.

During Wayson's study, each time a nonterminal operation was completed, a list of machines acceptable for the next operation of the job was prepared. (This list was made by examining the row of the matrix for the nominal machine, placing all machines with entry 1 on the list, and drawing random numbers to determine whether machines with fractional entries should be placed on the list.) Machines were selected from this list according to one of the following three disciplines:

1) Select at random.

2) Select the machine with the least amount of work in queue before it.

3) Select the machine with the least number of jobs in queue before it.

Ties which involved the nominal machine were always resolved in favor of the nominal machine. The job was placed in the queue selected in this way, and remained there until processed by the corresponding machine. It could not subsequently move to another queue; an irrevocable selection from among alternates was made at the time the preceding operation was completed. When a machine became available, jobs were selected from the queue for processing in the normal priority manner. Wayson's study was principally concerned with a comparison of FCFS and SPT under type (3) queue selection. Part of his results are given in Appendix C–5.

Figure 11–4 is a plot of the average number of jobs in queue under FCFS and SPT as a function of the degree of machine flexibility permitted. The SPT rule appears to be uniformly more powerful than FCFS, but the advantage diminishes with increasing flexibility. The actual frequency of alternate selection is also shown in Fig. 11–4.

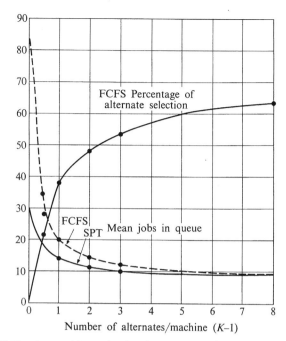

Fig. 11–4. Flexibility in machine selection in a symmetric random-routed job-shop of 9 machines. Sample size = 8880 jobs; utilization = 91.2% (from Appendix C–5).

The method of resolving ties in queue length in favor of the nominal machine causes the actual frequency to be less than the $(K - 1)/K$ that might be expected. For example, under FCFS with one alternate for each machine $(K = 2)$, the alternate is selected 39% rather than 50% of the time. With two alternates per machine, the selection is 48% rather than 67%.

The sensitivity of the system to flexibility of machine selection is very striking. Even a small amount of flexibility drastically reduces overall congestion. With alternate selection 20% of the time, FCFS yields the same number of jobs in queue as SPT without alternate selection. In terms of practical implementation of scheduling procedures, this effect is too important to be neglected. A sophisticated scheduling procedure (perhaps employing an expensive communication-computing system) that did not take advantage of this type of flexibility would risk being outperformed by a knowledgeable human scheduler.

11–4.2 Flexibility in Operation Sequence

Another type of flexibility present in real shops but excluded from most of the experimental work to date is the flexibility which enables one to depart from the strict ordering of operations as given on the routing. There are some precedence constraints that must be observed—for example, a hole has to be drilled before it can be tapped—but often these constraints do not impose a strict ordering on all the operations of a job.

Neimeier [148] undertook a preliminary exploration of this question, using Wayson's model and job set. (The control runs with FCFS and SPT, when flexibility was forbidden in both machine specification and operation sequence, are identical in these two studies.) Each of the jobs was partitioned into groups of operations such that the operations within a group could be performed in any order, but all the operations of a given group had to be completed before any operation of the following group could be performed. For example, a job of nine operations might be partitioned as follows.

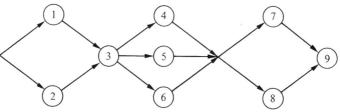

Processing could begin with either operation 1 or operation 2, but both 1 and 2 must be completed before 3, etc. The partitioning was accomplished in the experiment by specifying a probability, P, that an operation begins a new group. Neimeier took the operations in numbered order, and observed a random variable for each operation that either initiated a new group or added the operation to the preceding group.

He measured the degree of sequence flexibility by computing for each job the ratio of the number of permissible sequences to the number that would be possible if there were no precedence constraints. The result is the product of the factorials of the group sizes divided by the factorial of the total number of operations. In the example above, this would be

$$\frac{(2!)(1!)(3!)(2!)(1!)}{9!} = \frac{24}{36,288}.$$

A job set is characterized by the average of these ratios for the individual jobs. By varying P from one run to the next, Neimeier could systematically alter the group sizes and the routing ratio.

The problem of determining which of the operations of a group should be done next is comparable to Wayson's problem of determining which machine to use for a given operation. The same strategy was employed: perform next the operation that faces the queue containing the smallest number of jobs. Operations were selected from queue under either FCFS or SPT disciplines.

Partial results are given in Appendix C–6 and plotted in Fig. 11–5. The effect was similar to that of flexibility in machine specification (Fig. 11–4), although less pronounced. This would be expected, since the use of an alternate machine permits a job to avoid a congested queue altogether, whereas the use of an alternate sequence simply postpones the job's entry into a congested queue. During the interim the bypassed queue may become more rather than less congested, but the performance improvement clearly indicated that there is a tendency toward the latter.

Russo [177] carried out an investigation under very similar conditions, but concentrated on tardiness as a measure of performance. With some sequence flexibility

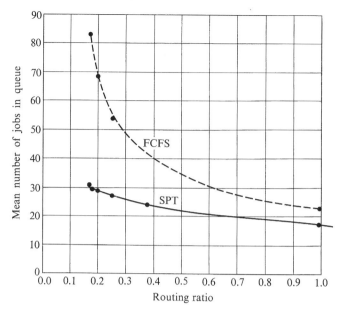

Fig. 11–5. Flexibility in operation sequence in a symmetric random-routed job-shop of 9 machines. Sample size = 4440 jobs; utilization = 91.2% (from Appendix C–6).

and well-designed priority rules he achieved extremely low levels of tardiness. He also demonstrated the value of postponing the decision as to what sequence to select to the point at which one of the possible operations would actually go onto a machine. Mechanically this involves placing the job in several queues simultaneously and requires a considerable increase in the complexity of the simulation program, but the performance is significantly improved.

11–4.3 Sequencing in an Assembly Shop

Assembly shop is a name for a job-shop in which tree-structured job routings are permitted. This is a generalization of the precedence constraints between operations, so that the following type of routing might be encountered:

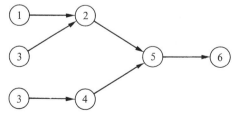

Such shops were considered explicitly in Section 4–3 and 7–1. They also provide the motivation for much of the concern with due-dates, since an assembly shop can be partitioned into simple job-shops and a system of due-dates used to coordinate these "independent" shops. However, this is at best a suboptimal procedure, and one should consider the entire shop and attempt to develop effective procedures. Work on this type of model has been much less extensive than for the simple job-shop, but a number of exploratory studies have been conducted. One such study took place at the RAND Corporation [130] under experimental conditions almost identical to the previous RAND job-shop study.

This study considered only a special case of tree-routings in which there was just a single level of branching, and this took place at the final operation of the job. Actually, this final operation was a dummy operation with zero processing-time. Typical routings would look like the following:

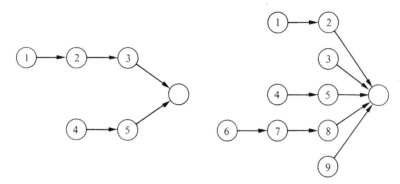

Each job consisted of a number of *branches*, each branch being equivalent to a job of the previous RAND study. The number of branches per job was a program parameter. With this equal to one, the simple job-shop process was obtained; values greater than one produced an assembly shop where all jobs had the same number of branches per job. The interarrival interval between jobs was exponentially distributed; its mean was set to yield a nominal shop utilization of 90%. The assembly time was assumed to be zero, so that the assembly operation consisted entirely of marshaling the branches of the job.

Due-dates for jobs were set by a variation of the NOP procedure. The due-date of a job was the time of arrival at the shop plus the maximum total amount of work on any branch of a job plus a due-date multiplier times the maximum total number of operations on any branch. The due-date multiplier was set so that approximately 50% of the jobs were late under FASFS sequencing.

FCFS, FASFS, SPT, DDATE, SLACK and S/OPN from the previous study were tested. Five additional rules were designed to use status information concerning the several branches of a job:

MAXRWD-NR   *Maximum remaining work difference;* no re-evaluation. The priority of an operation in queue equals the difference between the maximum remaining work over the branches of the job and the remaining work on the branch for this operation. This difference is computed at the time the operation enters queue and is not re-evaluated if work is completed on operations of other branches of the job.

MAXNRD-NR   *Maximum remaining number of operations difference;* no re-evaluation. The priority of an operation in queue equals the difference between the maximum number of remaining operations over the branches of the job and the number of operations remaining on the branch for this operation. This difference is computed at the time the operation enters queue and is not re-evaluated if work is completed on operations of other branches of the job.

NUB   Number of uncompleted branches. The priority of an operation is the total number of branches of the job for which not all the operations of the branch have been completed. This number is computed for each operation in queue every time an operation is to be selected from queue.

MAXNRD-SPT   The priority of an operation in queue equals the difference between the maximum number of remaining operations over the branches of the job and the number of operations remaining on the branch on the operation. This difference is computed for each operation in queue every time an operation is to be selected from queue. (SPT is used to break ties.)

NUB-SPT   The priority of an operation is the total number of branches of the job for which not all the operations of the branch have been completed. This number is computed for each operation in queue every time an operation is to be selected from queue. (SPT is used to break ties.)

A run consisted of a preload of 50 branches, a run-in period of 400 branches, a sample size of 9000 branches, followed by a run-out period of 600 branches. Three levels of number of branches per job were used. These were 2, 5, and 10, yielding sample sizes of 4500 jobs, 1800 jobs, and 900 jobs, respectively. Two additional measures of performance were introduced:

Branch flow-time mean, $\bar{B}$ the mean flow-time of all branches of the jobs. If the branches were in fact independent jobs in a job-shop process of the RAND study variety, then this would be the mean flow-time of these independent jobs.

Assembly wait-mean, $\bar{A}$ the mean time which is spent by the branches in waiting for the other branches of their job to be completed.

The mean job flow-time is related to these measures by the relation $\bar{F} = \bar{B} + \bar{A}$, since for any branch the sum of the time that it spends being worked on and waiting in queue plus the time that it spends waiting for the other branches of its job to be completed must equal the flow-time of its job.

MAXNRD—the first assembly rule tested—did not perform well for any measure of performance. It was apparent that it was not satisfactorily synchronizing the progress of the different branches of a job. Two modifications resulted in improved performance. One was re-evaluation of priority as an operation waited in queue

Table 11–3

Comparison of Priority Rules in an Assembly Shop with Two Branches for each Job*

Rule	Mean flow-time, $\overline{F}$	Mean branch flow-time, $\overline{B}$	Mean assembly wait-time, $\overline{A}$	Conditional mean tardiness, $\overline{T}_c$	Fraction tardy, f_t
FCFS	149.28	100.71	48.57	64.46	0.5789
SPT	73.80	45.01	28.79	139.65	0.1109
FASFS	121.13	98.66	22.47	50.66	0.5162
DDATE	116.91	95.59	21.02	40.71	0.3446
SLACK	114.40	96.11	18.29	35.79	0.3324
S/OPN	108.90	95.34	12.75	12.55	0.0993
MAXNRD-SPT	69.67	62.67	7.00	127.63	0.0957
NUB-SPT	72.31	61.99	10.32	92.82	0.1150

* Sample size = 4500 jobs; nominal utilization = 90%; RAND study conditions for branches as jobs (from Appendix C–7).

whenever work on other branches of the same job was completed. The second was the use of SPT rather than FCFS to break the frequent ties that occurred under this rule.

The first series of runs, with two branches for each job, is summarized in Table 11–3. The results for the branch flow-time means are consistent with the results of the previous RAND study (Section 11–2). The performance of SPT was very good with respect to both flow-time and fraction of jobs tardy. However, except for FCFS, SPT had the worst performance in getting the two branches of a job assembled without delay. The SPT performance was simply a consequence of its ability to keep jobs flowing quickly through the processing operations. The two rules which used status information, MAXNRD-SPT and NUB-SPT, achieved low values for average flow-time only by relying heavily on the superior performance of SPT on the branch flow-times. By contrast, the reduction in $\overline{A}$ from 12.75 for S/OPN to zero would in itself be insufficient to make S/OPN performance comparable to SPT performance.

The MAXNRD-SPT and NUB-SPT rules performed well for most measures, as compared with the due-date-related rules, the only exception being their performance in relation to conditional mean tardiness. It appears that these rules have the same characteristics as SPT: a lateness distribution with a small mean but highly skewed to the right. These two new rules require much more information for their implementation and operation. Status information on the branches of a job must be cross-referenced and updated on a real-time basis.

The effect of different levels of number of branches per job is shown in Table 11–4. The performance of SPT deteriorated as the number of branches per job increased. We can understand this by noting that SPT performance in an assembly shop is roughly comparable to selecting an appropriate number of observations from the flow-time distribution of a simple job-shop and using the maximum of the sampled

Table 11–4

Interaction Between Rules and Number of Branches per Job
in an Assembly Shop*

| Rule† | | Number of branches per job | | |
		2 (4500 jobs)	5 (1800 jobs)	10 (900 jobs)
SPT	(a)	73.80	112.01	264.36
	(b)	28.79	70.60	198.26
	(c)	64.46	129.90	279.43
	(d)	0.1109	0.1438	0.3761
FASFS	(a)	121.13	122.12	171.69
	(b)	22.47	38.34	54.42
	(c)	50.66	33.39	58.14
	(d)	0.5162	0.2167	0.2622
S/OPN	(a)	108.09	135.43	198.27
	(b)	12.75	33.31	39.49
	(c)	12.55	4.35	51.15
	(d)	0.0993	0.0044	0.1193
MAXNRD-SPT	(a)	69.67	91.22	164.47
	(b)	7.00	16.37	27.46
	(c)	127.63	108.14	182.37
	(d)	0.0957	0.0776	0.1128
NUB-SPT	(a)	72.31	96.86	144.90
	(b)	10.32	27.48	42.28
	(c)	92.82	73.99	129.60
	(d)	0.1150	0.0968	0.1479

* Sample size = 9000 branches; nominal utilization = 90%; RAND study conditions for branches as jobs (from Appendix C–7).
† Table entries: (a) $\bar{F}$, mean flow-time, (b) $\bar{A}$, mean assembly wait-time, (c) $\bar{T}_c$, conditional tardiness mean, (d) f_t, fraction of jobs tardy.

values. Since the flow-time distribution under SPT is highly skewed to the right, as the sample size increases, the observed maximum of the sample will increase rapidly. In contrast, S/OPN results do not increase as rapidly, since the branches are coupled with a common due-date. The same is true of MAXNRD-SPT and NUB-SPT, since the status information couples the branches for these rules.

There is a reversal in the comparison between MAXNRD-SPT and NUB-SPT. With two or five branches per job, MAXNRD-SPT was superior to NUB-SPT with respect to flow-time, but became substantially worse with ten branches per job. It may well be that with a larger number of branches NUB-SPT is more discriminating on the primary priority classification and relies less on SPT than does MAXNRD-SPT.

BIBLIOGRAPHY

1. S. S. ACKERMAN, "Even-Flow, A Scheduling Method for Reducing Lateness in Job Shops," *Management Technology* **3**, No. 1, May 1963

2. M. A. ACZEL, "The Effect of Introducing Priorities," *Operations Research* **8**, No. 5, September 1960

3. S. B. AKERS, JR., "A Graphical Approach to Production Scheduling Problems," *Operations Research* **4**, No. 2, April 1956

4. S. B. AKERS, JR., and J. FRIEDMAN, "A Non-Numerical Approach to Production Scheduling Problems," *Operations Research* **3**, No. 4, November 1955

5. B. AVI-ITZHAK, "Preemptive Repeat Priority Queues as a Special Case of the Multi-purpose Server Problem—I and II," *Operations Research* **11**, No. 4, July 1963

6. B. AVI-ITZHAK, "A Sequence of Service Stations with Arbitrary Input and Regular Service Times," *Management Science* **11**, No. 5, March 1965

7. B. AVI-ITZHAK, W. L. MAXWELL, and L. W. MILLER, "Queuing with Alternating Priorities," *Operations Research* **13**, No. 2, 1965

8. B. AVI-ITZHAK and P. NAOR, "Some Queuing Problems with the Service Station Subject to Breakdown," *Operations Research* **11**, No. 3, May 1963

9. B. AVI-ITZHAK and M. YADIN, "A Sequence of Two Servers with No Intermediate Queue," *Management Science* **11**, No. 5, March 1965

10. C. T. BAKER and B. P. DZIELINSKI, "Simulation of a Simplified Job Shop," *IBM Business Systems Research Memorandium*, August 1, 1958; also *Management Science* **6**, No. 3, April 1960

11. C. T. BAKER, B. P. DZIELINSKI, and A. S. MANNE, "Simulation Tests of Lot Size Programming," *Management Science* **9**, No. 2, January 1963

12. W. BAKKER, "De Levertijd in Machinefabrieken," thesis, Technische Hogeschool at Delft, Holland, June 1965

13. A. N. BAKHRU and M. R. RAO, "An Experimental Investigation of Job-Shop Scheduling," research report, Department of Industrial Engineering, Cornell University, 1964

14. B. P. BANERJEE, "Single Facility Sequencing with Random Execution Times," *Operations Research* **13**, No. 3, May 1965

15. L. L. BARACHET, "Graphic Solution of the Traveling Salesman Problem," *Operations Research* **5**, No. 6, December 1957

16. E. W. BARANKIN, "The Scheduling Problem as an Algebraic Generalization of Ordinary Linear Programming," UCLA Discussion Paper No. 9, August 28, 1952

17. J. Y. BARRY, "A Priority Queuing Problem," *Operations Research* **4**, No. 3, June 1956

18. R. BELLMAN, "Some Mathematical Aspects of Scheduling Theory," *J. Soc. Ind. and Appl. Math.* **4**, No. 3, September 1956

19. R. BELLMAN, "Dynamic Programming Treatment of the Travelling Salesman Problem," *J. Assn. for Computing Machinery* **9**, No. 1, January 1962

20. D. H. BESSEL, "List Processing FORTRAN," master's thesis, Cornell University, September 1964

21. E. H. BOWMAN, "The Schedule-Sequencing Problem," *Operations Research* **7**, No. 5, September 1959

22. G. H. BROOKS and C. R. WHITE, "An Algorithm for Finding Optimal or Near-Optimal Solutions to the Production Scheduling Problem," *J. Ind. Eng.* **16**, No. 1, January 1965

23. P. J. BURKE, "Equilibrium Delay Distribution for One Channel with Constant Holding Time, Poisson Input, and Random Service," *Bell System Technical Journal* **38**, No. 4, July 1959

24. P. J. BURKE, "The Output of Queuing Systems," *Operations Research* **4**, No. 6, December 1956

25. D. C. CARROLL, "Heuristic Sequencing of Single and Multiple Component Jobs," Ph.D. thesis, MIT, June 1965

26. A. CHARNES and W. W. COOPER, "A Network Interpretation and a Directed Sub-Dual Algorithm for Critical-Path Scheduling," *J. Ind. Eng.* **13**, No. 4, July 1962

27. W. CLARK, *The Gantt Chart* (3rd edition). London: Pitman and Sons, 1952

28. A. COBHAM, "Priority Assignment in Waiting Line Problems," *Operations Research* **2**, No. 1, February 1954

29. R. W. CONWAY, "Some Tactical Problems in Digital Simulation," *Management Science* **10**, No. 1, October 1963

30. R. W. CONWAY, "Priority Dispatching and Work-in-Process Inventory in a Job Shop," *J. Ind. Eng.* **16**, No. 2, March 1965

31. R. W. CONWAY, "Priority Dispatching and Job Lateness in a Job Shop," *J. Ind. Eng.* **16**, No. 4, July 1965

32. R. W. CONWAY, "An Experimental Investigation of Priority Assignment in a Job Shop," RAND Corporation Memorandum RM-3789-PR, February 1964

33. R. W. CONWAY and W. L. MAXWELL, "Network Dispatching by the Shortest Operation Discipline," *Operations Research* **10**, No. 1, February 1962, pages 51–73. Also *Industrial Scheduling*, J. F. Muth and G. L. Thompson, eds. Englewood Cliffs, N. J.: Prentice-Hall, 1963, Chapter 17

34. R. W. CONWAY, B. M. JOHNSON and W. L. MAXWELL, "An Experimental Investigation of Priority Dispatching," *J. Ind. Eng.* **11**, No. 3, May 1960

35. R. W. CONWAY, W. L. MAXWELL, J. DELFAUSSE, and W. WALKER, "CLP—The Cornell List Processor," *Communications of the ACM*, **8**, No. 4, April 1965

36. R. W. CONWAY, W. L. MAXWELL, and J. W. OLDZIEY, "Sequencing Against Due-Dates," *Proceedings of IFORS Conference*, Cambridge, Mass., September 1966

37. D. R. COX and W. L. SMITH, "On the Superposition of Renewal Processes," *Biometrica* **41**, Parts 1 and 2, June 1954

38. D. R. COX and W. L. SMITH, *Queues*. New York: John Wiley, 1961

39. D. R. COX, *Renewal Theory*. London: Methuen, 1962

40. T. B. CRABILL, "A Lower-Bound Approach to the Scheduling Problem," research report, Department of Industrial Engineering, Cornell University, 1964

41. G. A. CROES, "A Method for Solving Traveling-Salesman Problems," *Operations Research* **6**, No. 6, November 1958

42. M. F. DACEY, "Selection of an Initial Solution for the Traveling-Salesman Problem," *Operations Research* **8**, No. 1, January 1960

43. G. B. DANTZIG, "A Machine-Job Scheduling Model," *Management Science* **6**, No. 2, January 1960

44. G. B. DANTZIG, D. R. FULKERSON, and S. M. JOHNSON, "On a Linear-Programming Combinatorial Approach to the Traveling-Salesman Problem," *Operations Research* **7**, No. 1, January 1959

45. R. A. DUDEK and O. F. TEUTON, JR., "Development of *M*-Stage Decision Rule for Scheduling *n* Jobs Through *m* Machines," *Operations Research* **12**, No. 3, May 1964

46. W. L. EASTMAN, S. EVEN, and I. M. ISAACS, "Bounds for the Optimal Scheduling of *n* Jobs on *m* Processors," *Management Science* **11**, No. 2, November 1964

47. S. E. ELMAGHRABY and R. T. COLE, "On the Control of Production in Small Job Shops," *J. Ind. Eng.* **14**, No. 4, July 1963

48. S. E. ELMAGHRABY and A. S. GINSBERG, "A Dynamic Model of the Optimal Loading of Linear Multi-Operation Shops," *Management Technology* **4**, No. 1, June 1964

49. J. P. EVANS, "A Comparison of Two Queue Disciplines for Multi-Channel Queuing Systems," master's thesis, Cornell University, September 1962

50. W. J. FABRYCKY and J. E. SHAMBLIN, "A Probability-Based Sequencing Algorithm," *J. Ind. Eng.* **18**, No. 6, June 1966

51. D. W. FIFE, "Scheduling with Random Arrivals and Linear Loss Functions," *Management Science* **11**, No. 3, January 1965

52. H. FISCHER and G. L. THOMPSON, "Probabilistic Learning Combinations of Local Job-Shop Scheduling Rules," *Industrial Scheduling.* J. F. Muth and G. L. Thompson, eds., Englewood Cliffs, N. J.; Prentice-Hall, 1963, Chapter 15

53. M. FLOOD, "The Traveling-Salesman Problem," *Operations Research* **4**, No. 1, February 1956

54. L. R. FORD, JR., and D. R. FULKERSON, *Flows in Networks.* Princeton, N. J.: Princeton University Press, 1962

55. H. D. FRIEDMAN, "Reduction Methods for Tandem Queuing Systems," *Operations Research* **13**, No. 1, January 1965

56. W. GAPP, P. S. MANKEKAR, and L. G. MITTEN, "Sequencing Operations to Minimize In-Process Inventory Costs," *Management Science* **11**, No. 3, January 1965

57. D. P. GAVER, JR., "A Waiting Line with Interrupted Service, Including Priorities," *J. Royal Stat. Soc. Series B*, **24**, No. 1, June 1962

58. D. P. GAVER, JR., "Accommodation of Second-Class Traffic," *Operations Research* **11**, No. 1, January 1963

59. D. P. GAVER, JR., "A Comparison of Queue Disciplines when Service Orientation Times Occur," *Nav. Res. Log. Quart.* **10**, No. 3, September 1963

60. J. W. GAVETT, "Three Heuristic Rules for Sequencing Jobs to a Single Production Facility," *Management Science* **11**, No. 8, June 1965

61. W. GERE, "A Heuristic Approach to Job-Shop Scheduling," Ph.D. thesis, Carnegie Institute of Technology, 1962

62. B. GIFFLER, "Schedule Algebras and Their Use in Formulating General Systems Simulations," *Industrial Scheduling*, J. F. Muth and G. L. Thompson, eds., Englewood Cliffs, N. J.: Prentice-Hall, 1963, Chapter 4

63. B. GIFFLER, "Scheduling General Production Systems Using Schedule Algebra," *Nav. Res. Log. Quart.* **10**, No. 3, September 1963

64. B. GIFFLER and G. L. THOMPSON, "Algorithms for Solving Production-Scheduling Problems," *Operations Research* **8**, No. 4, July 1960

65. B. GIFFLER, G. L. THOMPSON, and V. VAN NESS, "Numerical Experience with the Linear and Monte Carlo Algorithms for Solving Production Scheduling Problems," *Industrial Scheduling*. J. F. Muth and G. L. Thompson, eds., Englewood Cliffs, N. J.: Prentice-Hall, 1963, Chapter 3

66. R. J. GIGLIO and H. M. WAGNER, "Approximate Solutions to the Three-Machine Scheduling Problem," *Operations Research* **12**, No. 2, March 1964

67. P. C. GILMORE, "Optimal and Suboptimal Algorithms for the Quadratic Assignment Problem," *J. Soc. Ind. and Appl. Math.* **10**, No. 2, June 1962

68. P. C. GILMORE and R. E. GOMORY, "Sequencing a One State-Variable Machine: A Solvable Case of the Traveling-Salesman Problem," *Operations Research* **12**, No. 5, September 1964

69. R. E. GOMORY, "An Algorithm for Integer Solutions to Linear Programs," Princeton-IBM Mathematics Research Project, Technical Report No. 1, November 17, 1958

70. M. H. GOTTERER, "Scheduling with Deadlines, Priorities, and Nonlinear Loss Functions," research report, Department of Industrial Engineering, Georgia Institute of Technology, Atlanta, Ga.

71. A. A. GRINDLAY, "Tandem Queues with Dynamic Priorities," *Operations Research* **11**, No. 2, March 1963

72. W. W. HARDGRAVE and G. NEMHAUSER, "A Geometric Model and Graphical Algorithm for a Sequencing Problem," *Operations Research* **11**, No. 6, November 1963

73. A. G. HAWKES, "Queuing at Traffic Intersections," *Proc. Second Symposium on Theory of Traffic Flow*, London, 1963

74. C. R. HEATHCOTE, "A Simple Queue with Several Preemptive Priority Classes," *Operations Research* **8**, No. 5, September 1960

75. M. HELD and R. M. KARP, "A Dynamic Programming Approach to Sequencing Problems," *J. Soc. Ind. and Appl. Math.* **10**, No. 2, March 1962

76. J. HELLER, "Combinatorial, Probabilistic, and Statistical Aspects of an $M \times J$ Scheduling Problem," NYO-2540, AEC Research and Development Report, February 1959

77. J. HELLER, "Some Problems in Linear Graph Theory that Arise in the Analysis of the Sequencing of Jobs through Machines," Report NYO-9487, AEC Computing and Applied Mathematics Center, New York, October 15, 1960

78. J. HELLER, "Some Numerical Experiments for an $M \times J$ Flow Shop and its Decision-Theoretical Aspects," *Operations Research* **8**, No. 2, March 1960

79. J. HELLER and G. LOGEMANN, "An Algorithm for the Construction and Evaluation of Feasible Schedules," *Management Science* **8**, No. 3, January 1962

80. W. A. HEUSER, JR., and B. E. WYNNE, JR., "An Application of the Critical Path Method to Job Shop Scheduling—A Case Study," *Management Technology* **3,** No. 2, December 1963

81. C. C. HOLT, "Priority Rules for Minimizing the Cost of Queues in Machine Scheduling," *Industrial Scheduling*, J. F. Muth and G. L. Thompson, eds., Englewood Cliffs, N. J.: Prentice-Hall, 1963, Chapter 6

82. C. C. HOLT, MODIGLIANI, MUTH, and SIMON, *Planning Production, Inventories and Work Force.* Englewood Cliffs, N. J.: Prentice-Hall, 1960

83. T. C. HU, "Parallel Sequencing and Assembly Line Problems," *Operations Research* **9,** No. 6, November 1961

84. G. C. HUNT, "Sequential Arrays of Waiting Lines," *Operations Research* **4,** No. 6, December 1956

85. IBM Mathematics and Applications Department, *The Job Shop Simulator.* 1271 Avenue of the Americas, New York, 1960

86. E. J. IGNALL, "A Review of Assembly-Line Balancing," *J. Ind. Eng.* **16,** No. 4, July–August, 1965

87. E. IGNALL and L. SCHRAGE, "Application of the Branch-and-Bound Technique to Some Flow-Shop Scheduling Problems," *Operations Research* **13,** No. 3, May 1965

88. J. R. JACKSON, "Notes on Some Scheduling Problems," Research Report No. 35, *Management Sciences Research Project*, UCLA, October 1, 1954

89. J. R. JACKSON, "Two One-Machine Scheduling Problems," Research Report 47, *Management Sciences Research Project*, UCLA, December 1954

90. J. R. JACKSON, "Scheduling a Production Line to Minimize Maximum Tardiness," Research Report 43, *Management Sciences Research Project*, UCLA, January 1955

91. J. R. JACKSON, "An Extension of Johnson's Results on Job-Lot Scheduling," *Nav. Res. Log. Quart.* **3,** No. 3, September 1956

92. J. R. JACKSON, "Networks of Waiting Lines," *Operations Research* **5,** No. 4, August 1957

93. J. R. JACKSON, "Simulation Research on Job-Shop Production," *Nav. Res. Log. Quart.* **4,** No. 4, December 1957

94. J. R. JACKSON, "Some Problems in Queueing with Dynamic Properties," *Nav. Res. Log. Quart.* **7,** No. 3, September 1960

95. J. R. JACKSON, "Sample Distribution of Mean Waiting Time in Queues," Research Report 67, *Management Sciences Research Project*, UCLA, September 20, 1960

96. J. R. JACKSON, "Simulation of Queues with Dynamic Priorities." Research Report 71, *Management Sciences Research Project*, UCLA, March 20, 1961

97. J. R. JACKSON, "Queues with Dynamic Priority Discipline," *Management Science* **8,** No. 1, October 1961

98. J. R. JACKSON, "Simulation as Experimental Mathematics." Research Report 72, *Management Sciences Research Project*, UCLA, June 22, 1961

99. J. R. JACKSON, "Waiting-Time Distributions for Queues with Dynamic Priorities," *Nav. Res. Log. Quart.* **9,** No. 1, March 1962

100. J. R. JACKSON, "Jobshop-Like Queuing Systems," Research Report 81, *Management Sciences Research Project*, UCLA, January 1963

101. J. R. JACKSON, and R. T. NELSON, "SWAC Computations for Some $m \times n$ Scheduling Problems," *J. Assn. for Computing Machinery* **4**, No. 4, 1957

102. B. JEREMIAH, A. LALCHANDANI, and L. SCHRAGE, "Heuristic Rules Toward Optimal Scheduling," research report, Department of Industrial Engineering, Cornell University, 1964

103. S. M. JOHNSON, "Optimal Two- and Three-Stage Production Schedules with Setup Times Included," *Nav. Res. Log. Quart.* **1**, No. 1, March 1954; also *Industrial Scheduling*, J. F. Muth and G. L. Thompson, eds., Englewood Cliffs, N. J.: Prentice-Hall, 1963, Chapter 2

104. S. M. JOHNSON, "Discussion: Sequencing n Jobs on Two Machines with Arbitrary Time Lags," *Management Science* **5**, No. 3, April 1959

105. W. KARUSH, "A Counterexample to a Proposed Algorithm for Optimal Sequencing of Jobs," *Operations Research* **13**, No. 2, March 1965

106. J. E. KELLEY, JR., "Critical-Path Planning and Scheduling: Mathematical Basis," *Operations Research* **9**, No. 3, May 1961

107. D. G. KENDALL, "Some Problems in the Theory of Queues," *J. Royal Stat. Soc.*, Series B **13**, No. 2, 1951

108. A. Y. KHINTCHINE, *Mathematical Methods in the Theory of Queuing.* New York: Hafner, 1960

109. H. KESTEN and J. TH. RUNNENBERG, "Priority in Waiting Line Problems," *Proc. Koninklijke Nederlandse Akademie van Wetenschappen*, A-60, No. 3 1957

110. J. F. C. KINGMAN, "Two Similar Queues in Parallel," *Ann. Math. Stat.* **32**, No. 4, December 1961

111. J. F. C. KINGMAN, "On Queues in Which Customers Are Served in Random Order," *Proc. Cambridge Phil. Soc.* **58**, Part 1, 1962

112. L. KLEINROCK, "A Delay Dependent Queue Discipline," *Nav. Res. Log. Quart.* **11**, No. 4, December 1964

113. E. KOENIGSBERG, "On Jockeying in Queues," *Management Science* **12**, No. 5, January 1966

114. H. S. KRASNOW and R. A. MERIKALLIO, "The Past, Present, and Future of General Simulation Languages," *Management Science* **11**, No. 2, November 1964

115. Y. KURATANI and J. L. McKENNEY, "A Preliminary Report on Job-Shop Simulation Research," Research Report 65, *Management Sciences Research Project*, UCLA, March 16, 1958

116. Z. A. LOMNICKI, "A Branch-and-Bound Algorithm for the Exact Solution of the Three-Machine Scheduling Problem," *Operational Research Quarterly* **16**, 1965

117. E. L. LAWLER, "The Quadratic Assignment Problem," *Management Science* **9**, No. 4, July 1963

118. E. L. LAWLER, "On Scheduling Problems with Deferral Costs," *Management Science* **11**, No. 2, November 1964

119. E. LeGRANDE, "The Development of a Factory Simulation Using Actual Operating Data," *Management Technology* **3**, No. 1, May 1963

120. F. K. LEVY, G. L. THOMPSON, and J. D. WEIST, "Multiship, Multishop, Workload-Smoothing Program," *Nav. Res. Log. Quart.* **9**, No. 1, March 1962

121. E. LIGTENBERG, "Minimal Cost Sequencing of n Grouped and Ordered Jobs on m Machines," *J. Ind. Eng.* **17**, No. 4, April 1966

122. J. D. C. LITTLE, "A Proof for the Queuing Formula $L = \lambda W$," *Operations Research* **9**, No. 3, May 1961

123. J. D. C. LITTLE, K. G. MURTY, D. W. SWEENY, and C. KAREL, "An Algorithm for the Traveling-Salesman Problem," *Operations Research* **11**, No. 6, November 1963

124. D. G. MALCOLM, J. H. ROSEBOOM, C. E. CLARK, and W. FAZAR, "Applications of a Technique for Research and Development Program Evaluation," *Operations Research* **7**, No. 5, September 1959

125. P. S. MANKEKAR and L. G. MITTEN, "The Constrained Least-Cost Testing Sequence Problem," *J. Ind. Eng.* **16**, No. 2, March 1965

126. A. S. MANNE, "On the Job-Shop Scheduling Problem," *Operations Research* **8**, No. 2, March 1960

127. H. M. MARKOWITZ, B. HAUSNER, and H. W. KARR, "SIMSCRIPT: A Simulation Programming Language," RM-3310-PR, RAND Corporation, November 1962 (published by Prentice-Hall, 1963)

128. W. L. MAXWELL, "An Investigation of Multi-Product, Single-Machine Scheduling and Inventory Problems," Ph.D. thesis, Cornell University, September 1961

129. W. L. MAXWELL, "The Scheduling of Economic Lot Sizes," *Nav. Res. Log. Quart.* **11**, No. 2, June 1964

130. W. L. MAXWELL, "Priority Dispatching and Assembly Operations in a Job Shop," RM-5370-PR, RAND Corporation, Santa Monica, Cal.

131. J. MAKINO, "On a Scheduling Problem," *J. Operations Research Soc. of Japan* **8**, No. 1, September 1965

132. J. O. MAYHUGH, "On the Mathematical Theory of Schedules," *Management Science* **11**, No. 2, November 1964

133. R. MCNAUGHTON, "Scheduling with Deadlines and Loss Functions," *Management Science* **6**, No. 1, October 1959

134. M. MEHRA, "An Experimental Investigation of Job-Shop Scheduling with Assembly Constraints," master's thesis, Cornell University, June 1967

135. L. W. MILLER, "Alternating Priorities in Multiclass Queues," Ph.D thesis, Cornell University, September 1964

136. L. W. MILLER, "Selection Disciplines in a Single-Server Queueing System," Research Memorandum RM-4693-PR, RAND Corporation, Santa Monica, Cal., October 1966

137. L. W. MILLER and L. SCHRAGE, "The Queue $M/G/1$ with the Shortest Remaining Processing Time Discipline," RAND Paper P-3263, November 1965

138. R. G. MILLER, JR., "Priority Queues," *Ann. Math. Stat.* **31**, No. 1, 1960

139. G. J. MINTY, "A Comment on the Shortest-Route Problem," *Operations Research* **5**, No. 5, October 1957

140. L. G. MITTEN, "Sequencing n Jobs on Two Machines with Arbitrary Time Lags," *Management Science* **5**, No. 3, April 1959

141. L. G. MITTEN, "A Scheduling Problem," *J. Ind. Eng.* **10**, No. 2, March 1959

142. J. J. MODER and C. R. PHILLIPS, *Project Management with CPM and PERT.* New York: Reinhold, 1964

143. P. M. Morse, *Queues, Inventories and Maintenance.* New York: John Wiley, 1958

144. J. F. Muth and G. L. Thompson, eds., *Industrial Scheduling,* Englewood Cliffs, N. J.: Prentice-Hall, 1963

145. I. Nabeshima, "The Order of n Items Processed on m Machines," *J. Operations Research Soc. of Japan,* 1961 (2 papers)

146. M. D. Naik, "m by n Job-Shop Scheduling," master's thesis, Cornell University, June 1967

147. Y. R. Nanot, "An Experimental Investigation and Comparative Evaluation of Priority Disciplines in Job-Shop-Like Queuing Networks," Research Report 87, *Management Sciences Research Project,* UCLA, December 1963

148. H. A. Neimeier, "An Investigation of Alternative Routing in a Job Shop," master's thesis, Cornell University, June 1967

149. R. T. Nelson, "Enumeration of a Three Job, Three Machine Scheduling Problem on SWAC," Discussion Paper No. 50, *Management Sciences Research Project,* UCLA, January 11, 1955

150. R. T. Nelson, "Priority Function Methods for Job-Lot Scheduling," Research Report 51, *Management Sciences Research Project,* UCLA, February 24, 1955

151. R. T. Nelson, "Waiting-Time Distributions for Application to a Series of Service Centers," *Operations Research* **6,** No. 6, November 1958

152. R. T. Nelson, "An Extension of Queuing Theory Results to a Series of Service Centers," Discussion Paper 68, *Management Sciences Research Project,* UCLA, April 29, 1958

153. R. T. Nelson, "An Empirical Study of Arrival, Service Time, and Waiting Time Distributions of a Job Shop Production Process," Research Report 60, *Management Sciences Research Project,* UCLA, June 19, 1959

154. R. T. Nelson, "A Simulation Study and Analysis of a Two-Station, Waiting-Line Network Model," Research Report 91, *Management Sciences Research Project,* January 5, 1965

155. R. T. Nelson, "Labor and Machine Limited Production Systems," *Management Sciences Research Project,* Paper 90, January 1966

156. R. T. Nelson, "Labor Assignment as a Dynamic Control Problem," *Operations Research* **14,** No. 3, May 1966

157. C. E. Nugent, "On Sampling Approaches to the Solution of the n-by-m Static Sequencing Problem," Ph.D. thesis, Cornell University, September 1964

158. J. W. Oldziey, "Dispatching Rules and Job Tardiness in a Simulated Job Shop," master's thesis, Cornell University, February 1966

159. G. L. Orkin, "An Experimental Investigation of Shop Loading for Setting Operation Due-Dates," master's thesis, Cornell University, February 1966

160. E. S. Page, "An Approach to the Scheduling of Jobs on Machines," *J. Royal Stat. Soc.,* Series B, **23,** No. 2, 1961

161. D. S. Palmer, "Sequencing Jobs Through a Multi-Stage Process in the Minimum Total Time—A Quick Method of Obtaining a Near Optimum," *Operational Research Quarterly* **16,** No. 1, 1965

162. E. Parzen, *Stochastic Processes.* San Francisco: Holden Day, 1962

163. T. E. Phipps, Jr., "Machine Repair as a Priority Waiting-Line Problem," *Operations Research* **4,** No. 1, February 1956

164. S. POLLACK and W. WIEBENSON, "Solutions of the Shortest Route Problem—A Review," *Operations Research* **8**, No. 2, March 1960

165. N. U. PRABHU, *Queues and Inventories*. New York: John Wiley, 1965

166. A. A. B. PRITSKER and W. W. HAPP, "GERT: Graphical Evaluation and Review Technique," *J. Ind. Eng.* **17**, No. 5, May 1966

167. E. REICH, "Waiting Times when Queues are in Tandem," *Ann. Math. Stat.* **28**, No. 3, 1957

168. R. C. REINITZ, "An Integrated Job-Shop Scheduling Problem," Ph.D. thesis, Case Institute of Technology, 1961

169. R. C. REINITZ, "On the Job Shop Scheduling Problem," *Industrial Scheduling*, J. F. Muth and G. L. Thompson, eds. Englewood Cliffs, N. J.: Prentice-Hall, 1963, Chapter 5

170. J. RIORDAN, *Stochastic Service Systems*. New York: John Wiley, 1962

171. J. G. ROOT, "Scheduling with Deadlines and Loss Functions on k Parallel Machines," *Management Science* **11**, No. 3, January 1965

172. M. H. ROTHKOPF, "Scheduling Independent Tasks on One or More Processors," Ph.D. thesis, MIT, January 1964

173. M. H. ROTHKOPF, "Scheduling Independent Tasks on Parallel Processors," *Management Science* **12**, No. 5, January 1966

174. M. H. ROTHKOPF, "Scheduling with Random Service Times," *Management Science* **12**, No. 9, May 1966

175. A. J. ROWE, "Sequential Decision Rules in Production Scheduling," Ph.D. thesis, UCLA, August 1958

176. A. J. ROWE, "Toward a Theory of Scheduling," *J. Ind. Eng.* **11**, No. 2, March 1960

177. F. J. RUSSO, "A Heuristic Approach to Alternate Routing in a Job Shop," master's thesis, MIT, June 1965

178. T. L. SAATY, *Elements of Queueing Theory*. New York: McGraw-Hill, 1961

179. P. SANDEMAN, "Empirical Design of Priority Waiting Times for Jobbing Shop Control," *Operations Research* **9**, No. 4, July 1961

180. M. SASIENI, A. YASPAN, and L. FRIEDMAN, *Operations Research: Methods and Problems*. New York: John Wiley, 1959, Chapter 9

181. A. SCHILD and I. J. FREDMAN, "Scheduling Tasks with Linear Loss Functions," *Management Science* **7**, No. 3, April 1961

182. A. SCHILD and I. J. FREDMAN, "Scheduling Tasks with Deadlines and Nonlinear Loss Functions," *Management Science* **9**, No. 1, October 1962

183. L. E. SCHRAGE, "A Survey of Priority Queueing," master's thesis, Cornell University, February 1965

184. L. E. SCHRAGE, "Some Queuing Models for a Time-Shared Facility," Ph.D. thesis, Cornell University, February 1966

185. M. SEGAL, "On the Behavior of Queues Subject to Job-Shop Conditions," D. Eng. thesis, John Hopkins University, June 1961

186. R. L. SISSON, "Sequencing Theory," *Progress in Operations Research*, Vol. 1. R. L. Ackoff, ed. New York: John Wiley, 1961, Chapter 7

187. W. E. SMITH, "Various Optimizers for Single-State Production," *Nav. Res. Log. Quart.* **3**, No. 1, March 1956

188. W. L. SMITH, "Renewal Theory and Its Ramifications," *J. Royal Stat. Soc.*, *Series B*, **20**, No. 2, 1958

189. A. E. STORY and H. M. WAGNER, "Computational Experience with Integer Programming for Job-Shop Scheduling," *Industrial Scheduling*, J. F. Muth and G. L. Thompson, eds. Englewood Cliffs, N. J.: Prentice-Hall, 1963, Chapter 14

190. W. SZWARC, "Solution of the Akers-Friedman Scheduling Problem," *Operations Research* **8**, No. 6, November 1960

191. L. TAKACS, *Introduction to the Theory of Queues*. New York: Oxford University Press, 1962

192. J. C. TANNER, "A Problem of Interference Between Two Queues," *Biometrica* **40**, Parts 1 and 2, June 1953

193. G. L. THOMPSON, "Recent Developments in the Job-Shop Scheduling Problem," *Nav. Res. Log. Quart.* **7**, No. 4, December 1960

194. G. L. THOMPSON and R. L. KARG, "A Heuristic Approach to Solving Travelling-Salesman Problems," *Management Science* **10**, No. 2, January 1964

195. D. R. TRILLING, "Job-Shop Simulation of Orders That Are Networks," *J. Ind. Eng.* **17**, No. 2, February 1966

196. S. P. VAN DER ZEE and H. THEIL, "Priority Assignment in Waiting-Line Problems Under Conditions of Misclassification," *Operations Research* **9**, No. 6, November 1961

197. H. M. WAGNER, "An Integer Linear-Programming Model for Machine Scheduling," *Nav. Res. Log. Quart.* **6**, No. 2, June 1959

198. R. D. WAYSON, "The Effects of Alternate Machines on Two Priority Dispatching Disciplines in the General Job Shops," master's thesis, Cornell University, February 1965

199. J. D. WEIST, "Some Properties of Schedules for Large Projects with Limited Resources," *Operations Research* **12**, No. 3, May 1964

200. P. D. WELCH, "Some Contributions to the Theory of Priority Queues," Ph.D. thesis, Columbia University, April 1963

201. H. WHITE and L. S. CHRISTIE, "Queuing with Preemptive Priorities or Breakdown," *Operations Research* **6**, No. 1, January 1958

202. D. M. G. WISHART, "Queuing Systems in Which the Discipline is Last-Come, First-Served," *Operations Research* **8**, No. 5, September 1960

APPENDIX A

THE LAPLACE-STIELTJES TRANSFORM
OF A DISTRIBUTION FUNCTION

Given that $G(t)$ is the distribution function of the random variable, T, defined by

$$G(t) = \text{Prob } (T \leq t),$$

the corresponding Laplace-Stieltjes transform is

$$\gamma(z) = E(e^{-zT}) = \int_0^\infty e^{-zt} \, dG(t). \tag{A-1}$$

The use of the Stieltjes integral allows this definition to be applied to distribution functions that are continuous, discrete, or a mixture of the two types. For example, given that $G(t)$ has a derivative, $g(t)$, everywhere except for a jump representing a concentration of probability mass at some point x, then

$$\gamma(z) = \int_0^{x^-} e^{-zt} g(t) \, dt + e^{-zx} \left(G(x) - G(x^-) \right) + \int_x^\infty e^{-zt} g(t) \, dt.$$

The properties of Laplace transforms associated with distributions which are used in this book are stated below without proof:

a) For $z \geq 0$ (or the real part of $z \geq 0$ if z is complex), $\gamma(z)$ exists, that is, $\gamma(z)$ is finite.

b) There is a one-to-one correspondence between a distribution function and its associated Laplace transform so that one uniquely determines the other. It is often possible to invert a Laplace transform to recover its associated distribution function by means of tables or inversion formulas.

c) If the kth moment of T, denoted by $E(T^k)$, exists, it is given by

$$E(T^k) = (-1)^k \frac{d^k \gamma(z)}{dz^k} \bigg|_{z=0}.$$

This is the "moment-generating" property.

d) If T_1 and T_2 are independently distributed random variables and $T = T_1 + T_2$, the distribution of T is given by

$$G(t) = \text{Prob } (T_1 + T_2 \leq t) = \int_{x=0}^t G_1(t - x) \, dG_2(x),$$

and

$$\gamma(z) = \gamma_1(z)\gamma_2(z).$$

This is the "convolution theorem" for Laplace transforms. It may be applied recursively so that the Laplace transform associated with the sum of n independently and identically distributed random variables, with Laplace transform $\gamma(z)$, would be $(\gamma(z))^n$. The integral above, known as a convolution, is sometimes abbreviated by the symbol: $G_1 * G_2(t)$.

259

When the distribution is discrete, with all probability mass concentrated at nonnegative integer points, it is convenient to define the probability-generating function, $\Pi(\xi)$, by letting $\xi = e^{-z}$ in Eq. (A-1). Then if $p_n = \text{Prob } (T = n)$, the probability-generating function would be defined by

$$\Pi(\xi) = \sum_{n=0}^{\infty} \xi^n p_n,$$

from which the factorial moments may be obtained by the formula

$$E(T(T - 1)(T - 2) \cdots (T - k + 1)) = \left. \frac{d^k \Pi(\xi)}{d\xi^k} \right|_{\xi=1},$$

and the individual probabilities are recoverable from the relation

$$p_n = \frac{1}{n!} \left. \frac{d^n \Pi(\xi)}{d\xi^n} \right|_{\xi=0}.$$

EXPERIMENTAL RESULTS:
n/m JOB-SHOP PROBLEM

Table B–1 Mean Flow-Time

Problem number	n	m	Routing	Replication	Type schedule	RANDOM
12	10	4	Random	1	Active Nondelay	*29.0*
13				2	Active Nondelay	*29.1*
14				3	Active Nondelay	*59.2*
15				4	Active Nondelay	*46.2*
16				5	Active Nondelay	*68.5*
17				6	Active Nondelay	*68.2*
18				7	Active Nondelay	*48.4*
19				8	Active Nondelay	*53.5*
20				9	Active Nondelay	*54.7*
21				10	Active Nondelay	43.2
22	10	4	Random	11	Active Nondelay	28.5
23				12	Active Nondelay	40.4
24				13	Active Nondelay	*68.1*
25				14	Active Nondelay	*40.8*
26	20	4	Random	1	Active	54.3
27				2	Active	*53.5*
28				3	Active	91.0

MOPNR	MWKR −P	MWKR/P	SPT	MWKR	LWKR	LPT	FCFS
32.0	33.0	33.0	29.9	32.4			
			16.3	26.2	17.4	*11.6*	15.2
33.7	33.8	33.8	29.7	33.4			
			22.3	29.3	*18.9*	21.5	22.0
72.0	78.6	71.6	61.6	75.4			
			47.2	*42.7*	42.8	54.0	46.1
54.8	53.4	50.0	49.0	52.7			
			36.9	44.8	*35.9*	42.5	40.5
94.5	88.8	86.1	70.4	85.6			
			44.0	61.4	*39.1*	59.6	50.4
89.2	85.4	82.5	68.7	84.0			
			53.7	73.1	53.8	65.8	67.5
64.9	62.5	59.2	52.2	61.7			
			37.4	44.9	*37.0*	48.6	40.2
69.0	64.9	69.0	57.9	66.6			
			39.0	42.4	*35.1*	48.0	46.7
70.4	72.2	72.2	59.8	72.6			
			37.1	52.5	37.3	43.7	38.1
51.7	58.0	57.5	*43.1*	57.6			
			37.9	46.7	*32.7*	39.5	39.1
29.8	31.0	30.1	*25.7*	34.2			
			24.1	33.2	*18.7*	28.9	26.1
42.9	43.5	46.4	*36.9*	42.2			
			24.8	36.7	25.0	31.6	30.4
91.5	94.1	102.7	75.9	96.2			
			43.9	70.6	49.0	63.8	46.4
48.0	44.4	50.3	43.8	49.8			
			31.0	47.5	39.0	48.0	31.7
64.9	70.6	64.3	*47.5*	74.2			
75.8	77.8	73.2	54.3	74.5			
128.4	126.3	129.0	*89.4*	131.0			

(*continued*)

Table B–1 (*continued*)

Problem number	n	m	Routing	Replication	Type schedule	RANDOM
29				4	Active	71.4
30				5	Active	74.7
31				6	Active	*67.4*
32				7	Active	87.9
33				8	Active	*51.1*
34				9	Active	109.8
35				10	Active	44.4
36	20	6	Random	1	Active	100.0
37				2	Active	95.5
38	20	6	Random	3	Active	109.3
39				4	Active	*114.5*
40				5	Active	82.5
41				6	Active	*106.2*
42				7	Active	121.7
43				8	Active	*67.3*
44				9	Active	*103.7*
45				10	Active	*107.4*
46	10	4	FixR	1	Active	
47				2	Active	
48				3	Active	
49				4	Active	
50				5	Active	
51				6	Active	
52				7	Active	
53	10	4	FixR	8	Active	
54				9	Active	

MOPNR	MWKR −P	MWKR P	SPT	MWKR	LWKR	LPT	FCFS
103.2	101.8	95.9	*64.3*	103.0			
97.5	104.1	97.5	*65.7*	105.5			
92.9	88.9	99.5	68.3	96.7			
121.6	119.1	119.9	*83.9*	124.0			
75.9	78.6	82.5	54.5	79.0			
146.4	151.4	156.7	*102.3*	150.4			
54.7	52.9	48.6	*43.9*	54.0			
120.2	125.1	118.3	*94.6*	126.8			
118.8	111.6	109.8	*88.1*	113.1			
125.3	132.1	126.8	*101.2*	126.4			
147.5	152.7	148.0	117.1	150.5			
90.4	87.0	89.7	*74.6*	90.6			
133.5	139.1	133.2	115.1	136.4			
153.8	155.2	157.8	*116.4*	159.2			
94.2	95.6	88.9	71.3	94.5			
146.7	148.1	145.2	109.3	148.2			
139.5	133.8	137.7	113.4	135.5			
			133.2	146.2	*123.5*		
			146.5	187.0	*140.1*		
			158.0	170.9	*150.5*		
			131.7	143.4	*128.9*		
			123.2	145.8	*110.2*		
			121.4	152.0	130.7		
			148.0	152.4	*126.3*		
			150.7	204.6	*134.7*		
			159.8	179.8	*151.3*		

(continued)

Table B–1 (*continued*)

Problem number	n	m	Routing	Replication	Type schedule	RANDOM
55				10	Active	
56	20	6	FixR	1	Active	138.3
57				2	Active	165.2
58				3	Active	163.1
59				4	Active	149.6
60				5	Active	174.4
61				6	Active	207.8
62				7	Active	162.0
63				8	Active	148.3
64				9	Active	163.8
65				10	Active	117.3
66				11	Active	176.0
67				12	Active	213.3
68				13	Active	180.7
69				14	Active	131.0
70	20	6	FixR	15	Active	*191.7*
71				16	Active	139.3
72				17	Active	162.0
73				18	Active	174.9
74				19	Active	141.9
75				20	Active	140.5
76	10	4	Permutation	1	Active Nondelay	113.3
77				2	Active Nondelay	158.1
78				3	Active Nondelay	162.7

MOPNR	MWKR −P	MWKR P	SPT	MWKR	LWKR	LPT	FCFS
			150.1	181.7	*129.4*		
130.9	132.0	142.0	133.2	156.0			
168.4	172.8	176.5	*146.5*	187.8			
161.5	174.8	164.9	*160.9*	195.3			
153.5	152.0	163.7	*145.2*	157.4			
146.0	164.7	170.5	163.8	181.9			
203.7	221.8	201.3	*168.2*	230.0			
162.0	150.5	146.0	*132.7*	171.7			
151.8	153.5	153.4	*139.4*	149.1			
168.2	161.9	172.0	*142.2*	176.7			
126.0	125.9	120.3	*113.3*	135.9			
202.0	193.3	182.3	*171.7*	202.4			
197.2	224.9	214.9	*187.2*	231.9			
189.1	189.6	173.1	*165.9*	172.7			
124.0	140.1	129.7	*110.0*	163.2			
211.2	203.9	209.4	192.8	211.4			
142.3	147.0	143.9	*125.5*	137.8			
179.7	185.2	154.4	*140.6*	172.6			
194.5	182.4	181.8	*162.4*	177.4			
161.6	177.9	164.0	*132.0*	171.7			
142.2	138.2	138.1	*106.3*	132.8			
			130.3 80.3	101.8	101.3	125.4	115.3
			157.6 119.0	174.6	123.6	161.1	160.0
			151.3 129.0	170.0	*121.4*	184.9	169.2

(continued)

Table **B–1** (*continued*)

Problem number	*n*	*m*	Routing	Replication	Type schedule	RANDOM
79				4	Active Nondelay	138.1
80				5	Active Nondelay	134.5
81				6	Active Nondelay	165.7
82				7	Active Nondelay	135.6
83				8	Active Nondelay	139.1
84	10	4	Permutation	9	Active Nondelay	139.4
85				10	Active Nondelay	132.2
86				11	Active Nondelay	141.6
87				12	Active Nondelay	160.9
88				13	Active Nondelay	151.7
89				14	Active Nondelay	101.8
90				15	Active Nondelay	*155.4* 174.1
91				16	Active Nondelay	127.7
92				17	Active Nondelay	148.2
93				18	Active Nondelay	145.3
94				19	Active Nondelay	174.1
95				20	Active Nondelay	97.1

MOPNR	MWKR −P	MWKR P	SPT	MWKR	LWKR	LPT	FCFS
		136.7					
		113.1	146.8	131.6	135.6		152.2
		145.1					
		121.9	142.1	129.0	152.9		166.8
		154.4					
		129.1	195.3	*109.3*	191.1		167.6
		127.7					
		105.2	158.9	*99.5*	155.0		159.6
		137.1					
		102.4	138.4	103.3	127.6		138.4
		167.3					
		118.9	167.5	120.6	176.2		159.9
		116.9					
		97.0	134.3	102.8	151.9		124.3
		162.4					
		140.4	170.9	*129.0*	178.0		168.6
		156.0					
		127.5	189.8	135.4	182.6		172.5
		156.6					
		132.1	170.6	159.8	172.2		170.8
		103.8					
		98.2	120.6	103.1	129.1		122.7
			137.3	189.2	*120.5*	179.8	170.9
		130.3					
		113.0	158.8	*111.9*	158.5		127.1
		148.6					
		112.6	158.5	120.4	164.5		160.2
		167.3					
		113.0	169.1	126.8	169.6		170.1
		123.4					
		121.7	176.7	125.6	171.2		146.8
		101.6					
		96.9	130.9	*85.8*	136.9		110.0

Table B-2 Maximum Flow-Time

Problem number	n	m	Routing	Replication	Type schedule	RANDOM
12	10	4	Random	1	Active Nondelay	67
13				2	Active Nondelay	92
14				3	Active Nondelay	118
15				4	Active Nondelay	88
16				5	Active Nondelay	156
17				6	Active Nondelay	155
18				7	Active Nondelay	116
19				8	Active Nondelay	190
20				9	Active Nondelay	119
21				10	Active Nondelay	97
22	10	4	Random	11	Active Nondelay	*53*
23				12	Active Nondelay	78
24				13	Active Nondelay	190
25				14	Active Nondelay	81
26	20	4	Random	1	Active	120
27				2	Active	136
28				3	Active	185
29				4	Active	162

MOPNR	MWKR −P	MWKR / P	SPT	MWKR	LWKR	LPT	FCFS
52	*52*	54	66	*52*			
			52	*52*	59	58	60
63	*63*	*63*	83	*63*			
			80	*70*	80	75	75
96	94	104	118	*93*			
			113	*98*	125	122	123
92	*76*	81	103	77			
			84	*78*	91	81	79
136	*136*	*136*	199	*136*			
			186	*167*	192	173	186
107	106	111	155	*101*			
			111	*99*	113	109	115
86	92	94	123	92			
			118	*113*	116	120	119
117	*110*	124	180	115			
			158	*109*	150	140	141
110	*110*	*110*	123	*110*			
			116	*107*	134	135	120
81	81	87	91	78			
			89	*83*	103	93	100
53	*53*	*53*	54	*53*			
			53	*53*	64	58	*53*
62	*62*	66	88	*62*			
			88	*67*	81	74	79
138	*122*	129	154	124			
			118	*111*	172	138	128
78	*77*	81	88	77			
			73	*73*	91	80	78
110	*110*	111	128	*110*			
113	115	116	126	*109*			
162	163	170	173	163			
132	132	*128*	170	129			

(continued)

Table B–2 (*continued*)

Problem number	*n*	*m*	Routing	Replication	Type schedule	RANDOM
30				5	Active	164
31				6	Active	156
32				7	Active	225
33				8	Active	145
34				9	Active	252
35				10	Active	102
36	20	6	Random	1	Active	192
37				2	Active	184
38	20	6	Random	3	Active	226
39				4	Active	216
40				5	Active	267
41				6	Active	237
42				7	Active	247
43				8	Active	168
44				9	Active	249
45				10	Active	250
46	10	4	FixR	1	Active	
47				2	Active	
48				3	Active	
49				4	Active	
50				5	Active	
51				6	Active Active	*170*
52				7	Active Active	191
53	10	4	FixR	8	Active	
54				9	Active	

MOPNR	MWKR −P	MWKR P	SPT	MWKR	LWKR	LPT	FCFS
139	139	143	205	*138*			
127	126	126	132	*124*			
167	168	169	202	*167*			
111	113	117	132	113			
196	198	213	230	201			
86	84	78	97	*79*			
177	178	182	220	184			
149	146	147	208	*144*			
176	185	188	248	*174*			
186	*185*	198	261	186			
151	155	157	199	*142*			
187	191	191	249	*186*			
187	*187*	193	238	194			
121	*116*	125	146	119			
181	187	185	284	187			
185	*177*	188	245	178			
			206	*164*	222		
			253	*219*	279		
			219	*205*	263		
			247	*164*	273		
			248	*195*	229		
			214	*193*	225		
			234	*193*	222		
			373	*243*	333		
			300	*210*	263		

(*continued*)

Table B–2 (*continued*)

Problem number	*n*	*m*	Routing	Replication	Type schedule	RANDOM
55				10	Active	
56	20	6	FixR	1	Active	199
57				2	Active	219
58				3	Active	247
59				4	Active	247
60				5	Active	253
61				6	Active	339
62				7	Active	228
63				8	Active	225
64				9	Active	251
65				10	Active	197
66				11	Active	254
67				12	Active	310
68				13	Active	310
69				14	Active	226
70				15	Active	324
71	20	6	FixR	16	Active	219
72				17	Active	298
73				18	Active	223
74				19	Active	209
75				20	Active	189
76	10	4	Permutation	1	Active Nondelay	189
77				2	Active Nondelay	254
78				3	Active Nondelay	280

MOPNR	MWKR −P	MWKR P	SPT	MWKR	LWKR	LPT	FCFS
			258	248	239		
218	183	233	206	249			
204	205	217	253	219			
214	229	231	280	241			
189	177	186	305	180			
242	253	248	270	247			
259	258	264	352	292			
207	194	198	299	200			
162	177	172	210	167			
198	195	233	259	201			
163	168	167	245	171			
235	235	246	327	241			
279	315	287	334	298			
251	271	247	292	236			
214	236	220	214	225			
255	251	268	277	251			
178	182	175	255	165			
219	212	231	331	194			
227	204	241	266	196			
201	212	218	278	205			
165	169	187	194	154			
			198				
			161	168	219	238	162
		223					
			228	231	226	255	238
				251			
			270	251	270	274	251

(*continued*)

Table B–2 (*continued*)

Problem number	*n*	*m*	Routing	Replication	Type schedule	RANDOM
79				4	Active Nondelay	218
80				5	Active Nondelay	206
81				6	Active Nondelay	266
82				7	Active Nondelay	*195*
83				8	Active Nondelay	216
84				9	Active Nondelay	*225*
85	10	4	Permutation	10 10	Active Nondelay	192
86				11	Active Nondelay	264
87				12	Active Nondelay	*223*
88				13	Active Nondelay	238
89				14	Active Nondelay	145
90				15	Active Nondelay	310
91				16	Active Nondelay	*171*
92				17	Active Nondelay	227
93				18	Active Nondelay	230
94				19	Active Nondelay	287
95				20	Active Nondelay	194

MOPNR	MWKR −P	MWKR P	SPT	MWKR	LWKR	LPT	FCFS
203							
			183	*168*	234	233	187
			177	*203* 196	204	258	228
			291	*254* *241*	265	277	271
	193						
			207	202	229	214	202
178							
			167	172	232	204	177
			225	*225* 225	239	236	*225*
170							
			204	178	207	216	*170*
	245						
			262	*241*	269	267	264
	236						
			223	248	225	241	237
			206	*214* 214	275	251	214
135							
			148	139	158	182	*132*
		257					
			241	268	267	293	290
		171					
			186	199	213	199	197
			209	*221* 202	310	240	221
217							
			231	226	272	230	*216*
		235					
			232	235	255	254	244
		171					
			144	150	188	181	167

Table B–3 Mean Machine Finish-Time

Problem number	n	m	Routing	Replication	Type schedule	RANDOM
12	10	4	Random	1	Active	53.8
13				2	Active Nondelay	78.5
14				3	Active Nondelay	109.0
15				4	Active Nondelay	74.5
16				5	Active Nondelay	149.5
17				6	Active Nondelay	145.3
18				7	Active Nondelay	108.8
19				8	Active Nondelay	174.3
20				9	Active Nondelay	105.0
21				10	Active Nondelay	82.0
22	10	4	Random	11	Active Nondelay	36.5
23				12	Active Nondelay	59.8
24				13	Active Nondelay	181.3
25				14	Active Nondelay	77.8
26	20	4	Random	1	Active	115.8
27				2	Active	127.3
28				3	Active	174.8

MOPNR	MWKR −P	MWKR P	SPT	MWKR	LWKR	LPT	FCFS
42.0	41.8	*41.5*	53.5	*41.5*			
				43.3			
55.0	*54.3*	*54.3*	69.5	*54.3*			
				57.8			
93.3	91.3	94.8	111.3	*87.8*			
				88.0			
79.0	72.3	*71.3*	83.5	74.0			
					72.3		
132.0	*129.5*	*129.5*	192.5	*129.5*			
				160.5			
100.3	100.3	100.5	145.3	*97.3*			
				91.0			
78.8	82.0	88.8	113.8	83.0			
				103.8			
115.0	*108.0*	119.5	167.0	110.5			
				114.5			
96.5	96.8	98.0	110.5	99.5			
				96.3			
72.8	72.5	78.8	77.3	*71.0*			
				69.0			
35.0	35.7	*33.0*	37.8	45.8			
							38.8
53.5	*53.5*	58.0	72.8	54.0			
				56.5			
120.5	*117.5*	122.8	135.3	119.3			
				107.8			
56.3	59.8	59.0	71.8	64.8			
				62.3			
81.3	*78.5*	80.5	114.3	88.5			
95.3	97.3	97.5	116.3	97.3			
145.0	*141.3*	153.0	163.5	149.8			

(*continued*)

Table B–3 (*continued*)

Problem number	n	m	Routing	Replication	Type schedule	RANDOM
29				4	Active	149.0
30				5	Active	153.5
31				6	Active	143.8
32				7	Active	205.5
33				8	Active	126.8
34				9	Active	240.0
35				10	Active	89.3
36	20	6	Random	1	Active	179.3
37				2	Active	169.2
38				3	Active	215.0
39	20	6	Random	4	Active	200.7
40				5	Active	248.0
41				6	Active	222.5
42				7	Active	224.7
43				8	Active	140.8
44				9	Active	228.3
45				10	Active	216.0

MOPNR	MWKR −P	MWKR P	SPT	MWKR	LWKR	LPT	FCFS
119.3	119.3	115.8	152.5	117.0			
127.8	129.0	131.5	190.8	127.8			
119.3	116.3	118.8	129.3	111.5			
142.8	146.5	151.0	196.3	139.5			
104.0	102.8	107.5	117.8	107.8			
167.3	172.8	182.8	219.5	175.0			
75.0	70.3	68.5	86.5	70.0			
156.2	157.7	156.2	191.5	170.3			
141.5	137.5	137.3	175.8	135.8			
151.7	163.8	166.2	227.5	154.0			
160.7	162.5	181.8	240.8	167.8			
132.7	128.7	133.5	177.2	125.3			
176.7	183.0	184.3	231.8	181.2			
165.3	163.2	172.5	212.5	172.0			
107.3	102.7	104.7	133.3	109.2			
165.0	165.7	169.8	254.3	170.7			
168.2	161.5	174.2	217.7	159.0			

EXPERIMENTAL RESULTS: CONTINUOUS-PROCESS JOB-SHOP PROBLEM

C-1 NOTATION USED IN PRIORITY RULES

t Time at which a selection for machine assignment is to be made

i Index over the jobs to be processed by the shop

j Index over the sequence of operations on a job

J Specific value of j; the imminent operation for which a job is in queue

g_i The total number of operations on job i; $1 \leqq j \leqq g_i$

$R_{i,j}$ The time at which job i became ready for its jth operation—i.e., the time at which the $(j-1)$th operation was completed; $R_{i,1} = r_i$, the time the job arrived at the shop

d_i The due-date of job i

$p_{i,j}$ The processing-time for the jth operation of job i

$p'_{i,j}$ An *a priori* estimate of $p_{i,j}$

$X_{i,j}$ Particular value of a random variable, uniformly distributed between 0 and 1, assigned to the jth operation of job i

$Y_{i,j}(t)$ The total work at time t, in the queue containing the jth operation of job i (total work is the sum of the imminent-operation processing-times of the jobs in that queue)

$Y'_{i,j}(t)$ The total work, including work that will "soon" arrive at time t in the queue containing the jth operation of job i (a job is expected to arrive "soon" if, at time t, its preceding operation is being performed)

$Z_i(t)$ The priority value of job i at time t

C-2 DEFINITION OF PRIORITY RULES FOR THE RAND STUDY

Symbol	Definition $Z_i(t) =$	Description Job selected which:
RANDOM	$X_{i,J}$	Has the smallest value of a *random* priority assigned at time of arrival at queue
FCFS	$R_{i,J}$	Arrived at queue first; *first come first served*
FASFS	r_i	Arrived at shop first; *first arrival at shop first served*
SPT	$p_{i,J}$	Has the *shortest processing-time*
LPT	$-p_{i,J}$	Has the *longest processing-time*
LWKR	$\sum_{j=J}^{g_i} p_{i,j}$	Has the *least work remaining* to be performed
MWKR	$-\sum_{j=J}^{g_i} p_{i,j}$	Has the *most work remaining* to be performed
FOPNR	$g_i - J + 1$	Has the *fewest operations remaining* to be performed
TWORK	$\sum_{j=1}^{g_i} p_{i,j}$	Has the greatest *total work* (all operations)
WINQ	$Y_{i,J+1}(t)$	Will go on for its next operation to the queue with the least work
XWINQ	$Y'_{i,J+1}(t)$	Will go on for its next operation to the queue with the least work, both present and expected
$\dfrac{\text{SPT}}{(\text{WKR})^u}$	$\dfrac{p_{i,J}}{\left(\sum_{j=J}^{g_i} p_{i,j}\right)^{-u}}$	Has the smallest "weighted" ratio of processing-time to work remaining
$\dfrac{\text{SPT}}{\text{TWORK}}$	$\dfrac{p_{i,j}}{\sum_{j=1}^{g_i} p_{i,j}}$	Has the smallest ratio of processing-time to total work
$(u)\,\text{SPT}$ $+ (1 - u)\,\text{WINQ}$	$u \cdot P_{i,J} + (1 - u)Y_{i,J+1}(t)$	Has the smallest weighted sum of processing-time and work in the following queue

(continued)

Table C–2 (*continued*)

Symbol	Definition $Z_i(t) =$	Description Job selected which:
(u) SPT + (1 − u) XWINQ	$u \cdot p_{i,J} + (1 - u) Y'_{i,J+1}(t)$	Has the smallest weighted sum of processing-time and work (including expected work) in the following queue
SPT | Trunc (u)	$p_{i,J}$ if $(t - R_{i,J}) < u$; 0 otherwise	Has the shortest processing-time, unless a job has been held in this queue longer than time u; a *truncated shortest-processing-time* rule
FCFS (u) | SPT	$R_{i,J}$ if $N_{i,J}(t) < u$; $p_{i,J}$ otherwise	Arrived at queue earliest, if the queue is short; the job with the shortest processing-time, if the queue is long
2-Class SPT (u)	$p_{i,J}$ if $k < u$; $(p_{i,J} + 10.0)$ if $k \geqq u$, where k is low-order digit of i, the job number	Has the shortest processing-time among the preferred class of jobs; if no preferred jobs in queue, then the shortest processing-time among other jobs
Est.-SPT (u)	$p'_{i,J} = $ $p_{t,J}(1 + 0.002u(X_{i,J} - 0.5))$	Has the smallest estimated processing-time (% errors of estimate assumed uniformly distributed between ±u)
2-Class (p)	$Z_i(t) = 0$ or 1; Prob $(Z_i(t) = 0 \mid p_{i,J} < M)$ $= p$; Prob $(Z_i(t) = 0 \mid p_{i,J} \geqq M)$ $= 1 - p$	Arrived first among the "short" jobs; if no "short" job in queue, then the "long" job which arrived first (probability that job is correctly classified short or long is p); M = mean processing-time
DDATE	d_i	Has the earliest due-date
OPNDD	$r_i + (d_i - r_i)(J/g_i)$	Has the earliest *operation due-date*; allowed flow-time divided equally among operations
SLACK	$d_i - t - \sum_{j=J}^{g_i} p_{i,j}$	Has the least *slack-time* remaining
S/OPN	$\dfrac{d_i - t - \sum_{j=J}^{g_i} p_{i,j}}{g_i - J + 1}$	Has the smallest ratio of slack-time to number of operations remaining
(u) SPT + (1 − u) S/OPN	$u \cdot p_{i,J} + (1 - u)$ $\times \dfrac{d_i - t - \sum_{j=J}^{g_i} p_{i,j}}{g_i - J + 1}$	Has the smallest weighted sum of processing-time and slack-operation

C–3 EXPERIMENTAL INVESTIGATION OF PRIORITY ASSIGNMENT IN A SIMPLE, SYMMETRIC, RANDOM-ROUTED JOB-SHOP OF 9 MACHINES*

Part I. Work-in-process inventory: Job set 1, utilization 88.4%. Based on measurements for 26 periods, each 400 time units long, representing the arrival of approximately 9360 jobs.

Rule	Mean number of jobs in queue, $\overline{N}_q$	Variance of period means	Mean total work, $\overline{p}$	Mean work remaining, $\overline{p}_r$	Mean imminent-operation work-content, $\overline{p}_q$
RANDOM	59.42	373	1080	554	58.39
FCFS	58.87	336	1078	559	58.29
SPT	23.25	41	545	297	62.54
LPT	167.44	1022	2709	1369	59.08
LWKR	47.52	85		989	51.81
MWKR	109.97	3178		276	69.07
FOPNR	52.23	96		1019	51.83
TWORK	82.91	1445	587	323	65.16
WINQ	40.43	170	801	421	39.16
XWINQ	34.03	105	709	379	41.26
$\dfrac{\text{SPT}}{\text{WKR}^{0.05}}$	23.48	42	539	290	63.80
$\dfrac{\text{SPT}}{\text{WKR}^{0.50}}$	29.63	94	540	235	70.58
$\dfrac{\text{SPT}}{\text{WKR}^{0.25}}$	24.85	55	524	260	65.95
$\dfrac{\text{SPT}}{\text{WKR}^{0.75}}$	37.99	195	597	224	73.77
$\dfrac{\text{SPT}}{\text{WKR}}$	62.94	879		220	73.79
$\dfrac{\text{SPT}}{\text{WKR}^{1.2}}$	83.23	1979	941	222	74.21
$\dfrac{\text{SPT}}{\text{WKR}^{2.0}}$	90.59	2226	1010	229	72.95

* From RAND study [32].   (*continued*)

Table C–3 (*continued*)

Rule	Mean number of jobs in queue, $\overline{N}_q$	Variance of period means	Mean total work, $\overline{p}$	Mean work remaining, $\overline{p}_r$	Mean imminent-operation work-content, $\overline{p}_q$
$\dfrac{\text{SPT}}{\text{TWORK}}$	42.39	2999	435	244	68.60
0.5 (SPT) + 0.5 (WINQ)	30.14	103	649	346	43.35
0.9 (SPT) + 0.1 (WINQ)	23.76	49	552	303	55.22
0.95 (SPT) + 0.05 (WINQ)	23.00	42	545	299	58.49
0.97 (SPT) + 0.03 (WINQ)	22.83	40	539	294	59.98
0.94 (SPT) + 0.06 (XWINQ)	23.26	46	552	303	57.89
0.96 (SPT) + 0.04 (XWINQ)	22.67	39	536	293	58.40
0.98 (SPT) + 0.02 (XWINQ)	22.74	38	541	295	59.96
SPT | Trunc (4)	55.67	408		534	57.77
SPT | Trunc (8)	53.50	310		520	60.32
SPT | Trunc (16)	44.20	334	863	452	60.54
SPT | Trunc (32)	32.85	197	664	368	61.36
FCFS (5) | SPT	29.49	52		342	61.59
FCFS (9) | SPT	38.67	90		407	59.08
2-Class SPT (0.1)	24.46	45	561	306	62.32
2-Class SPT (0.3)	27.29	52	602	326	62.09
2-Class SPT (0.5)	31.17	71	652	352	62.12
2-Class SPT (0.7)	34.57	103	696	375	59.74
2-Class SPT (0.9)	36.11	148	705	384	59.92
Est.-SPT (10)	23.23	41	546	297	62.50
Est.-SPT (100)	27.13	60	606	327	62.40
2-Class (1.00)	35.29	147	725	385	60.95
2-Class (0.75)	44.99	216	873	458	58.98

Table C–3 (*continued*)

Part II. Lateness: Job set 1, utilization = 88.4%; job set 2, utilization = 90.4; job set 3, utilization = 91.9. Sample size = 8700 jobs. Average allowance for queuing = 8 times total processing-time.

Rule	Job set	Type of due-date	Flow-time Mean, $\overline{F}$	Variance	Lateness Mean, $\overline{L}$	Variance	Fraction tardy, f_t
RANDOM	1	TWK	74.7	10822	−4.2	6914	0.3075
FCFS	1	TWK	74.4	5739	−4.5	1686	0.4479
FCFS	1	NOP	74.4	5739	−3.9	1126	0.3993
FCFS	1	CON	74.4	5739	−4.4	5739	0.3376
FCFS	1	RDM	74.4	5739	−4.9	7729	0.4121
FCFS	2	TWK	90.6	8165	11.27	2122	0.5767
FCFS	3	TWK	112.9	13259	31.39	4543	0.6733
FASFS	1	TWK	72.5	1565	−6.4	3359	0.5286
DDATE	1	TWK	63.7	6780	−15.5	432	0.1775
DDATE	1	NOP	70.6	6838	−7.7	529	0.2669
DDATE	1	CON	72.5	1565	−6.3	1565	0.4390
DDATE	1	RDM	72.9	2903	−6.4	2101	0.4878
OPNDD	1	TWK	69.0	28820	−9.9	14560	0.1036
SLACK	1	TWK	65.8	6524	−13.1	433	0.2202
S/OPN	1	TWK	66.1	5460	−12.8	226	0.0371
S/OPN	1	NOP	72.9	5548	−5.4	226	0.2159
S/OPN	1	CON	73.7	1118	−5.1	1118	0.4808
S/OPN	1	RDM	74.0	2590	−5.3	1718	0.5256
S/OPN	2	TWK	76.2	6274	−3.1	205	0.3093
S/OPN	3	TWK	93.5	9238	11.9	1029	0.5475
0.3 (SPT) + 0.7 (S/OPN)	1	TWK	67.7	5127	−17.3	368	0.0182
0.5 (SPT) + 0.5 (S/OPN)	1	TWK	57.6	4758	−21.3	518	0.0152
0.75 (SPT) + 0.25 (S/OPN)	1	TWK	52.2	4291	−26.7	696	0.0198
SPT	1	TWK	34.0	2318	−44.9	2878	0.0502
SPT	1	NOP	34.0	2318	−44.3	2930	0.0617
SPT	1	CON	34.0	2318	−44.8	2318	0.1100
SPT	1	RDM	34.0	2318	−45.3	4416	0.1984
SPT	2	TWK	39.8	5124	−39.6	4649	0.0609
SPT	3	TWK	46.3	13839	−35.3	12187	0.0726

C-4 EXPERIMENTAL RESULTS FOR A STATE-DEPENDENT DUE-DATE PROCEDURE IN A SIMPLE, SYMMETRIC, RANDOM-ROUTED JOB-SHOP OF 8 MACHINES*

Rule			Mean tardiness, $\overline{T}$	Fraction tardy, f_t	Conditional mean tardiness, $\overline{T}_c$	Mean lateness, $\overline{L}$	Std. dev. lateness	Mean flow-time, $\overline{F}$	Std. dev. flow-time
Simple rules:									
SPT			12.1	0.108	112.1	−49.0	106	107.8	147
S/OPN			24.0	0.543	44.2	11.8	62.2	168.6	183
DDATE			35.2	0.498	70.7	18.2	72.5	175.0	199
FASFS			59.1	0.629	94.0	29.6	115	186.4	117
Composite rules:									
b	r	h							
0	0	0	21.9	0.422	51.9	−1.2	60.5	155.6	160
5	0	0	11.1	0.266	41.8	−22.3	55.4	134.5	152
10	0	0	6.73	0.183	36.8	−34.2	55.0	122.6	147
15	0	0	6.52	0.152	42.8	−38.7	58.7	118.0	145
20	0	0	6.62	0.133	49.9	−41.8	62.1	115.0	143
0.0750	1	0	11.8	0.296	40.0	−17.6	53.2	139.2	154
0.1125	1	0	9.32	0.249	37.4	−23.2	52.0	133.6	153
0.1500	1	0	6.95	0.191	36.5	−29.5	51.3	127.3	150
0.2250	1	0	6.47	0.149	43.4	−34.0	54.0	122.8	150
0.3000	1	0	6.11	0.118	51.9	−37.3	56.2	119.5	148
0.3375	1	0	6.72	0.124	54.3	−37.4	58.4	119.4	149
0.3750	1	0	7.08	0.119	59.4	−38.3	60.6	118.5	149
0.4500	1	0	7.25	0.112	65.0	−39.5	62.8	117.3	148
0.3000	1	80	5.81	0.117	49.8	−38.8	56.4	118.0	147
0.3000	1	160	5.43	0.120	45.3	−39.4	56.3	117.4	145
0.3000	1	240	5.61	0.128	43.8	−39.2	57.1	117.6	146
0.1500	1	80	6.55	0.184	35.7	−31.8	52.1	124.9	149
0.1500	1	160	6.39	0.187	34.3	−32.3	52.4	124.5	148

C–5 FLEXIBILITY IN MACHINE SELECTION IN A SYMMETRIC RANDOM-ROUTED JOB-SHOP OF 9 MACHINES*

Number of alternate machines $(K - 1)$	Priority rule	Mean number of jobs in queue, $\bar{N}_q$	Mean flow-time, $\bar{F}$	Percentage of alternate selection
0	FCFS	81.77	97.41	0
0.5	FCFS	28.23	39.47	21.0
1	FCFS	19.72	30.25	38.7
2	FCFS	13.97	24.05	48.5
3	FCFS	11.77	21.67	53.7
8	FCFS	9.54	19.24	63.2
0	SPT	29.78	41.14	0
0.5	SPT	16.96	27.26	18.5
1	SPT	13.88	23.93	35.0
2	SPT	11.42	21.28	46.0
3	SPT	10.50	20.28	52.1
8	SPT	9.05	18.71	63.0

* Sample size = 8880 jobs; utilization = 91.2% (from Wayson [198]).

C–6 FLEXIBILITY IN OPERATION SEQUENCE IN A SYMMETRIC RANDOM-ROUTED JOB-SHOP OF 9 MACHINES*†

Routing ratio	Priority rule	Sample size, jobs	Mean number of jobs in queue, $\bar{N}_q$		Mean flow-time, $\bar{F}$		Percentage of alternate selection
0.1743	FCFS	8880	81.77		97.41		0
0.1743	FCFS	4440‡		82.7		98.6	0
0.1984	FCFS	4440		68.48		84.06	11.0
0.2502	FCFS	4440		53.53		67.26	23.9
1.0000	FCFS	8880	20.99		31.72		61.4
1.0000	FCFS	4440‡		22.3		33.2	
0.1743	SPT	8880	29.78		41.14		0
0.1743	SPT	4440‡		30.1		41.8	0
0.1823	SPT	4440		28.91		40.24	4.9
0.1984	SPT	4440		28.28		39.46	10.4
0.2502	SPT	4440		26.52		37.56	22.8
0.3731	SPT	4440		23.83		34.64	38.1
1.0000	SPT	8880	15.31		25.58		60.3
1.0000	SPT	4440‡		16.4		27.0	

* Utilization = 91.2% (for long runs) (from Neimeier [148]).
† The entries in columns 4 and 5 which are offset to the left indicate that a different sample size was used.
‡ Results for first half of longer run.

C–7 SEQUENCING IN AN ASSEMBLY SHOP*

Rule	Number of branches per job	Mean flow-time, $\bar{F}$	Mean branch flow-time, $\bar{B}$	Mean assembly wait-time, $\bar{A}$	Variance flow-time	Variance lateness	Conditional mean tardiness, $\bar{T}_c$	Fraction tardy, f_t	Jobs completed
FCFS	2	149.28	100.71	48.57	14750	5527	64.46	0.5789	4495
SPT	2	73.80	45.01	28.79	11808	11132	139.65	0.1109	4490
FASFS	2	121.13	98.66	22.47	2623	5787	50.66	0.5162	4500
DDATE	2	116.91	95.59	21.02	11031	2034	40.71	0.3446	4500
SLACK	2	114.40	96.11	18.29	9867	1745	35.79	0.3324	4500
S/OPN	2	108.09	95.34	12.75	7123	0796	12.55	0.0993	4500
NUB	2	154.13	150.18	03.95	29262	31380	168.28	0.4265	4431
MAXRWD-NR	2	156.58	95.10	61.48	39902	31791	186.16	0.3605	4435
MAXNRD-NR	2	202.63	141.78	60.85	48077	50592	216.29	0.5172	4356
MAXNRD-SPT	2	69.67	62.67	07.00	9310	9048	127.63	0.0957	4495
NUB-SPT	2	72.31	61.99	10.32	8844	8370	92.82	0.1150	4494
SPT	5	112.01	41.41	70.60	13417	14009	129.90	0.1438	1773
FASFS	5	122.12	83.78	18.25	1825	6233	33.39	0.2167	1800
S/OPN	5	135.43	102.12	33.31	6939	2577	04.35	0.0044	1800
MAXRND-SPT	5	91.22	74.85	16.37	7338	8819	108.14	0.0776	1791
NUB-SPT	5	96.86	69.38	27.48	5710	7659	73.99	0.0968	1796
SPT	10	264.36	66.10	198.26	67903	67142	279.43	0.3761	880
FASFS	10	171.69	117.27	54.42	4903	9218	58.14	0.2622	900
S/OPN	10	198.27	158.78	39.49	9138	4176	51.15	0.1193	900
MAXNRD-SPT	10	164.47	137.01	27.46	22692	22680	182.37	0.1128	897
NUB-SPT	10	144.90	102.62	42.28	12922	15193	129.60	0.1479	899

INDEX